Mom the Chemistry Professor

Kimberly Woznack • Amber Charlebois •
Renée Cole • Cecilia Marzabadi • Gail Webster
Editors

Mom the Chemistry Professor

Personal Accounts and Advice from
Chemistry Professors who are Mothers

Second Edition

Editors
Kimberly Woznack
Department of Chemistry & Physics
California University of Pennsylvania
California, Pennsylvania, USA

Amber Charlebois
Chemistry & Biochemistry Department
Nazareth College
Rochester, New York, USA

Renée Cole
Department of Chemistry
The University of Iowa
Iowa City, Iowa, USA

Cecilia Marzabadi
Department of Chemistry & Biochemistry
Seton Hall University
South Orange, New Jersey, USA

Gail Webster
Chemistry Department
Guilford College
Greensboro, North Carolina, USA

ISBN 978-3-319-78971-2   ISBN 978-3-319-78972-9   (eBook)
https://doi.org/10.1007/978-3-319-78972-9

Library of Congress Control Number: 2018945541

© Springer International Publishing AG, part of Springer Nature 2014, 2018
This work is subject to copyright. All rights are reserved by the Publisher, whether the whole or part of the material is concerned, specifically the rights of translation, reprinting, reuse of illustrations, recitation, broadcasting, reproduction on microfilms or in any other physical way, and transmission or information storage and retrieval, electronic adaptation, computer software, or by similar or dissimilar methodology now known or hereafter developed.
The use of general descriptive names, registered names, trademarks, service marks, etc. in this publication does not imply, even in the absence of a specific statement, that such names are exempt from the relevant protective laws and regulations and therefore free for general use.
The publisher, the authors, and the editors are safe to assume that the advice and information in this book are believed to be true and accurate at the date of publication. Neither the publisher nor the authors or the editors give a warranty, express or implied, with respect to the material contained herein or for any errors or omissions that may have been made. The publisher remains neutral with regard to jurisdictional claims in published maps and institutional affiliations.

Cover Photo: Mosaic Design: Doug Rickert
Photo credits: Individual Author Photo Credits have been provided in each author's chapter.

Printed on acid-free paper

This Springer imprint is published by the registered company Springer International Publishing AG part of Springer Nature.
The registered company address is: Gewerbestrasse 11, 6330 Cham, Switzerland

Foreword

In a faculty meeting I was at many years ago, a young professor and parent described the academic work–life balance mechanism as "a system of rotating neglect." The comment drew laughter, as one might expect, as many of us, even those of us without children, can identify with this sentiment.

The issue of how one can balance home life, which for some includes the fulfilling yet difficult task of raising children while working in an academic setting, is extraordinarily important to men and women alike.

The Women Chemists Committee of the American Chemical Society, which I have had the opportunity to lead as Chair, is tasked with attracting, retaining, developing, promoting, and advocating for women in the chemical sciences. Reasons for this mission include the harsh reality that the pipeline for women in our field has been and remains leaky. While nearly half of all bachelor's degrees are earned by women in our field, only 41% go on to earn a PhD, according to the National Center for Education Statistics. And as Madeleine Jacobs, the former CEO and Executive Director of ACS, once said (and I paraphrase), even if those women make it through the pipeline, they can find themselves dumped into toxic waters with currents taking them places they do not wish to go. According to "Women Faculty Positions Edge Up" in the April 7, 2014, issue of *Chemical & Engineering News*, "At the top 50 schools in terms of chemical R&D spending, women held just 18% of the tenured and tenure-track positions in the 2012–13 academic year." Although robust through the undergraduate years, the pipeline still leaks for women from graduate school and beyond.

One source of the leaks is the difficulty of harmonizing family life with a career in academia. Undergraduate students, graduate students, and postdoctoral fellows are observing the dynamics of this delicate balance in the home and work lives of women faculty, drawing their own conclusions about the possibilities of this career path for them. According to University of California Berkeley Professor Mary Ann Mason, who is one of the authors of the book *Do Babies Matter: Gender and Family in the Ivory Tower*, 55% of tenured women professors have children compared to 74% of their male colleagues (page 80) at the 12-year mark beyond earning a doctoral degree. Female professors are less likely to have children than their counterparts (women with terminal degrees) working in industry, government, law, or medicine. The same is true for male professors when compared against male doctors and lawyers. The environment in academia, while viewed by some as having tremendous flexibility, poses challenges for those who wish to have a family, and gender parity is only achieved when both professional and personal goals of raising children can be realized.

Dr. Jennifer Graves, the Australian geneticist, in a recent opinion piece for *Financial Review* (November 12, 2017), identifies these same issues and echoes the need to find ways to address the practicalities for women balancing the demands of motherhood and faculty careers. "It isn't that hard," she writes. "My suggestions are simple and not as expensive as losing half our trained workforce." Her ideas include providing technical help and formal, shorter work weeks.

The second edition of *Mom the Chemistry Professor* provides an expanded set of stories of women who have faced hurdles and setbacks along with joy and fulfillment in both their work and home life. Given the scarcity of role models, the compilation of the stories in the first and now this second edition is that much more important, not only to inform good policy but to provide guidance and inspiration to women far before they will ever need such policies. Annotations have also been included in the chapter focused on the reproductive safety challenges of working in chemistry laboratories while pregnant and breastfeeding and on the career challenges of not doing so.

I am grateful for the contributions of all these women who are sharing their personal stories. No matter how they might describe their "rotational systems" of focus between home and work, children and students, family members and colleagues, their stories can bring insight and inspiration to all of us.

Women Chemists Committee   Laura Sremaniak
Raleigh, NC, USA
November, 2017

Preface

The Women Chemists Committee (WCC) of the American Chemical Society (ACS) was first approached by Springer during the year 2012 and asked if we had interest in publishing an English-language, US-centric version of a book similar in nature to a book already published by Springer, *Professorin und Mutter- wie geht das?* (http://www.springer.com/de/book/9783827424310). They were interested in highlighting the challenges for women in the United States, where the maternity leave and family support policies are poor to nonexistent.

WCC was overjoyed to be approached with this opportunity, as it fit directly with the mission of the WCC. The mission of the WCC is to attract, retain, develop, promote, and advocate for women to positively impact diversity, equity, and inclusion in the Society and the profession. The WCC has four specific goals that address this mission:

- Increase engagement and retention of women
- Advocate and educate within the Society on issues of importance to women
- Enhance leadership and career development opportunities for women
- Highlight contributions of women

The book project addresses the first WCC goal of increasing participation and retention of women in the chemical sciences, specifically within academia. Additionally, it addresses the fourth WCC goal: to promote and recognize the professional accomplishments of women in the chemical sciences.

While the first edition of *Mom the Chemistry Professor* was not an official project of the WCC or ACS, the editors were all WCC members. It was released in 2014 and contained 16 personal accounts from chemistry professor moms. The first edition of the book also contained a chapter on perceptions with respect to chemical safety. The response of the chemistry community identified opportunities to highlight additional stories and to call for more reproductive health resources and policies.

This second edition is an official project of the WCC, containing accounts from mothers who have an even wider range of backgrounds and have followed a variety of pathways to become chemistry professors and mothers. The editorial team has

viii   Preface

chosen to include in this edition an introductory chapter highlighting some of the questions young women ask when considering both a family and a career in academia. This introduction calls attention to the various challenges faced by authors and different strategies they have used to counter these challenges. With 40 personal accounts in this edition, the introductory chapter may help the reader select which chapters they may most identify with or may want to read to focus on a specific challenge. This edition also addresses more specifically the challenges of chemical hazards and reproductive safety. The editors have added safety tips and key discussion questions into Megan Grunert Kowalske's research chapter and asked other authors to highlight some of the safety challenges they faced and choices they have made.

It is the hope of the editorial team and the WCC that this book will continue to not only inspire all women considering an academic career and motherhood but also to set the stage for more advocacy on behalf of women chemists.

Editorial Team
Mom the Chemistry Professor, Second Edition

California, PA   Kimberly Woznack
Rochester, NY   Amber Charlebois
Iowa City, IA   Renée Cole
South Orange, NJ   Cecilia Marzabadi
Greensboro, NC   Gail Webster

Acknowledgements

In addition to thanking the courageous women who were willing to share their very personal stories, we, the editorial team for the second edition of *Mom the Chemistry Professor*, have many people we would like to thank for their very important contributions to this work. We thank the Women Chemists Committee (WCC) members (both past and present) who provided suggestions, guidance, and recommendation regarding the book project.

For the second edition, the team was very grateful for the opportunity to assemble an advisory board. The members of the advisory board for the second edition and their ACS volunteer affiliation(s):

- Karl Booksh, ACS Diversity and Inclusion Advisory Board
- Amanda Bryant-Friedrich, Association for Women in Science, ACS Multidisciplinary Program Planning Group (MPPG)
- Nora Fredstron, ACS Committee on Chemical Safety
- Teri Quinn Gray, ACS Diversity and Inclusion Advisory Board
- Robin Izzo, ACS Committee on Chemical Safety
- Mary Jane Shultz, ACS Women Chemists Committee
- Weslene Tallmadge, ACS Committee on Chemical Safety
- Gloria Thomas, ACS Division of Professional Relations (PROF)
- Paula Christopher (*ex officio*), American Chemical Society, Staff Liaison to the ACS Diversity and Inclusion Advisory Board
- Jodi Wesemann (*ex officio*), American Chemical Society, Staff Liaison to the Women Chemists Committee (2016–2017)

The editorial team is grateful for the time and efforts of all of the members of the advisory board.

In addition to the advisory board members, the editorial team would like to thank Laura Sremaniak (WCC Chair 2017) for her assistance in keeping the book project moving forward and helping us to make the second edition into an official ACS WCC project. Many thanks also go out to 2015–2017 Committee on Chemical

Safety (CCS) Chair Betty Ann Howson and CCS ACS Staff Liaison, Marta Gmurczyk, for their discussions on chemical safety. The editorial team would also like to thank our new ACS WCC staff liaison Victoria Fuentes. Victoria began her position as the second edition was nearing completion, and the team is thankful for her support in the final stages.

Doug Rickert also provided assistance in the final stages. His graphic design work on the cover photo mosaic, which helped convey the diversity among our authors.

The editorial team will be eternally grateful for the limitless advocacy and assistance provided by Jodi Wesemann. This advocacy made it possible for the advisory board to hold our face-to-face meeting, which was a pivotal time in shaping the direction of the second edition. Without Jodi's input and tireless efforts, this book project would not have been finished.

Fig. 1 Group photo from the face-to-face advisory board meeting, which took place in December of 2016 at American Chemical Society headquarters in Washington, DC. Not pictured are Gloria Thomas and Karl Booksh, who participated via videoconference, and ACS WCC Staff Liaison Jodi Wesemann, who took the photo

Contents

Introduction 1
Amber F. Charlebois, Renée Cole, Cecilia H. Marzabadi, Gail H. Webster, and Kimberly A. Woznack

Motherhood and Academic Chemistry Laboratories: Safety and Career Challenges from a Qualitative Study 21
Megan Grunert Kowalske

The Best Job in the World 37
Stacey F. Bent

Equilibrium and Stress: Balancing One Marriage, a "Two-Body Problem," and Three Children 57
Stacey Lowery Bretz

If at First You Don't Succeed, Don't Give Up on Your Dreams 75
Pamela Ann McElroy Brown

Invictus 85
Amanda Bryant-Friedrich

My Circus: Please Note That I Have No Formal Training in Juggling ... 95
Amber Flynn Charlebois

Planned Serendipity 115
Renée Cole

Readymade Family 127
Mary Ann Crawford

Mother and Community College Professor 137
Elizabeth Dorland

From Premed to US Professor of the Year: My Personal Journey 155
Amina Khalifa El-Ashmawy

xi

Minha Vida e Minha Carreira e Minha Família (My Life and My Career and My Family) 177
Isabel C. Escobar

Chemistry in the Family 189
Cheryl B. Frech

"Are You Always This Enthusiastic?" 199
Jennifer M. Heemstra

How Motherhood Shaped My Professorship 209
Jani C. Ingram

Upward Bound to a PhD in Chemistry 223
Judith Iriarte-Gross

I'm.a.Gene: Destined for a Career in the Sciences 237
Margaret I. Kanipes

It Always Seems Impossible Until It Is Done 247
Sunghee Lee

On Breastfeeding, Supramolecular Chemistry, and Long Commutes: Life as an Associate Professor, Wife, and Busy Mother of Three 257
Mindy Levine

The Window of Opportunity 279
Nancy E. Levinger

Pieces of a Puzzle 297
Sherri R. Lovelace-Cameron

Wanting It All 307
Cecilia H. Marzabadi

On Our Own Terms 317
Sara E. Mason

Putting Family First Led to My Extraordinary Career 329
Saundra Yancy McGuire

Family Matters: How Family Influenced My Career 345
Jin Kim Montclare

Mother and Chemist: Every Pitfall is an Opportunity to Rise with a New Beginning 355
Ingrid Montes-González

Taking an Unconventional Route? 367
Janet R. Morrow

The Path to Academia and Motherhood: It Takes a Village 379
Emily Niemeyer

Contents

Elements to Successful Motherhood and the Professoriate 389
Sherine O. Obare

A Family Grows 401
Patricia Ann Redden

Teacher to Scientist and Back Again 417
Rachel E. Rigsby

A Divinely Ordered Path 425
Renã A. S. Robinson

Raising Three Children Across Three Continents 443
Omowunmi A. Sadik

Ebony and Women and Science. . . Oh My! 459
Darlene K. Taylor

From the Periodic Table to the Dinner Table 469
Danielle Tullman-Ercek

**The Bigger Picture: My Journey to a Purposeful Life and
Career in Academia** 485
Idalis Villanueva

Encounters of the Positive Kind 501
Michelle M. Ward

The Long and Winding Road 513
Gail Hartmann Webster

I Finally Know What I Want to Be When I Grow Up 525
Catherine O. Welder

Finding Rhythm 539
Leyte Winfield

Constant Refinement: Learning from Life's Experiments 553
Kimberly A. Woznack

**Remarkable, Delightful, Awesome: It Will Change Your Life,
Not Overnight But Over Time** 577
Sherry J. Yennello

Introduction

Amber F. Charlebois, Renée Cole, Cecilia H. Marzabadi, Gail H. Webster, and Kimberly A. Woznack

The second edition of "Mom the Chemistry Professor" features 40 women who have indeed had the privilege to identify as both mothers and chemistry professors. Many of the young chemists we have encountered through our daily work with students and our professional activities feel they are facing a choice between the professoriate and motherhood. Knowing that these roles are not diametrically opposed, we have assembled a collection of stories from which we hope readers can find inspiration to pursue a similar combination of career and family.

Each mother and chemistry professor navigates a unique set of circumstances and challenges. Though we often feel alone, wrestling with how to best reach our goals, others are asking similar questions. While editing this collection of stories, we have identified a series of commonly asked questions. The information provided in the introduction and the next chapter, combined with insights from others, will help readers develop their own answers.

A. F. Charlebois
Department of Chemistry and Biochemistry, Nazareth College, Rochester, NY, USA

R. Cole
Department of Chemistry, University of Iowa, Iowa City, IA, USA
e-mail: renee-cole@uiowa.edu

C. H. Marzabadi
Department of Chemistry & Biochemistry, Seton Hall University, South Orange, NJ, USA
e-mail: cecilia.marzabadi@shu.edu

G. H. Webster
Chemistry Department, Guilford College, Greensboro, NC, USA
e-mail: gwebster@guilford.edu

K. A. Woznack (✉)
Department of Chemistry & Physics, California University of Pennsylvania, California, PA, USA
e-mail: woznack@calu.edu

© Springer International Publishing AG, part of Springer Nature 2018
K. Woznack et al. (eds.), *Mom the Chemistry Professor*,
https://doi.org/10.1007/978-3-319-78972-9_1

The individual circumstances we navigate, and all the questions they raise, occur in the broader contexts of our departments, institutions, and disciplinary communities. As current and former members of the Women Chemists Committee (WCC) of the American Chemical Society (ACS), we also want to raise questions about the academic landscape, providing insights about the broader contexts and inspiring the collective work that needs to be done to support others to be mothers and professors.

How Many Women Pursue an Academic Career in Chemistry?

This simple question is not as easy to answer as it might seem. In addition to finding the data, there are the related questions associated with the educational pathways that are expected for academic careers.

First we turn to the number of women earning bachelor's degrees. The US Department of Education houses the National Center for Education Statistics (NCES). NCES administers the Integrated Postsecondary Education Data System (IPEDS). The American Physical Society (APS) Education program publishes selected IPEDS data and graphs on their website [1]. The data reveals that from 2001 to 2015, the percentage of undergraduate degrees in chemistry awarded women ranged from 48 to 52%. Some people might look at these numbers and say that there is no gender-based issue in chemistry, and it is just "a matter of time" until this population of female undergraduate students proceeds through the academic ranks and the lack of women among the ranks for full professors will disappear. Despite the elapsed time since women achieved 50% representation among earned chemistry bachelor's degrees, we don't continue to see women represented equally.

Next we consider the number of women earning PhD degrees. Each year the National Science Foundation (NSF) collects data on the number of doctorates earned in the United States. The Survey of Earned Doctorates (SED) results, including graduate demographics, are shared publicly on the NSF website [2]. According to the 2016 SED, there were 2704 doctorates awarded in the field of chemistry in 2016. Of these only 991 (or 36.6%) were awarded to women.

Determining the number of women in postdoctoral positions is more difficult. The SED asks participants if they have postgraduation employment commitments.

- Of the male respondents earning a PhD in chemistry, 55% (939 out of 1713) indicated that they had postgraduate employment commitments. Of those male chemistry PhDs with commitments, 61% ($n = 576$) indicated that they would be moving to a postdoctoral appointment or an academic appointment.
- Of the female respondents earning a PhD in chemistry, 50% (497 out of 991) indicated that they had postgraduate employment commitments. Of those female chemistry PhDs with commitments, 58% ($n = 289$) indicated that they would be moving to a postdoctoral appointment or an academic appointment.

Introduction

So, fewer women than men are earning doctoral degrees in chemistry. Of that smaller population, a slightly smaller percentage indicate by their commitments that they are interested in going into an academic career that in most cases requires the completion of a postdoctoral research appointment.

The question about the number of women pursuing academic careers in chemistry can only be partially answered, due to limited data sets. The Open Chemistry Collaborative in Diversity Equity (OXIDE) is a nationally funded initiative to advocate for fair and inclusive treatment of faculty by reaching out to administrators in the top research universities in the country. OXIDE collaborates with *Chemical & Engineering News* (C&EN) to collect data from these PhD-granting, research-focused departments to monitor the demographics of those faculty members [3]. According to the C&EN article "Women crack the academic glass ceiling," the percentage of women among the faculty ranks at the top research institutions is increasing across all ranks [4]. In the 2014–2015 academic year, women represent 26% of the assistant professors, 30% of associate professors, and 14% of full professors at the top 50 universities in terms of 2014 chemical R&D spending [4]. The higher percentage of women at the level of associate professor may suggest that they are not being promoted to the rank of full professor at the same rate as male faculty members.

What Are the Backgrounds of Chemistry Professors Who Are Mothers?

Chemistry professor mothers come from a variety of backgrounds. Each has her own combination of identities beyond gender—racial, ethnic, sexual orientation, socio-economic, geographical, religious, etc.—a combination conveyed in the feminist concept of *intersectionality*. This term was originally coined by legal scholar Kimberlé Crenshaw [5]. "Intersectionality references the two-fold idea that people's identities are complex, often not fitting easily into distinct social categories of gender, race, class, and sexuality..." [6]; while acknowledging all that makes each of us unique, intersectionality highlights the challenges associated with various identities. Intersectionality also indicates "that sexism, racism, classism, homophobia, ableism, religious persecution, and nationalism are interlocking systems of oppression that shape our lives and social institutions" [7].

As we celebrate and seek to increase the number of female chemistry professors from different backgrounds, each of us needs to consider how our own background influences our perspectives and behaviors. Even the most well-intentioned person sees the world from her or his own "privileged" standpoint, which includes implicit biases. For women who identify as coming from an underrepresented group, there are often additional career challenges in addition to those faced by women. Women from racial and ethnic minorities may encounter racism and stereotyping by students,

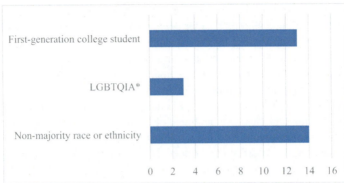

* Lesbian, gay, bisexual, transgender, queer/questioning, intersex, and asexual

Fig. 1 Distribution of authors who identify as first-generation college students; as lesbian, gay, bisexual, transgender, queer/questioning, intersex, or asexual; or as from a non-majority race/ethnicity

Table 1 Guide to chapters featuring authors who identify as coming from a non-majority race or ethnic background or identify as first-generation college students

Authors who identify as:	
Being from a non-majority race or ethnicity (14)	Amanda Bryant-Friedrich, Mary Crawford, Amina El-Ashmawy, Isabel Escobar, Jani Ingram, Sherri Lovelace-Cameron, Jin Montclare, Ingrid Montes, Sherine Obare, Renã Robinson, Omowunmi Sadik, Darlene Taylor, Idalis Villanueva, Leyte Winfield
LGBTQIA: lesbian, gay, bisexual, transgender, queer/questioning, intersex, and asexual (3)	Mary Crawford, Sara Mason and Emily Niemeyer
A first-generation college student (13)	Stacey Lowery Bretz, Amber Charlebois, Mary Crawford, Judith Iriarte-Gross, Sherri Lovelace-Cameron, Emily Niemeyer, Patricia Redden, Rachel Rigsby, Omowunmi Sadik, Michelle Ward, Gail Webster, Leyte Winfield

fellow faculty members, and administrators. Women may also experience discrimination on the basis of their gender expression or sexual orientation. It is truly important for us each to serve as campus advocates to embrace our differences and build a culture where we can work together to develop policies and practices on academic campuses to minimize the impact of biases.

A notable number of our authors identify as first-generation college students. They are the first person in their immediate family to attend and graduate from college. It is in no way a prerequisite that chemistry professors come from a family who has family members with advanced degrees (Fig. 1; Table 1).

Introduction

When Should I Start a Family?

Each woman should decide the time that's right for her to start a family. Women in this volume describe having or adopting children at all stages in their professional career: before graduate school, as graduate students, postdocs, pre-tenure faculty, and after receiving tenure.

Potential career impacts are among the aspects to consider. In the book *Do Babies Matter? Gender and Family in the Ivory Tower* [8], Mary Ann Mason states that "For women in academia, deciding to have a baby is a career decision." The timing of family formation is discussed, and the book reports that the impact of an "early baby" [baby arrives within 5 years of the completion of the PhD] is far greater when compared to women without children and men, whereas a "late baby" [baby arrives more than 5 years of the completion of the PhD] has a far less dramatic effect on one's career.

Many other aspects, personal and professional, must be considered along with the timing and nature of potential career impacts. What are your personal priorities? Is there a medical history or conditions that could impact your fertility? How comfortable are you with navigating the uncertainty of life and adjusting your anticipated trajectory? Along with these and other questions, there are a few practical aspects to consider.

One important consideration must be the ability to take some time off around the birth date or adoption date. A supportive mentor and departmental/university climate are helpful but may not always be present to support this leave request.

Also, careful considerations should be made about whether or not your laboratory activities can continue during pregnancy and breastfeeding. This is the focus of the next chapter "Motherhood and Academic Chemistry Laboratories: Safety and Career Challenges from a Qualitative Study," by Megan Grunert Kowalske. Besides consulting with your physician, be sure to notify your environmental safety officer and/or supervisor before continuing work in the lab once you have found out you are pregnant. If possible, initiate a hazard/risk assessment process before you become pregnant.

Many women describe the importance of a support network for helping them manage childcare and work obligations. It is often a spouse but also includes family and community members as well (Fig. 2; Table 2).

Will I be Able to Plan My Pregnancies, Childbirth, or Adoptions to Occur at Just the Right Time?

Many women, not just chemistry professors, hope to have their children during a certain stage of their life. As highlighted above, many academic women may want to have or adopt a child during a certain career stage.

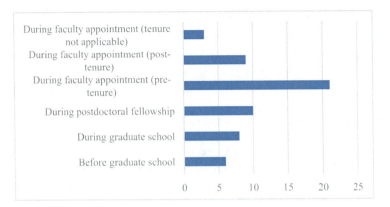

Fig. 2 Stage in their careers during which the authors had/adopted a child (the sum of the responses is greater than the number of authors because some authors have more than one child)

Table 2 Guide to chapters based upon the career stages at which authors had/adopted a child

Authors who have/had adopted a child:	
Before graduate school (6)	Pamela Brown, Mary Crawford, Judith Iriarte-Gross, Cecilia Marzabadi, Omowunmi Sadik, Michelle Ward
During graduate school (8)	Pamela Brown, Renée Cole, Ingrid Montes, Sunghee Lee, Saundra Yancy McGuire, Sherine Obare, Gail Webster, Leyte Winfield
During a postdoctoral appointment (10)	Stacey Bent, Stacey Lowery Bretz, Amber Charlebois, Jennifer Heemstra, Mindy Levine, Nancy Levinger, Omowunmi Sadik, Darlene Taylor, Danielle Tullman-Ercek, Idalis Villanueva
During faculty appointment (pre-tenure) (21)	Stacey Bent, Stacey Lowery Bretz, Amanda Bryant-Friedrich, Amber Charlebois, Elizabeth Dorland, Amina El-Ashmawy, Cheryl Frech, Jennifer Heemstra, Margaret Kanipes, Mindy Levine, Nancy Levinger, Sheri Lovelace-Cameron, Sara Mason, Jin Montclare, Janet Morrow, Sherine Obare, Rachel Rigsby, Renā Robinson, Danielle Tullman-Ercek, Idalis Vilanueva, Kim Woznack
During faculty appointment (post-tenure) (9)	Amber Charlebois, Isabel Escobar, Margaret Kanipes, Jin Montclare, Janet Morrow, Emily Niemeyer, Patricia Redden, Kim Woznack, Sherry Yennello
During faculty appointment (tenure not applicable) (3)	Elizabeth Dorland, Amina El-Ashmawy, and Catherine Welder

Unfortunately, most institutions of higher education in the USA do not offer paid parental leave or paid maternity leave to their faculty members. While most professors meet the requirements of the Family and Medical Leave Act (FMLA), this only entitles the mother to unpaid leave. Thus it is of interest to some chemistry professors to give birth or adopt at the end of an academic year so that they might have the summer, with a reduced or non-existing teaching load, to spend more time with their infant or new child. For biological mothers, it is not always possible to

Introduction

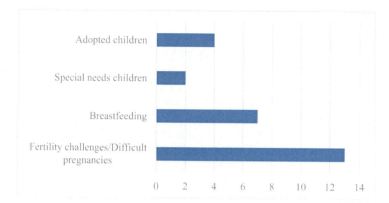

Fig. 3 Distribution of authors who report having adopted children, children with special needs, experience breastfeeding, or experience with fertility challenges or difficult pregnancies

Table 3 Guide to chapters featuring authors who experienced fertility challenges/difficult pregnancies, or adoptions

Authors who have experienced:	
Fertility challenges/difficult pregnancies[a] (14)	Stacey Lowery Bretz, Amber Charlebois, Amanda Bryant-Friedrich, Cheryl Frech, Margaret Kanipes, Mindy Levine, Cecilia Marzabadi, Sara Mason, Emily Niemeyer, Rachel Rigsby, Danielle Tullman-Ercek, Gail Webster, Catherine Welder, Kim Woznack
Adoption (4)	Isabel Escobar, Margaret Kanipes, Patricia Redden, Rachel Rigsby

[a]Trigger warning: these chapters potentially include stories of miscarriage or stillbirth

become pregnant or remain pregnant in exactly the perfect menstrual cycle to yield an end of spring semester birth.

Some of the authors have shared the challenges of becoming pregnant and experiencing miscarriage and stillbirth. In her article, "When Chemists Struggle with Infertility and Miscarriage," Linda Wang opens up a dialog that is crucial to have [9]. It can be challenging for women experiencing the loss of a pregnancy to communicate this grief, and the impact it may have on their work, when miscarriage is not always openly discussed.

For some women or families who experience fertility challenges, adoption has allowed them to start or expand their families. Fertility challenges are not the only reason to pursue adoption. Four authors have shared stories of adoption. In some cases of adoption, there is even less advanced notice for the arrival of a child. When a successful adoption match is made, the new parent(s) may have very little notice to prepare for the very sudden arrival of a child. This can be a challenge for women who may have a fixed teaching schedule and may need to adjust to childcare or school schedules (Fig. 3; Table 3).

How Do I Minimize the Risks of Working in the Lab?

Chemical safety with respect to pregnancy in the laboratory is a very important issue. A number of the authors in this work comment candidly on the issues of safety in the laboratory, and each author approached and handled their own situation the best way they knew how. We would like to note that in no way does this collection of stories claim to provide anything more than examples of how some women maneuvered this space. Once this collection of stories became complete, it was even more apparent that work and energy is needed in this area. WCC will advocate more proactively in this effort to create resources for women who are, or plan to become, pregnant and work in a chemical laboratory.

In order to manage risks, use a concept for science safety, developed by Robert H. Hill and David C. Finster in *Laboratory Safety for Chemistry Students* [10].

- Recognize hazards
- Assess the risks of hazards
- Minimize the risks of hazards
- Prepare for emergencies

By following these four key principles of safety, women who are pregnant and/or breastfeeding—and all those working with them—can make their laboratories safer.

In a previous section, the process of hazard/risk assessment was mentioned. We have also included below a contribution from Robin M. Izzo, an environmental health and safety professional for an academic institution to share her recommendations.

Reflections from an Environmental Health and Safety Professional Robin M. Izzo

In my more than 30 years as a laboratory safety professional in academia, I have consulted with scores of pregnant laboratory workers or scientists planning pregnancy. They are often just as nervous as they are excited about their pregnancy and are eager to know if it is safe to continue working in the lab. They usually want to keep the information confidential, not wanting to share the news of their pregnancy with their supervisor or others in the laboratory for the first trimester or longer. I always recommend sharing the news sooner rather than later, especially if they continue to work in the lab. Many find that their lab mates are more careful around them when they know that they are pregnant.

Most of the time, these women choose to continue their lab work through-out their pregnancy and go on to have healthy babies. Good work practices, engineering controls (like fume hoods, gloveboxes, etc.), and personal protective equipment are intended to prevent exposure to harmful chemicals, thereby limiting any risk to a developing fetus.

(continued)

My advice to women who are pregnant or are planning pregnancy is this:

- Make a list of the hazardous chemicals and operations that you conduct in your lab. Consider the work that others do around you.
- Consult with the health and safety staff at your institution, sharing the information you collected. If your health and safety staff are not able to assist you, use some of the resources listed here to research the potential health effects to you or your unborn child. The main things you want to know are:
  - What are the routes of exposure? Is it a powder or a volatile liquid that you can easily inhale? Is skin absorption a significant route of exposure? Does it irritate the skin?
  - Are there good warning properties? Can you smell it or detect it at concentrations lower than the odor threshold?
  - Is it a known or suspected carcinogen? Carcinogens can also be mutagens.
  - Are there studies or is there information about potential harm to the fetus? If so, is it at levels that are also harmful to the mother or can it harmful even before then?
  - Is it known to leach into breast milk?
  - What is the best way to protect from exposure (type of gloves, fume hood, etc.)? It might seem like a good idea to wear a respirator, but a respirator puts a bit of strain on your breathing and can become uncomfortable quickly.

- Take this information to your healthcare provider for recommendations of what type of work to avoid, if anything.
- If you work with radioactive materials or radiation-producing equipment, the US Nuclear Regulatory Commission (NRC) recommends lower exposure limits. In order to apply these limits (and, in some cases, be eligible for more frequent monitoring), you must make a formal declaration of pregnancy. Contact your institution's radiation safety officer for more information.
- When you are comfortable, review the recommendations and concerns with your supervisor or a health and safety professional. Come up with a plan of action that makes you feel comfortable.
- Let others in your lab know you are pregnant. Advise them of any processes, procedures, or materials that you want to avoid.

Keep in mind that stress from worrying about harmful chemical exposure can be more of a risk to pregnancy than the actual risk of chemical exposure. If you are not comfortable with the plan of action that your supervisor or institution has for you, talk it over with your healthcare provider, your advisor, or others at your institution who may be able to help.

(continued)

Resources
- Safety data sheets for the specific chemical
- Developmental (Teratology) Abstracts—National Toxicology Program
  https://ntp.niehs.nih.gov/testing/types/dev/abstracts/index.html
- Drugs and Lactation Database (LactMed)
  https://toxnet.nlm.nih.gov/newtoxnet/lactmed.htm
- *Developing Reproductive Protection Programs in Industrial and Academic Settings*—American Chemical Society
  https://www.acs.org/content/dam/acsorg/about/governance/committees/chemicalsafety/developing-reproductive-protection-programs-in-industry-and-academic-settings.pdf
- *Reproductive Health and the Workplace*—US Centers for Disease Control and Prevention
  https://www.cdc.gov/niosh/topics/repro/

What Kinds of Positions Allow for Professor and Mom?

There is often a perception that some academic positions are better than others if you want children and a career as a chemistry professor. Each position presents its own rewards and challenges, but as the stories in this book (and others) illustrate, it is possible to have a successful career and children in many different types of positions and institutions.

While there are differences in the demands and expectations among various types of institutions, there is also a significant degree of similarity in the experiences and strategies used to be productive and happy as a mother in an academic career. A more extensive analysis of the experiences of tenure-track women faculty at a variety of institutions is explored in the research of Kelly Ward and Lisa Wolf-Wendel. In *Academic Life and Motherhood*, they analyze the perspectives of women faculty at a variety of institutions who are also mothers of young children [11]. This work is extended in *Academic Motherhood*, where they present findings from longitudinal studies of motherhood at different types of institutions through different career stages [12]. Women have combined rewarding and successful careers as professors at community colleges, small liberal arts colleges, comprehensive universities, and research intensive PhD-granting institutions. Ward and Wolf-Wendel only studied tenure-track faculty; the stories in this collection include those of women in both non-tenure-track and tenure-track positions (Figs. 4 and 5; Tables 4 and 5).

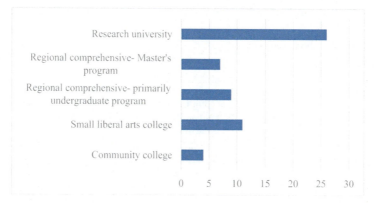

Fig. 4 Type of institutions at which the authors are/have been employed (the sum of the responses is greater than the number of authors because some authors have been at more than one institution)

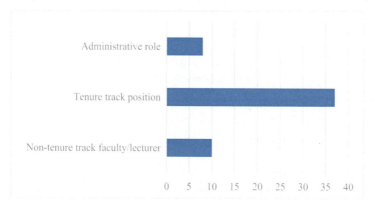

Fig. 5 Types of positions authors have held (the sum of the responses is greater than the number of authors because some authors have held more than one type of position or role)

How Can I Find a Position If My Partner or Spouse Is Also Looking for a Position?

This situation is sometimes referred to as the "two-body problem," an inelegant term describing the dilemma that couples have in finding desirable positions for each person that are geographically close enough that they can continue to live together. It is particularly prevalent when one or both people in the relationship are in academia. Given the shortage of full-time academic positions, couples are often put in a situation where they have to choose between serious underemployment for one of them and living separately. It is not limited to academics, however, but it does seem to affect couples/partners who are both specifically and highly trained. Of our authors, ten report experiencing a "two-body problem" including Stacey Lowery Bretz, Amber Charlebois, Elizabeth Dorland, Cheryl Frech, Jennifer Heemstra, Isabel Escobar, Cecilia Marzabadi, Sherine Obare, Gail Webster, and Catherine Welder.

Table 4 Guide to chapters based on of the types of institutions at which authors are/have been employed

Authors who have been employed at:	
Community colleges (4)	Elizabeth Dorland, Amina El-Ashmawy, Judith Iriarte-Gross, Sunghee Lee
Small liberal arts colleges (11)	Amber Charlebois, Mary Crawford, Elizabeth Dorland, Judith Iriarte-Gross, Sunghee Lee, Emily Niemeyer, Patricia Redden, Rachel Rigsby, Gail Webster, Catherine Welder, Leyte Winfield
Regional comprehensive—primarily undergraduate programs (9)	Pamela Brown, Amanda Bryant-Friedrich, Stacey Lowery Bretz, Amber Charlebois, Renée Cole, Cheryl Frech, Judith Iriarte-Gross, Darlene Taylor, Kim Woznack
Regional comprehensive—master's programs (7)	Stacey Lowery Bretz, Pamela Brown, Jani Ingram, Judith Iriarte-Gross, Sherri Lovelace-Cameron, Saundra Yancy McGuire, Gail Webster
Research universities (26)	Stacey Bent, Stacey Lowery Bretz, Amanda Bryant-Friedrich, Renée Cole, Mary Crawford, Isabel Escobar, Jennifer Heemstra, Judith Iriarte-Gross, Margaret Kanipes, Sunghee Lee, Mindy Levine, Nancy Levinger, Saundra Yancy McGuire, Jin Montclare, Ingrid Montes, Janet Morrow, Sherine Obare, Renã Robinson, Omowunmi Sadik, Danielle Tullman-Ercek, Idalis Villanueva, Michelle Ward, Gail Webster, Sherry Yennello

Table 5 Guide to chapters featuring authors who have held non-tenure-track positions or administrative roles

Type of positions held by authors:	
Non-tenure-track position (10)	Pamela Brown, Amber Charlebois, Elizabeth Dorland, Amina El-Ashmawy, Judith Iriarte-Gross, Saundra Yancy McGuire, Idalis Villanueva, Michelle Ward, Gail Webster, Catherine Welder
Administrative role (department chair, dean, or vice president) (17)	Stacey Bent, Pamela Brown, Amanda Bryant-Friedrich, Mary Crawford, Isabel Escobar, Cheryl Frech, Margaret Kanipes, Sunghee Lee, Cecilia Marzabadi, Saundra Yancy McGuire, Ingrid Montes, Emily Niemeyer, Sherine Obare, Patricia Redden, Gail Webster, Leyte Winfield, Kim Woznack

What Kinds of Family Structures Work for Being a Mom and Professor?

The diversity of family structures found in American life is also reflected in the stories collected in this book. While many women raise children with a husband, this is not the only structure that works, which is true in general and for academic careers. Women have combined motherhood and a successful career as a chemistry professor

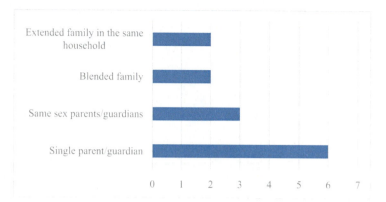

Fig. 6 Distribution of types of family structures among the authors

Table 6 Guide to chapters featuring authors with various family structures

Structure of the authors' families:	
Single parent/guardian (6)	Stacey Bent, Judith Iriarte-Gross, Ingrid Montes, Patricia Redden, Michelle Ward, Leyte Winfield
Same sex parent/guardians (3)	Mary Crawford, Sara Mason, Emily Niemeyer
Blended family (2)	Stacey Bent, Darlene Taylor
Extended family in same household (2)	Judith Iriarte-Gross, Sunghee Lee

with many different family structures. The stories in this collection include women who were single parents, part of same-sex couples, or lived with extended families. Some of these women have also experienced changes to their family structures over time (Fig. 6 and Table 6).

What Do I Do About Childcare?

Issues involving childcare are discussed by many of the mothers in this book. In 2001, the American Association of University Professors (AAUP) adopted a statement regarding "Principles on Family Responsibilities and Academic Work" [13]. The 2001 document states that, although many institutions realize that child care is a pressing issue facing faculty, few institutions offer on-site childcare or subsidize the cost of childcare. AAUP recommends that institutions commit to provide "quality childcare for the children of faculty and other academic professionals." Moreover, having "just-in-time" care for sick children or emergency situations is yet another tangible way that institutions can recruit and retain faculty.

Childcare costs, especially infant care, can often be one of the largest expenses for families and is often higher than the cost of a college education [14]. Because of this, AAUP further recommends institutional support for subsidized daycare. This is particularly important since women remain disproportionately employed as

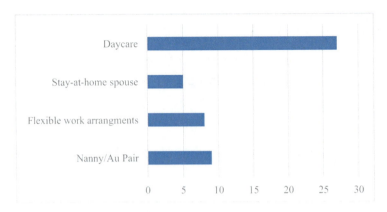

Fig. 7 Distribution of childcare strategies used by the authors

Table 7 Guide to chapters featuring authors using various childcare strategies

Childcare strategy used	
Nanny/Au Pair (9)	Pamela Brown, Amanda Bryant-Friedrich, Amber Charlebois, Cheryl Frech, Jani Ingram, Mindy Levine, Cecilia Marzabadi, Nancy Levinger, Renā Robinson
Flexible work arrangements (8)	Mary Crawford, Amina El-Ashmawy, Jin Montclare, Janet Morrow, Patricia Redden, Rachel Rigsby, Danielle Tullman-Ercek, Catherine Welder
Stay-at-home spouse (5)	Renée Cole, Amina El-Ashmawy, Rachel Rigsby, Catherine Welder, Kim Woznack

contingent faculty (57% of women are in lecturer, instructor, unranked positions) and at community colleges [13]. Subsidizing the high cost of childcare for these typically lower salaried positions, as well as for graduate students and postdocs, could be a way for women to remain in the academy.

Since it is often the case that academics take positions in locations far from the support of extended family, their need for other childcare strategies is amplified. Many authors navigated the difficult situations that can arise from being far away from their extended families (Fig. 7 and Table 7).

What If I Face a Hostile Work Environment?

Although we cannot provide legal advice, you should be aware of how a hostile work environment is defined and initial steps to take if you feel you are in one. "A hostile environment can result from the unwelcome conduct of supervisors, co-workers, customers, contractors, or anyone else with whom the victim interacts on the job, and the unwelcome conduct renders the workplace atmosphere intimidating, hostile, or offensive" [15].

Introduction

A hostile work environment could exist in the form of sexual harassment, as Linda Wang and Andrea Widener describe in their article, "Confronting sexual harassment in chemistry." They report that the chemistry community has the same discouraging sexual harassment problems as the larger science community and the nation [16]. In 2017, the National Academies Committee on Women in Science, Engineering, and Medicine formed an ad hoc committee to study the impact of sexual harassment on career advancement of women in academia [17].

An entirely distinct (nonsexual in nature) form of hostile work environment, as defined by the US Equal Employment Opportunity Commission, includes conduct that creates a work environment that would be intimidating, hostile, or offensive to a reasonable person. This is often difficult to prove, because in order to be considered illegal, the conduct must be so objectively offensive as to alter the conditions of the individual's employment [18]. It can be described as constant disregard, disrespect, and non-inclusivity with reoccurring degrading and derogatory comments and consistent assignment of menial tasks. Additionally, the concept of "any reasonable person" is quite vague and very difficult to measure.

If you feel you are in a hostile work environment and experiencing discrimination, you should take steps for your protection. Document each conversation, interaction, and experience you consider hostile and identify people who have witnessed the treatment. Utilize your network (mentors/coaches/champions/advocates, both inside and outside the university) for support, strength, and guidance during this challenging and frustrating time. Tap into your human resources department, university administration, Title IX representative, and/or legal counsel.

At least one of our authors has experienced a hostile work environment based on gender. Amber Charlebois shares some of her story in her chapter.

How Do I Find Work/Life Satisfaction?

The Wikipedia definition of work-life balance (noun), the more commonly used term, is the division of one's time and focus between working and family or leisure activities. This balancing act is ongoing and presents unique challenges for those in academic careers.

Parents add the challenges of childcare and family obligations. In the article, "Balancing Parenthood and Academia," the term work/family conflict is used [19]. The three types of conflict addressed in the article are time-based, strain-based, and behavior-based conflicts.

We have embraced the term work/life satisfaction. Even if you feel conflicted or off-balance, you can be satisfied. This collection of personal accounts provides numerous unique approaches to combining one's work and one's family life. The authors share times when there was a nice and stable balance, as well as their stories of when this became more of a conflict, where decisions were made and sometimes sacrifices were made. Work-life satisfaction is something evident in every story here. You will find unique, creative, and inspirational examples of how to find satisfaction with your work and life.

How Do I Find Mentors?

Given the importance that mentoring plays in professional and personal development, it is wise to have relationships that fulfill a range of needs. Mentoring helps identify the skills, knowledge, and experience you need to be successful. It helps ease transitions [20].

Consider what you need—inspiration, guidance, insightful questions—as you seek out those who can serve as role models, career coaches, and sources of socio-emotional support. You will also want to identify those who can serve as advocates and champions. Having clear expectations will help frame requests and make the most of your interactions.

Like other relationships, mentoring relationships develop in different ways and take time. Sometimes these relationships develop naturally. Other times you may have to initiate the initial contact. This may occur in the course of everyday activities, virtually, or at events, such as conferences, short courses, business meetings, or social gatherings. Those in your network may refer you to someone. Once you have made the initial contact, be sure to follow up with clear expectations [21].

Preston (2004) has shown that mentoring among female scientists appears to lead to a greater likelihood of completing graduate school and of successfully attaining employment. Preston's studies also suggest that mentoring does not seem to have the same beneficial effect for male scientists, whose success rates are similar regardless of the presence of a strong mentor [21]. In addition, women without mentors find it more difficult to make contacts with the recognized authorities in their fields and do not have critical advocates to push for their promotion, tenure, and nomination for awards [22].

Research on mentoring is investigating how "these complex and dynamic relationships form, evolve, and impact the lives and careers of the current and next generation" [23]. Many of the authors describe mentoring they sought and received at different stages of their careers and how their relationships played important roles in their lives.

What Are the Policies that Impact Motherhood in Academia?

It is recognized nationally that structural and systematic changes must be made to support women in academic positions in STEM field. The National Science Foundation (NSF) funds a program, "Increasing the Participation and Advancement of Women in Academic Science and Engineering Careers" (ADVANCE) [24].

The Virginia Tech ADVANCE portal website classifies funded ADVANCE projects or approaches into a few categories: individual or interpersonal (advice, leadership development, mentoring, negotiation, networking, on ramps, professional development, tenure, and promotion), departmental (recruitment, resources), institutional (climate and culture, glass ceiling, pipeline, policies, and retention), and general (initiatives, study, theater) [25].

There are many institutions that have already adopted policies designed to support academic mothers. The Center for WorkLife Law based at University of California Hastings College of the Law published a best practice guide titled, "Effective Policies and Programs for Retention and Advancement of Women in Academia" [26]. This guide mentions policies and programs such as:

- Stopping the clock
- Opting-out policies instead of opting-in
- Dual Career Support
- Mentoring
- Networking
- Convenient, affordable childcare
- Part-time tenure-track positions

Policies and programs focused on these and other areas, such as safety and work environment, are part of the collective work that needs to be done to support mothers and professors, current and prospective.

Making structural and systematic changes requires collective, persistent, and thoughtful efforts. Our experiences and insights should inform effective policies and programs. These policies need to be designed and implemented in ways that:

- Shift behaviors and develop departmental, institutional, and disciplinary cultures where our questions and needs can be addressed in broader contexts, rather than in isolation;
- Reflect our unique circumstances and help us determine appropriate steps, rather than dictate; and
- Provide the ongoing support needed for work–life satisfaction as we navigate individual, departmental, institutional, and disciplinary contexts.

Our efforts are a key part of a building a more effective, safer, and more diverse academic landscape. Sharing our stories and reflecting on our experiences, whether or not they are guided and supported by policies and programs, will help us move from individual to collective action. With its contributions from 40 women, the second edition of "Mom the Chemistry Professor" conveys that the roles of mother and professor are not diametrically opposed. It also positions us all to advocate for changes that facilitate the pursuit of our personal and professional goals.

Disclaimer The personal accounts contained in this manuscript have been contributed by a collection of individuals from various different backgrounds and institutions. This manuscript is intended to serve as an inspiration for those women who may be considering balancing a chemistry career in academia with motherhood and thus this book does not purport to specify career advice, safety advice, or legal advice or to represent the policy of the American Chemical Society. No warranty, guarantee, or representation is made by the American Chemical Society as to the accuracy or sufficiency of the information contained herein, and the Society assumes no responsibility in connection therewith. Registered names, trademarks, etc., used in this publication, even without specific indication thereof, are not to be considered unprotected by laws.

References

1. Physics Graphs and Statistics, excerpts from IPEDS (2015) American Physical Society, College Park, MD. https://www.aps.org/programs/education/statistics/womenmajors.cfm. Accessed 4 Feb 2018
2. Survey of Earned Doctorates (2016) National Science Foundation, Alexandria, VA. https://www.nsf.gov/statistics/2018/nsf18304/data.cfm. Accessed 4 Feb 2018
3. Demographics Data (2015) OXIDE, Baltimore, MD. http://oxide.jhu.edu/2/demographics. Accessed 4 Feb 2018
4. Wang L (2016) Women crack the academic glass ceiling. Chem Eng News 94(36):18–21
5. Crenshaw K (1991) Mapping the margins: intersectionality, identity politics, and violence against women of color. Stanford Law Rev 43(6):1241–1299
6. Saraswati LA, Shaw BL, Rellihan H (eds) (2018) Introduction to women's, gender & sexuality studies: interdisciplinary and intersectional approaches. Oxford University Press, New York
7. Collins PH, Bilge S (2016) Intersectionality. Wiley, Cambridge
8. Mason MA, Wolfinger NH, Goulden M (2013) Do babies matter? Gender and family in the ivory tower. Rutgers University Press, New Brunswick
9. Wang L (2016) When chemists struggle with infertility and miscarriage. Chem Eng News 94 (6):33–35
10. Hill RH, Finster DC (2016) Laboratory safety for chemistry students. Wiley, Hoboken
11. Wolf-Wendel LE, Ward K (2006) Academic life and motherhood: variations by institutional type. High Educ 52:487–521
12. Ward K, Wolf-Wendel LE (2012) Academic motherhood: how faculty manage work and family. Rutgers University Press, New Brunswick
13. American Association of University Professors (2001) Statement of principles on family responsibilities and academic work. https://www.aaup.org/AAUP/pubsres/policydocs/contents/workfam-stmt.htm. Accessed 17 Feb 2018
14. Childcare Aware America (2017) Parents and the high cost of childcare 2017 Report. https://usa.childcareaware.org/wp-content/uploads/2017/12/2017_CCA_High_Cost_Report_FINAL.pdf. Accessed 18 Feb 2018
15. United States Department of Labor (2011) What do I need to know about...Workplace harassment. https://www.dol.gov/oasam/programs/crc/2011-workplace-harassment.htm. Accessed 18 Feb 2018
16. Wang L, Widener A (2017) Confronting sexual harassment in chemistry. Chem Eng News 95 (37):28–37
17. National Academies of Sciences, Engineering & Medicine (2017) Impacts of sexual harassment in academia. http://sites.nationalacademies.org/pga/cwsem/shstudy/index.htm. Accessed 18 Feb 2018
18. US Equal Employment Opportunity Commission (2018) Harassment. https://www.eeoc.gov/laws/types/harassment.cfm. Accessed 12 Feb 2018
19. O'Laughlin EM, Bischoff LG (2005) Balancing parenthood and academia: work/family stress as influenced by gender and tenure status. J Fam Issues 26(1):79–106
20. Grant CS (2015) Mentoring: empowering your success. In: Pritchard PA, Grant CS (eds) Success strategies for women in STEM2nd edn. Elsevier, London, pp 63–96
21. Preston AE (2004) Leaving science: occupational exit from scientific careers. Russell Sage Foundation, New York
22. Farley J (1990) Women professors in the USA: where are they? In: Stiver Lie S, O'Leary VE (eds) Storming the tower: women in the academic world. Nichols/GP Publishing, New York, pp 194–207
23. National Academies of Sciences, Engineering, and Medicine 2017 Effective mentoring in STEMM: practice, research, and future directions: Proceedings of a workshop—in brief. The National Academies Press, . Washington, DC. https://www.nap.edu/catalog/24815/effective-mentoring-in-stemm-practice-research-and-future-directions-proceedings

24. National Science Foundation (2016) Increasing the participation and advancement of women in academic science and engineering careers (ADVANCE). https://www.nsf.gov/funding/pgm_summ.jsp?pims_id=5383. Accessed 18 Feb 2018
25. Virginia Tech (2018) ADVANCE portal. http://www.portal.advance.vt.edu. Accessed 18 Feb 2018
26. The Center for WorkLife Law (2013) Effective policies and programs for retention and advancement of women in academia. http://worklifelaw.org/publications/Effective-Policies-and-Programs-for-Retention-and-Advancement-of-Women-in-Academia.pdf. Accessed 18 Feb 2018

Motherhood and Academic Chemistry Laboratories: Safety and Career Challenges from a Qualitative Study

Megan Grunert Kowalske

Photo Credit: Western Michigan University

As more and more women enter the workforce, issues regarding childbearing and caretaking have become more prominent. Challenges women face with the timing of pregnancy, finding affordable childcare, and meeting the multitude of demands inherent with raising children are starting to be addressed by many employers. In the university setting, structural and cultural factors pose barriers yet also provide almost unheard-of flexibility.

M. G. Kowalske (✉)
Department of Chemistry & The Mallinson Institute for Science Education, Western Michigan University, Kalamazoo, MI, USA
e-mail: megan.kowalske@wmich.edu

While some of these challenges are being addressed by employers, funding agencies, and communities, one area of particular concern is safety for women chemists. There are hazards associated with working with chemicals, and the risks associated with them may be higher during pregnancy and breastfeeding. There is no easy way of assessing and minimizing the risks.

This chapter will review the safety concerns for women chemists, current policies and practices for dealing with safety challenges, experiences of women faculty, and considerations for departments and administrators.

Editors' Note

Megan's study, like the collection of stories in the chapters that follow, plays an important role in:

- Conveying the varied and complex circumstances that women chemists navigate
- Informing the choices we make and the approaches we consider
- Inspiring us to address the challenges associated with being a mother and pursuing an academic career

Since Megan collected and reported this data, the emphasis on safety and availability of safety resources has increased. We highlight some of these resources throughout the chapter along with safety tips and recommendations. We have also included some key questions for thought and discussion with others on your campus that are raised while reading some of the stories shared here.

Research Design

The data presented here are drawn from a larger qualitative study exploring women's career choices in chemistry after obtaining a doctorate [1]. This study, conducted in 2008–2009, investigated the perceptions and motivations that led women with doctorates in chemistry to select their careers. The study included interviews with senior graduate students, faculty members at research institutions, and faculty members at primarily undergraduate institutions.

One of the themes that emerged during the data analysis process was the challenge associated with working in an academic chemistry research laboratory. The women faculty members discussed potential exposure to chemicals while pregnant and breastfeeding, the lack of policies and procedures in place to help them navigate decisions related to chemical exposure in the laboratory, and the impact of their decisions on their careers and colleagues.

The data presented here are excerpts from the faculty interviews highlighting the concerns these women expressed about working in chemistry laboratories and approaches they used to navigate their roles as researchers, educators, and mothers. These data, the literature, and publicly available safety information indicate the need for departments and administrators to update policies and implement practices to support women in their decisions regarding chemical exposure while pregnant and breastfeeding.

Safety Concerns for Mothers in Chemistry

Chemists are familiar with safety training, and highly publicized cases, including a fatal accident at UCLA [2], have recently led to an increased examination of safety practices and appropriate training. Often not addressed in safety training are issues for pregnant and breastfeeding women. Obviously, many hazards are directly associated with the chemicals used in the laboratory setting. Many organic solvents, heavy metals, and other hazardous compounds pose risks to developing fetuses and infants. Safety data sheets (SDS)[1] provide information regarding hazards known at the time the SDS is published[2], but what about the hazards not yet known? Even when a hazard is known, SDS do not make recommendations on personal protection or exposure levels for pregnant mothers. Without a sense of the hazards, it is much more difficult to assess the risks and take appropriate actions.

> **Editors' Safety Tip: Differentiate Between Hazard and Risk**
> Understanding the differences and relationships between these two terms will help frame safety discussions and risk assessments. According to the American Chemical Society,[3]
>
> hazard is a potential source of danger or harm, and
> risk is the probability of suffering harm from exposure to an unsafe situation.
>
> The relationship, according to the Toxicology Education Foundation, is the following[4]:
>
> $$Risk = Hazard \times Exposure$$
>
> In other words, reducing the exposure to hazards, especially hazards that may not be yet known or well characterized, reduces risk.

---

[1]The Globally Harmonized System of Classification and Labeling of Chemicals (GHS) was developed by the United Nations (UN) to better align adopting countries with respect to chemical hazards. Countries that have adopted the GHS require safety data sheets (SDS) in step with GHS expectations. These SDS replace the former materials safety data sheets (MSDS) system. The SDS are required to provide information on the identity, composition, and the physical and chemical properties of the substance. Additionally, the SDS provide information on the hazards, measures for addressing fire or accidental release, exposure controls and personal protection, and toxicological information. For more information about the GHS SDS, see the US Occupational Safety and Health Administration Brief on SDS (https://www.osha.gov/Publications/OSHA3514.html).

[2]It is important to note that SDS are not guaranteed to be correct and each contains a disclaimer at the end. It is always wise to search for other reliable sources for additional chemical information, such as the National Institute for Occupational Safety and Health (NIOSH) pocket guide (https://www.cdc.gov/niosh/npg/search.html).

[3]https://www.acs.org/content/acs/en/chemical-safety/ramp.html

[4]http://toxedfoundation.org/hazard-vs-risk/

One of the greatest challenges for women chemists who are pregnant or breastfeeding is the lack of standardized information. The Centers for Disease Control (CDC), the National Institutes of Health (NIH), the Environmental Protection Agency (EPA), and some university environmental health and safety offices have publicly available information about hazards for pregnant workers; however, this information does not include all chemicals used in research. These databases cover common hazardous chemicals, medications, and illegal substances. The Agency for Toxic Substances and Disease Registry (www.atsdr.cdc.gov/toxfaqs/index.asp), within the CDC, maintains a list of fact sheets addressing frequently asked questions about many chemicals. As with many references, it is not overly helpful with regard to pregnancy and breastfeeding. Consider the entry for acetone, where the last paragraph under the heading "How can acetone affect my health?" says, "Health effects from long-term exposure are known mostly from animal studies. Kidney, liver, and nerve damage, increased birth defects, and lowered ability to reproduce (males only) occurred in animals exposed long-term." For a woman deciding whether it is safe to be around acetone while pregnant, this is so vague as to not be helpful. Similarly, the page for mercury includes the statement, "Exposure to high levels of metallic, inorganic, or organic mercury can permanently damage the brain, kidneys, and developing fetus." It is unclear what "exposure to high levels" means, although the outcomes are clearly severe. The National Institute for Occupational Safety and Health, part of the CDC, includes a statement in their document The Effects of Workplace Hazards on Female Reproductive Health, about exposure to hazardous chemicals during pregnancy, "Whether a woman or her baby is harmed depends on *how much* of the hazard they are exposed to, *when* they are exposed, *how long* they are exposed, and *how* they are exposed" (http://www.cdc.gov/niosh/docs/99-104/pdfs/99-104.pdf, p. 14). Again, it is a less than helpful warning for women chemists.

While working in a research laboratory, women have to look up possible hazards on the SDS or through other available reference lists for any chemicals they are using. Either working with a safety officer or on their own, they then need to consider how risks could be minimized and make a decision about whether they want to expose a developing fetus or breastfeeding infant to these chemicals. Often, these decisions are strained by a lack of information about what specifically poses risk, what levels are dangerous, how mobile chemicals are in relation to skin and cell barriers, and what personal protective equipment would help and how effective it is. This can effectively put women chemists in a position of choosing their career or the health and development of their child(ren).

Editors' Safety Tip: Initiate the Risk Assessment Process Early
If you are pregnant or trying to get pregnant, a variety of people can help identify hazards, assess the risks associated with them, and determine appropriate approaches for minimizing them.

(continued)

- Environmental health and safety (EHS)[5]
- Human resources
- Your research advisor or department chair
- Trusted colleagues

If you are early in your pregnancy and you have not shared this information publicly within your department, be sure to ask for privacy and confidentiality.

Current Policies and Practices for Dealing with Safety Challenges

Another major challenge for women chemists is the limited policies and established practices associated with working in the laboratory while pregnant or breastfeeding. Publicly available recommendations are few and far between, and those that are available are vague. They leave the determination of safety up to the individual, putting each person in the challenging position of conducting risk assessments and determining appropriate courses of action with incomplete or minimal information and guidance.

For example, Virginia Tech has a public statement on Safe Pregnancy for Laboratory Workers (http://www.chem.vt.edu/facilities/resources/safety-pregnancy-accommodation.pdf). It is just over a page long and includes the following paragraph:

> The Chemistry Department seeks to minimize the risks of working in its laboratories for all employees and students, especially for pregnant women because of the known sensitivity of the fetus to specific chemicals, in particular teratogens. All laboratory workers are expected to know the hazards of chemicals they are using, including the pregnant woman. Material Safety Data Sheets (MSDS) are essential, but may not provide a complete set of recommendations. Additional protective equipment may be available, but alternatives to laboratory work such as spectroscopic or computational studies, library work, writing, or seminar preparation may be requested by the pregnant laboratory worker. Each woman's situation will be different, so the Department can be creative and flexible. We encourage a pregnant woman to consider those accommodations that she might request for her well being, and for the well being of her fetus.

Also in this statement are recommendations for pregnant graduate students, encouraging them to consult with their research director or graduate program director regarding questions they may have. In many cases, research and graduate program directors are likely to have few answers and may not be able to provide answers that graduate students are seeking. This Virginia Tech statement mentions students, but it is contained within a statement for laboratory "workers." Students working in classroom laboratories may have even less guidance and a higher degree

---

[5]Depending on the size of your institution and the degree to which chemical research is emphasized, this may be an office of only a single person, or it could be a much larger department with several staff members.

of vulnerability based on their career stage. Students may want to consider seeking out a professor for assistance in navigating this landscape.

> **Editors' Safety Tip: Using Your Networks to Answer Your Questions**
> Turn to a variety of places and people for information and advice that will help you develop your own answers. Each campus has a person, team, or office responsible for environmental health and safety. Tap into them, as well as the networks of other women on campus and in your field. Collectively, they can help you determine the answers appropriate for your circumstances.

Also available online is a website from The University of California, San Francisco Office of Research: Environmental Health and Safety (http://ehs.ucsf.edu/faq-pregnant-workerstrainees-research-laboratory-setting). It provides the following statement regarding pregnant women working with chemicals:

> Pregnant workers should avoid unnecessary exposure to chemicals. Since the beginning of the 20th century, thousands of new synthetic chemicals have been developed, and only a small portion of these chemicals have been adequately studied to determine whether they pose a risk of cancer or birth defects. Therefore, it is advisable to limit any unnecessary chemical exposure during pregnancy. Some chemicals are well known to increase the risk of cancer or birth defects.

While these types of statements are meant to be helpful, the lack of concrete recommendations can be unsettling for women. There are few chemistry departments or universities that have a statement regarding pregnant workers at all, so it is positive that some statements exist to address the needs of pregnant women in chemistry. It is more common to find recommendations for accommodating graduate students rather than faculty members.

> **Editors' Safety Tip**
> For more information on limiting unnecessary chemical exposure and to read some advice from an environmental health and safety professional, please see the introductory chapter of this book.

Currently, the American Chemical Society does not have a specific published document or website regarding reproductive hazards in the chemical workplace. The ACS Committee on Chemical Safety website lists a few resources pertaining to reproductive hazards (https://www.acs.org/content/acs/en/about/governance/committees/chemicalsafety/safetypractices.html).

Editors' Safety Tip: Building an Academic Safety Culture

Conversations about pregnancy and breastfeeding may be facilitated when students, faculty, staff, and administrators incorporate chemical and laboratory safety into everyday practice. The report Creating Safety Cultures in Academic Institutions identifies seven key *elements of strong safety cultures*:

1. **Leadership and management**—Responsibilities and accountability for safety must be clearly defined from the highest levels of the administration, down through schools and departments, to individual faculty and staff. Leading by example is a requirement of all members of an institution, especially faculty.

2. **Teaching laboratory and chemical safety**—Every chemist needs an in-depth knowledge of laboratory and chemical safety. He or she should acquire this safety knowledge and education continually through years of educational process. The report suggests about 80 safety topics that should be taught, and it emphasizes teaching "critical thinking" skills in laboratory and chemical safety.

3. **Strong safety attitudes, awareness, and ethics**—Strong positive attitudes about safety require long-term efforts through continuous emphasis on safety. Teaching (and learning) safety is an ethical responsibility. The safety ethic reflects the proper attitude of valuing safety.

4. **Learning from laboratory incidents**—Studies of incidents capture interest and teach lessons about safety. Institutions should implement a system of reporting and investigating incidents.

5. **Establishing collaborative relationships**—Safety culture requires close, trusting collaborations among all members of the academic community—including faculty members, administrative staff, students, postdoctoral scholars, and environmental health and safety staff—as well as public emergency responders.

6. **Promoting and communicating safety**—Demonstrating safety practices through personal example and recognizing positive safety behaviors are important ways to promote safety. Safety should be reinforced through continuous and diverse efforts.

7. **Strong safety programs require funding**—All strong safety programs require investment of substantial effort with adequate and continuous funding by institutional administrations. Identifying responsibilities for safety is a critical step in determining budgets.

https://www.acs.org/content/acs/en/education/students/graduate/creating-safety-cultures-in-academic-institutions.html

It can be highly challenging to find information to support women trying to make decisions about safety during pregnancy and breastfeeding so that they can advocate for their own needs.

> **Editors' Safety Tip: Know You Are Protected**
> Faculty members or postdoctoral associates employed by a university are protected by the Pregnancy Discrimination Act (PDA). For more information on the PDA, please see the US Equal Employment Opportunity Commission (EEOC) website on Pregnancy Discrimination (https://www.eeoc.gov/eeoc/publications/fs-preg.cfm).
>
> Undergraduate and graduate students are legally protected from discrimination on the basis of pregnancy status under Title IX (https://www2.ed.gov/about/offices/list/ocr/docs/dcl-know-rights-201306-title-ix.html).
>
> If you feel that these policies are being violated on your campus, there are places you can go to find help. The office or title of the people to contact may vary from campus to campus, but some suggestions include the Office of Human Resources, Institutional Equity or Social Equity, the Ombudsman's Office, the Title IX officer, and the office for Equal Education and Employment Opportunities.

Approaches Used by Women Faculty to Address Safety Challenges

The impact of motherhood for women chemists is different depending on the type of institution, the teaching load, research expectations, and service responsibilities. The challenges are somewhat different at primarily undergraduate institutions (PUIs) and research-intensive institutions.

At research institutions, research productivity can be somewhat maintained by postdoctoral associates and graduate students in the absence of the research advisor/primary investigator (PI). The National Science Foundation (NSF) has also instituted policies allowing for grant funding to be used to hire a lab technician or for the grant to stop for a year.

At PUIs, faculty members may have to maintain an undergraduate research program. In the absence of graduate students or postdocs, it falls to the PI to train student researchers and work with them in the lab. Thus, a pregnancy can force PIs out of the lab, essentially halting research progress. This is a challenge after pregnancy as well, when the PI tries to restart her research program.

Experiences of Faculty Members at Predominantly Undergraduate Institutions

One PUI faculty member, Ellen, took 5 years off from undergraduate research during two pregnancies due to the use of a mercury bubbler. While her department adjusted her workload by removing the research component and adding instructional and service responsibilities, this did not address the stalled research progress or the difficulty in restarting her research program. When telling her story, she said:

> ...I'm back this year for the first time basically in. . .five years, because I had two kids. So I was out of the lab [for] more or less five years straight, [because of] pregnancies and then breast-feeding, and then trying to get the second one, and. . .so I'm finally back. I was a big believer and didn't go near [the lab] the whole time. . .I'm an organometallic chemist who uses a mercury bubbler! Academic institutions do not understand how to deal with female faculty members, especially in the sciences, who choose to have children. They're just confused with how to deal with that. . . .There's no road map. . .the difficulties of dealing with chemicals and the campus was very, very confused about . . .how to deal with that. That's probably been the biggest challenge. . . .I picked up other teaching duties, I took responsibility of making sure that all the general chemistry laboratories were [updated], the lab manual, that became my baby. Writing the whole thing, proofing the whole thing, dealing with our stockroom support on making sure the labs get prepped, that became my job, but then I also picked up additional teaching load, and that I worked out with the dean [to compensate for not being in the research or teaching labs].

As she points out, there is no "road map" for science departments to address the concerns of women chemists who are pregnant and/or breastfeeding. Luckily, she was able to work out a compromise with her overall workload, but she went on to discuss how challenging it was to return to the undergraduate research lab.

Key Discussion Questions

How do we maintain research momentum without being in the lab with students? How do we help women regain research momentum?

Another PUI faculty member, Laura, discussed challenges associated with teaching organic laboratory courses while pregnant. At PUIs, faculty members are often the laboratory instructors, rather than graduate teaching assistants as is standard practice at research universities. She discussed feeling uncomfortable being in the teaching lab while breastfeeding as well as being limited in the research lab:

> I think the number one thing is that if you're an organic chemist, and, really I would say a wet chemist, I don't think it applies to all fields, but it's a challenge, child-bearing. As a chemist, you know, you're doing wet chemistry or using things sometimes that you don't wanna expose your kids to. If you teach labs, then that's nine months that you're technically not teaching organic lab. In terms of, breast-feeding your child, I was in the lab when I was doing that, and every day I felt sick [for] my child. Subsequently the other female faculty did not teach with their second children, in the lab, while they were breastfeeding. So really that pretty much gives you two academic years almost where you are not teaching the lab, and it will limit the time you spend actually in the research lab also. I think time-wise, certainly at an undergraduate institution you don't have to work as long [in terms of] hours, but in terms of the impact child-bearing [has] on the ability to do your job, [it] is entirely different.

Her concerns about how pregnancy and breastfeeding impact female professors' careers echo those stated earlier. She discussed feeling an obligation to uphold her teaching responsibilities, but later wished she had worked out an alternate solution. Many women, especially graduate students, untenured faculty members, and temporary instructors, may have felt pressured to put themselves in positions where they felt uncomfortable because of the expectations or attitudes of others they work with. Women should know that they are not alone.

> **Key Discussion Questions**
>
> How do we learn from other's experiences?
> How do we help develop and select solutions appropriate for different women in different circumstances?
> What are alternatives to help pregnant or breastfeeding women whose regular duties are in the teaching laboratory?

> **Editors' Note**
> While this faculty member was honest with the chapter author about her experience, we want to emphasize that no person should continue to work in a laboratory situation in which they feel that the chemical exposure risk is too great. This would apply to any person regardless of their gender or pregnancy status. Those who are asked to work in an environment where they feel exposed to risk with which they are not comfortable should begin a dialog with the supervisors and administrators with whom they work.

While Laura's colleagues changed their post-pregnancy plans based on her experiences, another faculty member, Irene, worried about how her decision to not take maternity leave would affect her colleagues. She had no regrets about not taking time off, as her department was understaffed at the time and she worried about overburdening her colleagues. Irene did not express the same safety concerns as Laura and Ellen, but she is a plant biochemist and believed the chemicals she worked with safe to be handled during pregnancy. She described her experience, saying:

> I think being one of the first faculty women, there's also pressure that you do it [and] do it right, so that you set a good precedent. I'm somewhat concerned about the precedent I've set with having kids since I didn't really have maternity leaves for two of the three, we did not hire a replacement for me, so that put pressure on my colleagues. And I'm now in the position [where] younger women faculty ask for advice. Am I giving good advice? I don't know. Well, I can tell you what I did, [but] I don't know if it's the best answer, or what I would do differently.

As seen previously, she expressed doubt about her decisions. Lack of clear policies, recommendations, and precedents makes navigating pregnancy and breastfeeding challenging for women chemists.

> **Key Discussion Questions**
>
> How do we advocate for other women even if their choices are different from our own?
> How can we feel more comfortable being role models for other women on our campus?

Experiences of Faculty Members at Research-Intensive Institutions

At research-intensive institutions, women expressed more concern about the timing of pregnancy with regard to tenure rather than safety concerns with chemicals. They commented on a lack of structure or support within the university regarding maternity leave and arranging teaching assignments, concerns regarding postponing pregnancy until after tenure, and having fewer children than they would have liked due to their careers.

Petra, a professor at a research-intensive university, discussed some of the challenges she specifically had while pregnant. She commented on the university structure at large and a general lack of support for mothers. She believed the hurdles for mothers in the university setting were part of the reason women left research institutions. She said:

> ...teaching assignments when I was pregnant, arranging maternity leave, there were just lots of discussions and issues that made these arrangements not ideal, so I think there is a lot of work to be done, [and] this was a little bit a cause of grief...there is so much to be done to make this job for a woman more human and more balanced, such as, dealing properly with day care situations, not only when a woman gets pregnant, but also when the baby's born, some facilities and even time off...I think if the structures and the system and colleagues and the whole machinery were more conducive to really understanding the needs of women or that family, and make sure that they can spend time with kids, perhaps women would stay.

Key Discussion Questions

How can we advocate for better university policies and structures to support pregnant faculty members and new mothers?

What mechanisms are in place at our university to support faculty parents and families?

Danielle, an associate professor at a research-intensive university, regretted postponing starting a family. She talked about her decision, saying:

> I think we sacrificed quite a bit personally. Like right now, I'm pregnant, we're having a baby, it's really exciting, but, sometimes you feel like, I'm 36, right?, Why did I wait this long?...I think we did make a lot of sacrifices personally, and I think we're [at] a stage right now where, granted, we've got to keep everything going and moving forward in the lab, but, I think we just need to take a breath here and assess what's going on... 'cause, you can't get that time back and you have to sit and realize, I think I've gotta physically make time for things other than this [job].

She chose to wait until she was through the tenure process, which despite her regret, did allow her some flexibility in terms of exposure to chemicals. She felt her research group was at a place where it was relatively self-sustaining, allowing her to minimize her time in the lab while pregnant.

Marie, an assistant professor, expressed similar concerns regarding the timing of starting a family and the tenure process:

> ...by the time I get tenure, I'll probably be 36? And I feel, if I waited until I got tenure, and then I got tenure, and then I had kids, then everything would be great. But, if I waited until I got tenure, and then I didn't get tenure, then I would have put off something that was really, really important to me and then I wouldn't have [kids] and I wouldn't have a job, so then I think I would be even more upset.

Interestingly, Marie did not report that safety issues were her predominant concern, even though her research group works with very hazardous materials. It is believed that this is due to the structural difference at research institutions that will be discussed further.

Petra echoed the challenges with the tenure system and the university structure with regard to women getting pregnant and starting families. She advocated for more supportive and flexible pathways to tenure, saying:

> I don't think a woman should ever compromise [her] biological clock for her job, but that's why I [support] having better structures that would facilitate all aspects of having children, when you're expecting and after you have them. I think [this is] sorely needed, because, frankly, I think women are just discouraged from having kids when they are in [a] tenure-track situation. They are afraid about...the repercussions on their future. I had my second child [before tenure] because I wanted to have a second child and I was not young compared to American students, so it was just either you do it now or you won't do it, but if I had been a little younger, perhaps I would have waited. And some people may think, justly, that it's not fair that you have to wait to have kids because of your job. So I totally sympathize with that.

Catherine, a professor and department chair, discussed how her career and the challenge of finding jobs in the same location for her and her husband led to them having a smaller family than planned. She said:

> ...my husband and I lived apart for a number of years trying to get jobs in the same place, and I would definitely say I had fewer children than I would [liked to] have [had]. I have one daughter, I have fewer children than I would have [had], I think, if I weren't at a research institution...I had one fewer child than I wanted.

Again, the career and institutional structures have changed women's plans for starting and building a family.

> **Key Discussion Questions**
>
> How can we support women who do choose to have children pre-tenure and pre-promotion?
> What policies or practices are in place so that women don't feel discouraged from starting families pre-tenure?

While women at research institutions felt there were significant challenges with regard to being a professor and mother, they did not bring up the issues of safety that were highlighted by women at PUIs. While these safety challenges were not nonexistent, they were less salient for professors at research universities. One primary reason for this difference between faculty concerns is the structure of research labs at PUIs compared to research institutions. At PUIs, faculty maintain the research program continuously. They tend to work with inexperienced and transient undergraduates, which means they are in the research lab training student researchers and maintaining research progress when in between students. At research institutions, postdocs and experienced graduate students can help train new graduate students and undergraduate researchers. This lessens the need for PIs to be physically in the research lab, affording more flexibility during and post-pregnancy.

Additionally, instructional roles are very different between PUIs and research institutions. As noted previously, instructional labs at research universities are usually taught by graduate teaching assistants, particularly at the introductory levels. At PUIs, lab courses may or may not be taught by undergraduate teaching assistants at the introductory level. More likely than not, faculty members need to be present in the instructional laboratory. As a result of this structure, it is again more flexible for faculty at research institutions to avoid contact with hazardous chemicals during pregnancy and breastfeeding. The main concern for women at research institutions was the high research expectations for tenure and deciding to postpone starting a family, whereas for women at PUIs, safety in the presence of hazardous chemicals was the primary concern. If the tenure process at research institutions was seen as more family-friendly and supportive of women, it's likely that safety would become more of a focal point for women at these institutions.

Considerations for Departments and Administrators

As seen here, there are few universally concrete recommendations for women chemists regarding reproductive health, pregnancy, and breastfeeding. This poses a unique challenge for women chemists, as well as departmental and university administrators who must help find accommodations for them.

The hazards posed by many chemicals during pregnancy and breastfeeding are unknown and are likely to remain unknown. It is up to individuals to decide what they are comfortable with in terms of exposure to chemicals during pregnancy and breastfeeding. This is an uncomfortable position for women to be in, as they are trying to make difficult decisions with incomplete data.

From interviews with chemistry professors who are also mothers, it is clear that open communication between faculty members and department chairs is key to addressing safety concerns and making accommodations during pregnancy. It is also imperative to foster an environment of support and community, where women feel they can communicate with administrators and have support from colleagues. There needs to be recognition of the challenges inherent in pregnancy and motherhood across departments and universities so that adequate support systems and accommodations can be implemented.

As seen from faculty members' reports, the challenges to motherhood are partially systemic, due to structural issues with tenure. One of the major challenges is the disruption of research progress, especially during the pre-tenure years. If research progress stops for motherhood, it can be hard to get it restarted. Even stopping the tenure clock does not fix this. Timing is important in the sciences, and breaks from research pose significant challenges in research progress, publishing, and grant awards. This is a very real challenge outside of the safety concerns women chemists face.

Recommendations for Departments and Administrators
As part of building a safety culture, consider the times when women may be pregnant or breastfeeding.

- Highlight the importance of supporting women during pregnancy and breastfeeding.
- Discuss approaches for navigating various circumstances that women may face.
- Develop overarching policies.

When women are pregnant or breastfeeding, be supportive.

- Assist with risk assessments.
- Help determine appropriate approaches.
- Serve as advocates.
- Intervene if needed accommodations are not being made.

Conclusions

In conclusion, pregnancy and breastfeeding pose unique challenges to women chemists. These challenges are exacerbated by incomplete information about chemical hazards, a lack of policies and concrete recommendations, and the tenure system structure. Women have chosen accommodations based on their individual needs and concerns. While it is appropriate to be flexible and meet each individual's circumstances, it is problematic to not have overarching policies or recommendations. It leaves each woman to advocate for herself and be subject to colleagues, chairs, and administrators who may be less than accommodating.

As seen from the reports of women faculty, there are several strategies used to address the safety concerns of working in the laboratory during pregnancy. Finding alternate teaching or service responsibilities and avoiding chemical exposure was a popular choice, but it came at the expense of research progress. At research-intensive institutions, it was more common to postpone starting a family until after tenure based on the rigorous tenure expectations as well as the added flexibility that tenure brings. These women stressed the importance of communication and a supportive chair in finding accommodations they were comfortable with during pregnancy and breastfeeding. Rather than waiting for a faculty member to become pregnant and address it at that time, discussions at the department and university level are encouraged to develop possible strategies. It is recommended that departments include a statement about accommodating pregnancy in their policy statements. This would not only provide clearer strategies for handling pregnancy but would demonstrate a willingness to support female faculty members, a critical element in recruiting and retaining female chemists. Finally, a statement or policy from ACS or other professional organizations is encouraged as a way to recognize the challenges mothers in chemistry face and to open discussion about how departments can best accommodate and support female faculty.

> **Editors' Note**
> In 2017, the importance of safety was emphasized in the ACS Strategic Plan by specifying it in the following core value:
>
> ### Professionalism, Safety, and Ethics
> *We support and promote the safe, ethical, responsible, and sustainable practice of chemistry coupled with professional behavior and technical competence. We recognize a responsibility to safeguard the health of the planet through chemical stewardship.*
>
> ACS also established a new ACS staff position to integrate and expand safety resources, working with the Committee on Chemical Safety, the Division on Chemical Health and Safety, and others across the chemistry community.
>
> The Women Chemists Committee (WCC) also completed a strategic planning process in 2017. Advocacy for women's reproductive safety emerged as a priority for the WCC to conduct over the next 5 years. WCC looks forward to potential collaborations with the Committee on Chemical Safety (CCS) on developing resources for women seeking additional information regarding reproductive health.

Acknowledgments Thank you to all the women who shared their stories and made this work possible.

Disclaimer

The personal accounts contained in this chapter have been contributed by a collection of individuals from various different backgrounds and institutions. This chapter describes the results of a qualitative research study and aims to catalyze individual, departmental, and campus discussions on supporting women's reproductive safety in academic chemistry laboratories. This chapter does not purport to specify career advice, safety advice, or legal advice or to represent the policy of the American Chemical Society. No warranty, guarantee, or representation is made by the American Chemical Society as to the accuracy or sufficiency of the information contained herein, and the Society assumes no responsibility in connection therewith. Registered names and trademarks, etc., used in this publication, even without specific indication thereof, are not to be considered unprotected by laws.

About the Author

Education and Professional Career

2004	BS Chemistry, Biology, Spanish, University of Indianapolis, Indianapolis, IN
2008	MS Chemistry, Purdue University, West Lafayette, IN
2010	PhD Chemistry, Women's Studies Graduate Certificate, Purdue University, West Lafayette, IN
2010–2011	Research and Teaching Postdoc, Iowa State University, Ames, IA
2011–2016	Assistant Professor, Chemistry and Science Education, Western Michigan University, Kalamazoo, MI
2016–present	Associate Professor, Chemistry and Science Education, Western Michigan University, Kalamazoo, MI

Honors and Awards (Selected)

2017	Western Michigan University Excellence in Discovery: Research and External Funding over $1 Million Dollars for Five Years, Principal Investigator 2012–2016
2015	Western Michigan University Excellence in Discovery: Research and External Funding in 2014–2015
2012	Western Michigan University College of Arts and Sciences Teaching and Research Award
2011	The Michigan College Chemistry Teachers Association Stanley Kirschner Award
2009	University of Michigan National Center for Institutional Diversity Emerging Diversity Scholar Citation

Megan's current projects include exploring the identity development and experiences of graduate students from underrepresented minority groups, women chemists' academic and career choices, motivational theories in undergraduate and graduate STEM education, feminist critiques of academic science, experiences of LGBTQ+ scientists, and assessing online STEM undergraduate courses.

References

1. Grunert ML (2010) Women's career choices in chemistry: motivations, perceptions, and a conceptual model. Dissertation, Purdue University
2. Kemsley J (2009) Researcher dies after lab fire. Chem Eng News. American Chemical Society. Available: https://cen.acs.org/articles/87/web/2009/01/Researcher-Dies-Lab-Fire.html. Accessed 16 Feb 2018

The Best Job in the World

Stacey F. Bent

Best Job in the World

I often tell people I have the best job in the world. But more precisely, I have two of the best jobs in the world—being a mom and being a professor—both of which I am convinced work well together.

So why is this such a great combination? First, I get to do a job that I love. I pursue creative work that challenges and interests me. I have a position that allows me to use my talents and to grow and thrive. And at the same time, I have a loving family with children whom I adore and I cannot imagine a richer life. And these two parts of my life fit very closely together.

A big benefit of a job as a professor is the flexibility of the academic schedule. When my children were young, if I needed to take them to the doctor or attend a school performance or sports competition, I could plan my calendar around those activities. I didn't need to formally request time off with a boss, find a substitute, or punch a timecard. The work could be made up at any time, including evenings and weekends, and—importantly—from home. Of course, the tradeoff is that a faculty position can be an intensive career with seemingly never-ending work that can consume those same evenings and weekends.

Another highlight of my job is that as my children have grown older, I have been able to model for them the benefit of having a professional career as a scientist. From the "science shows" at my children's school, to family events at my university's school of engineering, to the invited talk or research group meeting that they occasionally attended, my children repeatedly saw their mother as a respected professional. As a result, it seems natural to them that women should be in positions of leadership. It also allowed them to see through my eyes the pleasure of having a job that one loves, and I believe that the power of that message should not be underestimated.

Early Years

I grew up in Orange County, California, in a middle-class family with two brothers and went to local public schools. I don't remember focusing on a career in science until I was in high school. In fact, when I was much younger, I had wanted to be a veterinarian to indulge my love for animals. I was never involved in science fairs nor did I play much with engineering or science toys as a child. As an older child and teen, I typically spent my free time going to the beach or the local shopping mall with family or friends.

My public high school was adequate but not particularly strong. Even though advanced placement (AP) tests existed, my high school still hadn't adopted them at that time. The concept of final exams hadn't even made it to my high school yet! The curriculum was limited in its course offerings. For example, my high school didn't offer a class in calculus. However, my high school did have one terrific advantage— it was located across the street from a well-respected community college. The lack of advanced math turned out to be a blessing in disguise because it allowed me to take calculus (and several other courses) at the community college while I was in high school.

My decision to major in chemical engineering in college was made when I was applying to universities as a high school senior. This was under direct instruction from my father, who wanted each of his children to start off in an engineering major, even if we might later change our major. As he always said, "it's much easier to transfer *out* of engineering than to transfer *in*." This advice stemmed from his own experience as a first-generation college student. He did not receive good guidance about what to study and always regretted his choice of major in college. So, despite the fact that neither of my parents nor any of our close relatives were scientists or engineers (my father was an accountant and my mother was a teacher and librarian), my parents ended up with three children who not only majored in engineering in college but went on to careers in engineering and science (one brother is a computer scientist and the other an aerospace engineer). My parents always remained supportive of my career in science and engineering as it progressed. My mother would often ask me to read to her the title of a talk I would be giving at an upcoming conference or an honor I had received; she would carefully write it down so that, although she might not understand what it meant, she could share the information with pride among her close friends.

My choice of chemical engineering was made by a combination of practicality and inspiration. With my father insisting that I choose an engineering major and application deadlines looming, I spent quite a bit of time staring at lists of engineering majors to decide which I would pick. The problem with this approach was that I really had no idea what a civil engineer, mechanical engineer, electrical engineer, or chemical engineer did. This was before the days of career nights, informational interviews, or shadowing of professionals. It was also before the rise of the internet, and so one could not simply google what various professions entailed. However, I did enjoy chemistry in high school, and with the faintest notion that chemical engineering might involve a lot of chemistry, I decided to major in chemical engineering. It also didn't hurt that in rankings of starting salaries among the different engineering majors, chemical engineering was always featured near the top!

I had been a good student in high school, graduating as valedictorian of my class, and was fortunate to be accepted to the University of California at Berkeley as a chemical engineering major. I experienced a bit of a shock in making the jump from my low-key high school to UC Berkeley, where the workload was tough and the class material was challenging. But I quickly adapted, and, although I worked hard there, I enjoyed my college experience tremendously.

Fortuitous Invitation

When I was in high school, I was convinced that I didn't want a career as a scientist and certainly not one as a research scientist. I had in mind the stereotype of a scientist in a white lab coat, working alone in a lab. This was definitely *not* what I wanted for a career. Once I was in college, I figured that after graduation I would get a job in industry as a BS-level chemical engineer. Also, when I was studying chemical engineering at Berkeley, there wasn't a strong culture of undergraduate research, and few of my friends were doing research.

So, research was not on my mind when, in my freshman year, my teaching assistant for Honors Chemistry asked me if I'd be interested in doing research in her advisor's laboratory. I'm not sure what had inspired my TA, Laurie Butler (now a Professor of Chemistry at the University of Chicago), to extend this invitation, but that one act of outreach ended up changing the path of my entire career. Although I don't remember the details, Laurie's faculty advisor must also have agreed and thus began my 3-year experience doing research in the laboratory of Prof. Yuan T. Lee of the Department of Chemistry. Y.T. Lee was an expert in molecular beam studies, and my various projects in his lab involved crossed laser-molecular beam research on state-selective photodissociation of alkyl halides. One of my clearest memories of that research experience was feeling that I was in over my head, and I didn't really understand a lot of the theory behind the research. But what I did enjoy was the excitement of doing research at the cutting edge of the field and sharing in those experiences with the graduate students and postdocs in the lab.

On the other hand, although I enjoyed working with the graduate students in Y.T. Lee's lab, I was not thinking of going to graduate school myself. No one in my family had a PhD, and it wasn't something I aspired to. I had a wonderful academic advisor for my 4 years at Berkeley, the late Prof. Charles Tobias, who was always pushing me to get a PhD. We had regular advising meetings, and frequently Prof. Tobias would refer to the future using phrases like "...when you're in graduate school." I always sat there and thought to myself, "I'm *never* going to graduate school." In part, I was also rebelling against taking the path that all my professors expected I would follow. But by my senior year at college, my thinking on this started changing, and I decided to apply to graduate schools. Around this time, a rare, exciting event occurred. When I was a senior in Prof. Lee's lab, he was awarded the Nobel Prize in Chemistry. That was a wonderful and fun time to be in his group. And it further piqued my interest in going to graduate school. The surprise to some of the faculty in chemical engineering was that I decided to apply to PhD programs in chemistry, not chemical engineering. This decision was largely inspired by the research experience I had in Y.T. Lee's group. Not everyone was positive about this switch. I remember one of my teaching assistants (a PhD student himself) saying "why would you want to get a PhD in chemistry? Chemistry PhDs are a dime-a-dozen; it's much more special to be a chemical engineering PhD." I'm glad I ignored the naysayers and followed my own interests.

Even though I applied to graduate schools, I still wasn't fully committed to taking that path, and so during my senior year I also applied for traditional bachelor-level engineering jobs at places like Dow Chemical, Procter & Gamble, and Exxon. This meant that much of my final year at Berkeley was spent doing on-site interviews at major companies while also visiting PhD programs around the country. I refused to commit to one path or the other until it was nearly the deadline for PhD decisions in the spring, and I finally turned down the last job offers after I had selected my choice of graduate program: Stanford University. I chose Stanford for several reasons, one of which was a specific research topic that I wanted to work on in the lab of Prof. Richard (Dick) Zare. Dick had a research effort combining molecular beams and quantum-state-resolved molecular detection (the topic of my undergraduate

The Best Job in the World   41

research) and solid surfaces (the direction I wanted to go in the future), and he agreed to let me work on that project.

One feature of my years at Berkeley that I find surprising today—but didn't even notice at the time—is that I had no female professors in my entire 4 years. I did have a female lecturer for a technical communications class, but she was not part of the regular professoriate. Every other one of my classes was taught by male professors. But what's more surprising than the lack of diversity was the fact that I didn't even notice it or think anything about it until a conversation I had after I graduated. One day, I was speaking with my younger brother, who was by then an engineering major in college, and he referred to one of his classes by saying, "My professor said. . .and then she. . .," and I was in shock. She?! Just hearing "professor" and "she" together made me do a double take. And that was the real surprise: the lack of diversity was so prevalent that I didn't even recognize it.

Auspicious Meeting

Before heading off to Stanford to start the PhD program, I spent the summer working at Bell Laboratories in Murray Hill, New Jersey. This was definitely not how I had imagined spending my last summer before graduate school, and I resisted taking the internship. I had lived my entire life in California, and New Jersey seemed very far away and isolated. However, I had received a grant from AT&T Bell Laboratories (the Graduate Research Program for Women) for my graduate studies, and the terms of the grant required that I spend the summer before starting graduate school working in one of their laboratories on a research project. I tried to talk my way out of this requirement with the fellowship coordinators, but to no avail.

So, off I went to New Jersey shortly after graduating college. Despite my initial misgivings, it turned out to be a wonderful experience, both professionally and personally. Many things in life are this way, and I've learned that it can pay dividends to embrace the random-walk nature of a career. At Bell Labs, I worked with two terrific research scientists, Dr. John Tully and Dr. Mark Cardillo, who continued to serve as mentors for many years. More importantly, Bell Laboratories is where I met my future husband, Brian Bent, who at the time was a postdoctoral fellow at Bell Labs. We met shortly after I arrived and enjoyed getting to know each other over those few short months. At the end of the summer internship, I was faced with heading back to Stanford for graduate school, which now didn't seem like quite the good idea it had at the start of the summer before I'd met Brian. However, I had committed to graduate school at Stanford, and Brian had just accepted a faculty position in the Chemistry Department of Columbia University in New York City, and our relationship felt too new to justify changing those life plans. So, I went to Stanford and Brian to Columbia, and we proceeded to have a long-distance relationship over the next 4 years.

With my serious boyfriend (and eventually husband) living 3000 miles away, I spent my entire graduate years in a hurry to finish my PhD. We traveled to see each other every month or two, typically for long weekends, and then we would focus on working hard in the periods between visits. But of course, research takes time, and the experimental project that I worked on was challenging. It eventually took about four and a half years for me to finish, which was still relatively short for a PhD in Chemistry at Stanford. About 2 years before I graduated, I arranged to spend a few months doing another internship at AT&T Bell Laboratories, both to take a break from the grind of graduate school and to spend time living in the same part of the country as Brian. Then a year before I graduated from Stanford, Brian and I got married. However, a few months before our wedding, Brian was diagnosed with testicular cancer, and underwent surgery and radiation therapy. Those were scary and difficult times, and my being away at Stanford for much of that time (minus several trips to New York) was very stressful. Thankfully, Brian recovered and was given a clean bill of health. Finally, I had my thesis defense, with both Brian and my beaming mother in attendance for the technical portion. It was a very proud moment!

Next Steps

While I was in my last year as a PhD student, my PhD advisor, Dick Zare, recommended me for a faculty opening in the Chemistry Department at New York University. Since my husband was at Columbia University, which is also in New York City, NYU was in a good location, and Dick was proactive in making this connection for me. I interviewed at NYU while I was still in graduate school. I remember the interview being something of a disaster. At that stage of my graduate career, I was not prepared to be looking for a faculty position; I wasn't even sure I wanted a faculty career. A mere week or two before my interview, I still hadn't come up with a research proposal for my job talk! I remember going on a ski trip to Lake Tahoe with several other graduate students from the Zare group, and one of them exclaiming in surprise, "You haven't prepared your research proposal yet?!" I ended up throwing some slides together based on work I had done during my second internship at Bell Labs, and in fact during that interview, one of the NYU faculty commented that some of the slides looked like something taken directly from a book (which they were).

At the end of my interview, several faculty members took me to dinner at a nearby restaurant in Greenwich Village. At that dinner, one of them told me that he knew I would have to take the NYU position because, with my husband at Columbia University, there weren't other viable faculty options for me in New York City, and he then proceeded to explain why each of the other schools were not options. That didn't exactly put me in a good negotiating position if I were to get the job offer! Nevertheless, I was hopeful about the position in the NYU Chemistry department, and somehow they were intrigued enough to say that they were interested in making me an offer. However, they weren't ready to make a firm commitment, and the process dragged on for many months.

In the meantime, I started looking for postdoctoral positions in the New York City area so that I could finally live in the same part of the country as my husband. After

The Best Job in the World 43

interviewing at a couple of universities and industrial research laboratories in the area, I accepted a postdoc position back at AT&T Bell Laboratories in Murray Hill, New Jersey, with Dr. William Wilson, an expert in ultrafast laser spectroscopy. During my postdoc fellowship, Brian and I lived in his Columbia University-owned apartment in Manhattan, and I commuted to Bell Labs (about a 45-minute drive) by car. It was the first time I was able to live with Brian. Although weekdays were long with the commute in New York area traffic, I got to work with terrific colleagues at Bell Labs on studies of new materials and ultrafast laser spectroscopy, and Brian and I greatly enjoyed our weekends in the city.

By the time I had moved out to New York City, Brian and I had already been a couple for over 4 years, and I became eager to have a child. This sense of urgency was intensified due to Brian's cancer diagnosis and treatment, with the attendant risks as well as introspection that come from facing cancer. Just a few months after I started my postdoc position, I became pregnant with my daughter. Due to morning sickness and the general distractedness of pregnancy, much of the rest of my postdoc experience was a blur. I'm sure I was not as effective or productive a postdoc as I should have been! Since I shared an office with other postdocs, I remember going into a seating area tucked away behind the darkened laser laboratory and propping myself in a chair for catnaps whenever I could get away with it. Because of issues with mild pregnancy-induced hypertension, I was advised by my obstetrician not to work for the last 6 weeks of my pregnancy. I had planned to take maternity leave from Bell Labs after the baby was born, then return to my postdoc position. So, about 1 year after starting my postdoc position, I went on medical/maternity leave.

Around this time, NYU reached out to me and told me that while they were still interested in potentially making me a faculty offer, they needed me to come in to reinterview. I told them that I was not supposed to travel (this was during my 6 weeks of home confinement), but they convinced me that they would make the interview easy for me: they would send a car service to pick me up, schedule my research and my proposal talks back-to-back, let me give the two talks while sitting in a chair, then have the car bring me home. So, I agreed to interview. That interview is a bit surreal because I was heavily pregnant, rushed through two talks, and then was whisked home. But somehow it went well enough for me to finally get a formal job offer from them.

First Faculty Position

Because of that offer, I never returned to my postdoc position at Bell Labs but instead arranged to start my faculty job at NYU in January, when my daughter Rachel was 9 months old. So, with an infant at home, I began my first academic appointment. I remember those early days as a new faculty member as a wonderful period in my career, during which I could really immerse myself in my research and in building up my independent research program in experimental physical chemistry, with a focus on gas-surface reactions in electronic materials processing. The teaching load was moderate at NYU, and every course I was assigned was team-taught with a more senior faculty member, so I was lucky enough to learn the ropes from more experienced faculty.

I became pregnant with my son during my second year as a faculty member at NYU. I worked up until nearly my due date, commuting from our apartment near Columbia University down to NYU on the subway, about a 45-minute trip. Although New Yorkers can be very nice, I do remember many crowded subway rides strap-hanging while very pregnant without anyone offering to give up their seat. I often thought that the dynamic of those interactions would make a good sociology exper-iment. One time, in a real "New York moment," a lady kindly gave up her seat for me, only to have it taken by a man who rushed past me to get into the open seat; fortunately, another passenger saw the whole interaction and gave me her seat instead.

During the time that I was starting my academic career at NYU, Brian's faculty career was going very well at Columbia University. He had been promoted to associate professor with tenure the year that I received my PhD, and he was promoted to full professor in 1996. I remember being home with a toddler and pregnant with my second child when a colleague threw him a promotion party at a local restaurant. Not realizing what big a deal that was, and also feeling a bit exhausted with the pregnancy and a toddler, I declined to attend, something I still regret. But soon afterward, our second child was born, and that was a very happy time (Fig. 1).

Because NYU in those days didn't have a formal maternity leave policy, my chair arranged to give me teaching relief the semester after my son, Drew, was born. That allowed me to take an unofficial 4-month maternity leave but with full pay. This time, with a research group to supervise, it wasn't a complete leave, since I continued running my research program and advising students, occasionally taking my son in with me to NYU. We had hired a nanny when I first started at NYU because in considering childcare options, Brian and I decided that having a nanny would offer the most flexibility. So, we hired our first nanny to help with Rachel, and after he was born she watched Drew, as well.

During this time, I built up a group of three PhD students and a postdoc, and my group and I had begun publishing our first papers. Things were busy with two very young children at home and a faculty position, and life was very fulfilling (Fig. 2).

Tragedy

When my daughter had just turned 3 years old and my son was 8 months old, we took a vacation to Brian's family's summer cabin in the Boundary Waters area of Northern Minnesota. This was a relatively remote site on a small lake, in a cabin without running water. A few days after we arrived, Brian went on a short bike ride with his brother-in-law. While on that ride on a dirt road near the cabin, Brian collapsed and died. Brian was only 35 years old, and his sudden death from an underlying heart condition was unexpected and devastating. This tragedy plunged the entire family into mourning and grief. We somehow dragged ourselves back to

Fig. 1 Stacey with Brian, Rachel, and a newborn Drew (1995)

Fig. 2 Stacey with her early research group at New York University (1997)

New York, arranged a funeral, and tried to pick up the pieces. It was a tremendously difficult period of my life.

After a few weeks of bereavement leave, I returned to work at NYU. Work gave me something to think about besides my grief and worry and in that way was helpful to my healing. After my children were born, I had occasionally entertained idle thoughts about taking a leave or quitting my job to raise my children, but not with any real degree of seriousness. However, after my husband's death, I was extremely glad that I had not, among other reasons because he died without life insurance and my family with two small children needed my salary. Even with my continued salary, those first years after his death were financially challenging. I was fortunate that our children qualified for social security benefits, which allowed us to keep our nanny so that Rachel and Drew could have some semblance of continuity and I could still work.

With an infant and young child, a full-time job as an assistant professor of chemistry, and the pain of grief, I needed a lot of help. Fortunately, my family and Brian's family rose to the occasion. Over that first year, I had help nearly all the time by a combination of Brian's parents, my parents, and our brothers and sister. Although they lived in various cities across the country, they took turns staying with us in our apartment for a week or two at a time, especially my mother-in-law Anne, whose generosity I can never repay.

About a year after Brian died, I decided that I needed to make plans to move away from New York City. Life in New York City was challenging enough when we were a two-parent household, but going it alone was too difficult. Dealing with childcare, decisions about public schools, commuting, shopping—all while pushing a double stroller on and off buses and subways—it was overwhelming. Colleagues at NYU tried to be supportive, although one of them once told me in what was perhaps a misguided attempt at humor that what I needed as a young widow was a "wife." One Christmas Eve day, I was at work at NYU and stepped out to get some takeout lunch at a nearby Chinese food restaurant. While I was walking to the restaurant, a homeless woman stopped me and asked for money. In the spirit of the holiday, I told her I would buy her lunch, which she gratefully accepted. While we were both waiting for our food to be ready, we started talking and she found out what had happened to my family. When she kept repeating that she was going to have to tell all her friends my story, I realized that I didn't want to live my life as an object of pity and that the time had come to make a change.

With this resolve, I began applying for jobs in the West Coast to be closer to my parents and brothers, who still lived in California. Because I had been in a faculty position for under 4 years, I applied for positions at the untenured assistant professor level. The interview process and the travel that it required really taxed my family network for childcare, and I used to joke that I practically had to rely on strangers off the street to watch my children during this time (though it wasn't really true!). Over the 2-year period after Brian died, I gave nearly 25 department seminars and conference presentations all around the country. I was fortunate to receive a few offers in both chemistry and chemical engineering departments. I decided to accept an offer from the chemical engineering department at Stanford University.

The Best Job in the World   47

There were some poignant and also funny moments associated with my visits to Stanford at this time. Once, during a recruiting trip, one of my future faculty colleagues was kind enough to offer to babysit my children while I went house hunting with my mother, who had flown up from Southern California. During that same trip, I was swimming with my children in the hotel pool and I looked up to see a whole group of prospective PhD students looking at me. They were staying at the same hotel, and another one of my colleagues, noticing me in the pool, was pointing me out to them "Look! There's our newest faculty member!" It was not quite the professional image I had hoped to portray. Finally, during that same visit, my mother began to feel unwell and was diagnosed with a relapse of leukemia, which sadly would eventually take her life about a decade later.

I didn't make the choice to take the faculty position at Stanford without significant apprehension. I was concerned about the pressure of being a single parent in such a competitive program, and I had strongly considered going to a less demanding school. I asked my former PhD advisor for advice, and he convinced me that there was no question but that I should take the Stanford offer.

Starting Over at Stanford

So, 2 years after my husband's death, my children and I moved to Stanford to begin a new chapter in our lives. It was both a scary and a hopeful time. Rachel was now 5 years old and entering kindergarten, and Drew was 2 years old. I again searched for a nanny, and I was lucky enough to find Slavka, a warm, smart, creative, and talented young woman. Slavka was a wonderful nanny who became like a second mother to my children. My kids absolutely adored her, and our family became the envy of many neighborhood parents, since with Slavka at the helm, every weekday was like a summer camp, arts and crafts class, and professional tutoring session all-in-one for Rachel and Drew. She taught my kids everything from making handmade paper to making sushi. Importantly, she played a significant role in allowing me to successfully juggle all that I had to do during those early days at Stanford (Fig. 3).

At Stanford, I was focused on teaching new courses and expanding my research program. At NYU, I had taught courses in physical chemistry, general chemistry, and physical chemistry lab, but now I was teaching chemical engineering classes in kinetics and reactor design, microelectronics processing, and applied spectroscopy. My colleagues at Stanford were very supportive of me, although they couldn't really know how challenging my life was. Those first years were, admittedly, exhausting. On a typical day, I would get Rachel and Drew off to school in the morning, spend the full day on campus, return home in the evening in time to make dinner, spend time with the kids, and get them into bed so that I could then have a few hours to do work (often until 1 or 2 am). Many evenings, I would fall asleep while the children watched a video or two, until they finally woke me up so that I could put them to bed. Then I would stay up for several more hours working.

Fig. 3 Stacey doing a science demonstration in Rachel's first-grade class (2000)

Traveling to give talks still proved one of the most challenging aspects of my job, and before I could accept any speaking invitation, I first had to check to see if a family member could travel to Palo Alto to stay at our house and watch Rachel and Drew while I was away. I still marvel at the kindness of family and friends who stepped up and provided overnight childcare, over and over again, for me to pursue the kind of "external presence" required of my job. My travel also impacted my children's lives. For one thing, the trips were usually arranged several months in advance, and at times that meant that I missed important family events because they occurred when I already had travel planned. What made it especially difficult is that Rachel and Drew hated when I traveled, an emotion that never went away even as they got older. What made it tolerable was the recognition that the anticipation of Mom going on a trip was always worse than the reality. Once I was gone, the kids were fine, and in fact they developed close relationships with many of their extended family as a result of spending this time with them.

Four years after moving to Stanford (and 8 years after I began my first faculty appointment), I was promoted to associate professor with tenure and three years after that to full professor. Around that time, I started dating a professor whom I had met from another engineering department at Stanford. As a senior faculty member, Bruce understood the academic life, and, as the father of two young adult sons, he also understood parenting. After a brief courtship, we got married in a simple wedding ceremony at which Rachel and Drew provided the music on their cellos. Finally, 9 years after being widowed, I was no longer a single parent. Many aspects of my life got easier with a supportive partner and co-parent. But as Bruce and my children

The Best Job in the World

Fig. 4 At Stacey and Bruce's wedding, with her stepchildren Daniel and Andrew and children Rachel and Drew (2004)

would admit, being a stepparent or stepchild is difficult, and not everything went smoothly. Still, my kids and I acquired this wonderful new family member as well as his extended family. When Bruce and I got married, I gained two stepsons and my children gained two stepbrothers. Although they were already young adults living on their own by that point, they were and continue to be a big part of our lives. And Rachel and Drew gained more grandparents! We have happy photographs that I treasure to this day of my kids with their three pairs of grandparents. On the other side, Bruce was embraced by the family of my late husband. Bruce even traveled with us sometimes to visit Brian's family, and he always enjoyed people's confusion when he told them he was going to visit his wife's in-laws—and that *no*, that didn't mean his own family! (Fig. 4).

Although Slavka had moved on after 7 years with our family, we continued to have nannies, which simplified our lives. As the kids got older, the childcare role became less "nanny" and more "chauffeur," since we still needed someone to get them to their various after-school activities. At work, my research group grew to about 10–15 PhD students, and I very much enjoyed my work "family." And as an added benefit, as my children have gotten older and become college-aged themselves, I have found myself becoming more in tune with and sympathetic to my students in the classroom and laboratory.

Travel continued to be both a benefit and liability of my job. My previously arranged travel sometimes conflicted with major events in my children's lives. For example, I wasn't able to move Rachel into her dormitory during her freshman year at college due to a travel conflict (although Bruce was there and was much more helpful in lifting boxes and furniture, and in socializing with the roommate's family, than I would have been). I don't believe she's quite forgiven me for this. To make up for it, when Rachel started medical school, I made sure I was there for her move-in, orientation, and her white coat ceremony—all of it. Similarly, because I was traveling, I also missed the day Drew competed in (and won in his category) the state science fair in middle school. However, Bruce was again there to cheer him on, and it turned out to be a wonderful experience that the two of them were able to share.

The flip side of all the travel is that it has given my children opportunities that they wouldn't have had otherwise. For example, the whole family took a memorable trip to the Brazilian Amazon when I was invited to a conference there. We also spent 4 months on sabbatical in Sweden, where my kids were enrolled in an international school and had a life-enriching experience. As they got older (high school and college age), I discovered that it is a lot of fun to travel with adult children, and a few times I have been able to take them with me when I attended an international conference that allowed us to do some sightseeing afterward. In this way, I have taken Drew with me to Japan and Denmark and Rachel with me to Italy and Sweden.

In parallel with my job as a professor, I have always been very involved in my children's lives and I am especially close to them. Although I raised them alone for nearly 10 years, I tried to make their childhood as full as possible. That included music lessons, sports, religious school, etc. as well as developing and nurturing shared traditions with family and friends. I was the mom who planned sometimes elaborate birthday parties complete with art projects, scavenger hunts, and—always—homemade birthday cake. I can't say I went to *every* track meet, but I was there cheering on my children at many of them, as well as most cello lessons, robotics competitions, school performances, award ceremonies, and so on. We spent countless hours together as a family, time that I will always cherish.

Rachel and Drew are now in their early 20s and both studying in scientific fields (medicine and physics). I tried to encourage them to pursue their own interests and never explicitly said that they should study science in college. However, both of them did choose science majors, and Rachel says that even though I wasn't explicit, it was obvious to them that non-science fields were not going to be good enough for their mom. Although I regret having inadvertently pressured them in this way, I feel redeemed in that they both seem happy with their choices. And I am tremendously proud of the wonderful adults that they have become! (Fig. 5).

Career Progression

As my career at Stanford progressed, various opportunities presented themselves for me to take on different leadership roles in addition to my research, teaching, and service activities. About 8 years ago, I became a co-director of a DOE-funded Energy Frontier Research Center, which gave me experience leading a large group

Fig. 5 A spring break reunion in Palo Alto (2017)

of faculty in carrying out a broad research agenda. As often is the case, the visibility I received from this first role brought other opportunities my way. Soon after taking on that role, I was asked to direct a new endowed center on sustainable energy, a position that I still hold.

With the leadership roles came more responsibilities, and at times that was challenging with children still at home. However, I had been given advice years before that I should not become department chair until my youngest child was in college, and I held fast to that mantra. The year after Drew started college, I was asked to serve as department chair, and having run out of excuses, I accepted. Shortly afterward, I was appointed as the senior associate dean for faculty and academic affairs in the School of Engineering. I enjoy this role because it allows me to see what is happening across the school and the university, have an impact larger than that of my own research group, and meet people throughout the university.

At the same time, I'm still very active with my research program and run a group of 15 PhD students and postdoctoral scholars. My students and I continue to carry out research on surface and materials chemistry, and our projects range from fundamental studies of molecular bonding at surfaces to fabrication and testing of solar energy devices. I very much enjoy being involved in scientific discovery, and just as I did when I was an undergraduate in my first research experience, I find it exciting to uncover new knowledge about our physical world.

Reflections

- How I ended up where I am today, as a mom and a professor, was not preordained. My story, like that of so many others, is one of opportunities and challenges. There were fortuitous opportunities at critical points in my life that I took advantage of, either by design or by dumb luck. For example, I was provided with opportunities by virtue of the location of my high school, by having a college TA who gave me a chance at research and by meeting the right people at the right time. But some were opportunities only because I took a chance at them, despite being outside my comfort zone. I have also had serious challenges in my life, such as Brian's death and balancing career and family as a single parent. In many cases, I overcame these challenges by leveraging the help of others.
- My life is infinitely richer because of my children. A few years ago, I gave a speech at a reception to celebrate receiving an endowed chair at Stanford. In that speech, which was attended by family and many colleagues, I said "My children, Rachel and Drew, have been with me through thick and thin—the three of us moved out here from NY together— and I believe they are my greatest fans, just as I am their greatest fan." I can't imagine my life without them.

Reflections from Stacey's Children

Rachel Bent

Screams of excitement and wonderment filled the outdoor walkway of my elementary school as a stream of liquid nitrogen flowed outside my classroom creating a thick mist. In that moment, I remember being in awe of my mom. Had I known the word "badass" at the time, this surely would have come to mind as I watched my mom with her goggles on, leading this chemistry demonstration. Many of my strongest memories as a kid are a direct result of my mom's career as a chemical engineering professor. I remember how much fun I had when I got to hang out in her office while she worked. My brother, Drew, and I would spend hours drawing pictures on her whiteboard (obviously, only after my mom had split the whiteboard down the middle so we each had an equal amount of space to unleash our creativity). I remember her research group coming over to our house for barbeques. Some of her students would chase me and Drew around the backyard while others chatted with my mom over freshly grilled salmon. Even as a young kid, I could see how much my mom's students admired and respected her.

While my mom's work did sometimes interfere with certain family events, I never doubted that Drew and I were her priority. In high school, when my cross-country team made it to the state meet, I didn't expect my mom to come. Not because I thought she didn't support me, as she had been to several other races throughout the season, but because this meet was 3 hours away. Despite my

(continued)

reservations, my mom wasn't deterred. She was there, on the morning of the race, wishing me luck as I anxiously headed to the start gate. As I ran, I thought about the long list of work items that would be waiting for her when she got home. As I rounded the final turn and heard my mom's cheers from the sidelines, I felt especially grateful for her presence and kicked it into the finish line.

As I now embark on the path toward my own career, I sometimes think about the challenges that I might face when it comes time for me to have a family. And while I know that it will be difficult at times, I have never once doubted that I will be able to do both. And for that, I have my mom to thank.

Drew Bent

That my mom was a professor was a simple fact of life as a young child. It was clear to me that not everyone had a mom who worked or a mom who raised them singlehandedly for many years. But other than that, things felt normal.

I attended public lectures by my mom, built dye-sensitized solar cells with her for a science fair project, and even traveled to Sweden and Brazil with her, yet none of those seemed out of the ordinary.

As I grew older, this simplistic picture evolved. High school friends coming over now saw my mom as *their* professor, a respected authority figure. Thus, I too started to think more critically about my mom's profession, and I began to understand how her academic tendencies were the basis for much of what I had become.

My work ethic is a mirror image of my mom's habit of working late into the night with laser focus. My quantitative interests are a byproduct of her nonstop encouragement — to the point where she even helped organize a "math problem-solving party" for my friends and me over the summer before fifth grade. And ultimately, my desire to pursue a life's work, and not simply a career, is due to how much I admire my mom's passion for her research and teaching, transcending everything she does.

What my mom managed to accomplish is nothing short of remarkable: she instilled in my sister and me the virtues of an academic, while rarely letting us in on the behind-the-scenes show and countless hours needed to pull that off. So, although my life may have felt ordinary with our mom taking us to track meets, orchestra concerts, and Sunday school, she was actually doing much more: leading by example.

Acknowledgments It should already be abundantly clear from this chapter that I have *many* people to thank. First and foremost, I am indebted to my family, including my husband (Bruce), parents (Ed and Judy) and parents-in-law (Anne, Henry, Reba, and Leland), brothers, brother-in-law and sisters-in-law (Mike, Yumei, Russ, Julia, Libby, Rolf, Susan, Laura, and Heidi) who gave in so many ways to me and my children over the years and enabled me to succeed in the two "best jobs in the world." I am eternally grateful for the time I had with my late husband, Brian; it was far too brief, but it helped shape who I am both as a mother and as a scientist. At its core, this is a story about being a mom, so it would not be complete without expressing my gratitude to my children (Rachel and Drew) and stepchildren (Daniel and Andrew) who have always shown tremendous support of my career and have made it all worthwhile. Special thanks to my sister-in-law Susan Fletcher and colleague Matt Abrahams, who gave insightful and encouraging feedback on an early draft. I would also like to express deep appreciation to my many colleagues and mentors who have given support and encouragement to me throughout the years and to my students and postdocs who have shared this journey of scientific discovery with me.

About the Author

Education and Professional Career

1987	BS Chemical Engineering, University of California, Berkeley, CA
1992	PhD Chemistry, Stanford University, CA
1992–1993	Postdoctoral Fellow, AT&T Bell Laboratories, Murray Hill, NJ
1994–1998	Assistant Professor of Chemistry, New York University, NY
1998–2002	Assistant Professor of Chemical Engineering, Stanford University
2002–2005	Associate Professor of Chemical Engineering, Stanford University
2005–present	Professor, Department of Chemical Engineering, Stanford University
2012–present	Jagdeep and Roshni Singh Professor, Stanford University
2015–2016	Department Chair, Department of Chemical Engineering, Stanford University
2016–present	Senior Associate Dean for Faculty and Academic Affairs, School of Engineering

Honors and Awards (Selected)

2018	ACS Award in Surface Chemistry
2013	Bert and Candace Forbes University Fellow in Undergraduate Education
2013	Fellow of the American Chemical Society (ACS)
2013	Stanford University Medal for Faculty Excellence Fostering Undergraduate Research
2006	Fellow of the American Vacuum Society (AVS)
2006	Tau Beta Pi Award for Excellence in Undergraduate Teaching
2001	Coblentz Award for Molecular Spectroscopy, Coblentz Society
2000	Peter Mark Memorial Award of the American Vacuum Society
1998–2003	Camille Dreyfus Teacher-Scholar Award

1998	Research Corporation Cottrell Scholar
1997–2000	Beckman Young Investigator Award
1995–2000	National Science Foundation CAREER Award

Stacey is active in research on understanding and controlling surface and interfacial chemistry and applying this knowledge to problems in semiconductor processing, nanoelectronics, and sustainable energy. She holds courtesy faculty appointments in the Departments of Chemistry, Materials Science and Engineering, and Electrical Engineering at Stanford University.

Equilibrium and Stress: Balancing One Marriage, a "Two-Body Problem," and Three Children

Stacey Lowery Bretz

Introduction

As a young girl in the 1970s, I heard all about "superwomen." These were not women flying around in capes and tights, but they seemed just as powerful to me. Superwomen were so talented that they could have both a family *and* a career. No longer were young women expected to stay home with their children, nor forego children to pursue a career. Both my parents (Fig. 1) worked full time, and many nights my daddy was the one who put supper on the table. My parents were my role models. It was possible to "have it all," and I intended to do just that.

S. L. Bretz (✉)
Department of Chemistry & Biochemistry, Miami University, Oxford, OH, USA
e-mail: bretzsl@miamioh.edu

Fig. 1 Stacey and her parents, Jim and Ruth Lowery, in June 2016

As with many childhood dreams, they are visions borne out of endless optimism and naïvete. I had no idea what an impossible standard I envisioned for myself. Nothing less than perfection! As a young girl, everything seemed possible, including having everything you ever wanted—having it "all."

I've since come to learn that chasing the elusive "having it all" is folly and can create unrealistic pressures and disappointment for women and their families. No one can have it all. But, and this is a big BUT, you most definitely can have *both*. A family *and* a career. There is an important difference between having it all and having both. And that is the story I'd like to share with hopes that it inspires others.

Middle School and High School (1980–1985)

I can still remember the first time someone told me I would not "have it all." My middle school biology teacher told me I wasn't strong enough to be a doctor. I would want to have children and choose to give up my career for staying home once children started to "come along." He made me so mad. Who was he to decide my future? I remember thinking: I'll show him!

I did not end up in medicine, however, because in my sophomore year in high school, Mrs. Palmer introduced me to chemistry. She invited me to stay after school to join her team of students who conducted independent research projects. I had seen the 4 feet tall trophies these students won at the state research competitions. I doubted I could do original research as a 14-year-old. But Mrs. Palmer had confidence in me, and before I knew it, I had spent 3 years working in her classroom after

Fig. 2 Stacey and her high school sweetheart (now husband), Richard Bretz, in spring 1983

school to explore the catalysis of urea and factors that affected the function of the enzyme urease.

That single invitation, and my teacher's unfailing belief in me, shaped my life in ways I could have never imagined. I won the "Best Chemistry Project" award from Dow Chemical as a high school junior, which eventually led to a summer internship in college. My high school didn't offer AP (advanced placement) courses, but conducting independent research made a strong impression and helped me get into Cornell University. But the most profound impact of that research experience was not professional—it was personal. There wasn't enough time after school to make all my solutions and then run my experiments. So, Mrs. Palmer invited one of her senior honors students (who had been on the state, trophy-winning teams) to make my solutions during his study hall so that I could get straight to work at the end of the school day. That boy was named Richard (Fig. 2), and so began an incredible friendship that blossomed into young love. We were together for just 6 months as he left for college at Rensselaer Polytechnic Institute. Thirty-five years later, though, I am proud to say I am still married to my high school sweetheart, and we've made a wonderful family. What follows in the rest of this chapter is the story of how we, as two chemists, have tried to balance our marriage, family, and careers.

College (1985–1989)

My first college course in chemistry was co-taught by Barbara Baird and Roald Hoffmann. Professor Baird was the only woman chemist on the faculty at Cornell, and when the semester began, she was pregnant. Professor Hoffmann is a Nobel

Laureate. Looking back, I didn't realize then how exceptional my experience was. I just took it for granted then that of course you could be a woman professor and have a family, all while on the tenure-track. Of course other college freshmen were being taught chemistry by a Nobel Laureate. I was so naïve on both counts!

I went on to do undergraduate research in the lab of Héctor Abruña. All throughout college, I was still in a long-distance relationship with my high school sweetheart. And back in the days of no email, no texting, and no social media—this meant writing letters. Lots and lots of letters. In my senior year at Cornell, Richard came to begin his PhD in the Abruña group. For a while, we told no one we were dating (and had been for 6 years at that point!). We didn't know how that would be perceived. What if one of us would have to leave the group? After a while, though, we felt comfortable letting the group, and eventually Dr. Abruña, know. (As I write this, I realize now that was probably the first, and only, time I ever "hid" some part of my personal life from my colleagues.)

However, as I started looking toward graduate school, I learned that it was not OK to earn your PhD in the same department where you were an undergraduate. It just wasn't done. So, after 1 year of togetherness, it was time to be apart again from Rich. I set off to earn a PhD at Penn State (3 hours away). The goal: to earn my PhD and become a faculty member at a small college where I could focus on teaching—and to do so in a location where Rich could either do the same or put his PhD in analytical chemistry to work in industry.

Graduate School (1989–1994)

I had no idea how transformative graduate school would be. I joined a research group that had never (ever) had a female graduate student. I had been warned not to do so by many people, but I was really interested in the chemistry. And more importantly, I was a confident woman. There was no way anyone or anything was going to derail my plans. I won Honorable Mention from the National Science Foundation (NSF) Graduate Research Fellowship Program—clearly I had potential. I joined the group with two other guys, and they (and their projects) quickly outpaced me. I was assigned to an old lab by myself with a barely functional piece of equipment. It took me a while to realize what was going on. Discrimination can be subtle. And denial can take a while to recognize and to admit the anger and the pain. Long story short, I dug deep. I got help with my instrument and data collection from the Abruña group back at Cornell and successfully defended my MS thesis before two committee members. My advisor chose not to come that day, saying he would pass me if my committee did. Which they did, and I graduated.

Which is not to say that the 2 years I spent at Penn State were nothing but frustration. On the contrary, some wonderful things happened while I was there, and they set me on a new and exciting path that has shaped my career and my family. As a graduate student, I was assigned to be a TA (teaching assistant) in first semester general chemistry. This was a very large course at Penn State (several thousand

Fig. 3 Stacey and Rich on their wedding day

students), taught in lecture sections of 500 and small recitation sections of 25 or so. I loved being a TA. I was fascinated with the reasons why students had problems learning chemistry (and, to be blunt, why some professors could be so terrible at teaching it!). I wanted to investigate these problems and create solutions. I took a class on higher education from Mary Ellen Weimer, and it opened my world to an entire literature I did not even know existed. So, I decided to make a change. Instead of looking for another PhD program in chemistry or another research group at Penn State, I wanted to get a PhD in chemistry education research (CER). Earning a PhD is the last opportunity for formal research training, and I wanted to learn how to investigate problems in teaching, learning, and assessment.

However, there were only MS programs in CER in chemistry departments then, so, I decided to return to Cornell. And then, something truly wonderful happened next. Because I was at Cornell again, Rich and I finally decided to marry—after 8 1/2 years of dating. When Pastor Schieber welcomed our guests to our wedding, he started the ceremony by saying he knew what all who were present that day were thinking: "it's about time!" (Fig. 3).

Now, Cornell did not award PhDs in CER, but they did have a structure within the graduate school where a student could create a novel interdisciplinary program with the endorsement of the faculty. At the time, my husband was a TA for Roald Hoffmann who suggested I contact Joseph Novak in the Department of Education at Cornell. Rich had taken a course with Joe and found it very interesting for thinking about a career in academia. I reached out to Joe, applied to Cornell, and soon had a committee of Joe, Roald, and Joän Egner who specialized in adult education and higher education. They helped me round out my chemistry coursework. (Roald insisted I take biochemistry, which I am forever grateful he did. It fascinates me.) I took courses in curriculum and instruction, learning theory, and the history of science education in the Department of Education, survey design and program evaluation courses in the College of Human Ecology, and applied statistics in the School of Industrial and Labor Relations. I met colleagues who were embarking on careers in physics education research. One day at a seminar on physics education, I learned from a chemistry professor that Jerrold Meinwald, an organic chemist at Cornell, had just gotten a grant from the Mellon Foundation to develop a chemistry course for nonmajors and needed a research assistant (at this point I was filing papers in the Graduate School Admissions Office to pay the bills). I set off to schedule an appointment with Dr. Meinwald.

What I thought would be a short conversation with his secretary turned into a conversation to join him right then and there. An hour and a half later, he had offered me an RA (research assistantship) to help him design the course and was sending me to a brand new Gordon Conference on chemistry education to learn the latest and greatest goings on in the field. I had no idea how attending that one seminar would be such a pivotal point in my career.

I went to that Gordon Conference not even knowing what a Gordon Conference was. I had no idea how prestigious they were. I had no idea you must apply and be accepted to attend by the chair. I had no idea I was the only graduate student in the room. I had no idea who else was in the room. I didn't know any of these things. My naïvete was probably a good thing or I might have been too intimidated to say anything during discussion. I was simultaneously fascinated, and frustrated, by the first three talks. There were really novel ways of teaching being presented, but no data on how it impacted student learning. Most of the talks simply said "my students really seemed to like it." I raised my hand, noted that I found such claims to be odd because as chemists we would never accept liking something as a substitute for data and that there were most certainly methods for gathering and analyzing such data, and I hoped we would see some as the week went on. (In hindsight, I cannot believe I had such chutzpah to do that! I was a new PhD student who thought she was learning the keys to the universe in her methods courses. But one could easily see how my enthusiasm might have been confused for hubris!)

Well, my comment (it never really was a question) turned out to be the last before we broke for lunch. I never got to stand up. I was surrounded by more people than I could count, all who were excitedly telling me about the planning grants they had just gotten from NSF for something called the Systemic Initiatives. (I told you I had no idea about who else was at this conference!) The NSF program officers were there, as were representatives from 14 different university collaborations. Each had $50K planning grants to prepare the submission of their $5M proposals to make radical changes in the undergraduate chemistry curriculum. They all needed to prepare detailed assessment and evaluation plans for these proposals. And more than a few were interested in what I knew about how to do such things. It was overwhelming to say the least.

The irony is, that at a time when I was so very excited to be finding opportunities in this new discipline of CER, an unexpected event in my personal life happened that was completely beyond my control. A few weeks before the Gordon Conference, I found out that I was pregnant. This was not planned and definitely unexpected. We weren't sure how this was going to work with both of us still in graduate school. But while I was out in California at the Gordon Conference, I started to miscarry. And even though I hadn't planned on getting pregnant, I was devastated to lose our first baby. I hadn't told friends the good news yet. How could I explain my near constant tears and sadness? It was a difficult time.

Postdoc (1994–1995)

I threw myself into my dissertation and the excitement of getting multiple postdoc offers to work on the Systemic Initiatives. Ultimately, I accepted an offer from the University of California, Berkeley. Rich and I graduated, packed up our two-bedroom apartment, and drove a rented moving van across the country.

Because the NSF had not yet officially made the award to Berkeley, I had to work on an alternative project for a few months. I was a postdoc with Angy Stacy in the Department of Chemistry, and we worked together on designing a new course for nonmajors. The grant was funded, and I became the Director of Evaluation, Assessment, and Pedagogy for a $5M grant (as a postdoc!). I couldn't believe my good fortune to have secured such a position for the next 5 years.

Meanwhile, Rich was looking for employment, which everyone had assured us would be easy to find in the Bay Area for a chemist. It took a long time. He taught part time for a university and worked for an environmental company. In the give and take of a marriage, Rich was willing to support me in pursuing my "dream job" first.

And while I learned a good deal of chemistry working with Angy, I learned even more from her example of how to combine motherhood with professorhood. When I walked into Angy's office on the first day, I remember my utter surprise at seeing a pack-and-go portable crib behind her desk. There were toys, the drawings of young children—things I had never seen before in a chemistry building, and certainly not in a professor's office. And to see this at a place with the stature of Berkeley! It forever changed my thoughts on what was possible. Angy didn't ask permission to bring her kids into the building. She just did it. We had many long conversations over the next year. I was especially keen to learn all I could about how she balanced family and career—for Rich and I had decided to have a child, and I was pregnant!

We thought we would be in California for several years, but in November a job ad appeared for a tenure-track position in CER at the University of Michigan-Dearborn. We knew that CER positions in chemistry departments were not commonplace. And Dearborn was only 2 hours from our hometown—and the four grandparents awaiting their first grandchild. We had to take a chance. I applied, interviewed when I was

Fig. 4 Suzannah Bretz in May 1996 helping her parents grade final exams

6 months pregnant (which was obvious despite my carefully chosen suit), and was offered the job. Suzannah was born in May, and we drove back across the country in July (Fig. 4).

Assistant Professor (1995–2000)

I began my tenure-track appointment with a 3-month-old child. My husband took an adjunct teaching position in the same department, and the chair agreed to not assign us courses at the same time so that one of us could always be with the baby. The plan was for Rich to continue teaching part time until Suzannah was old enough to go to the Child Development Center on campus (she had to be walking). I asked for and was assigned an office big enough to put a pack-and-go in. By February, I was pregnant with our second child who was due in late October.

When I've shared this story with young undergraduates or graduate students, I often get asked something akin to "weren't you worried they wouldn't think you a serious scientist to show up with a baby and get pregnant again during your first year?" And the simple answer is no. I wasn't worried. And, what's more, I didn't really care. My philosophy about family and work is pretty simple; I've never hidden my children, my wedding ring, pictures, toys, or kids' artwork. My children are an integral part of my life. If I have to hide them or not talk about them in order to be taken seriously at work, then that's not a place I want to work. And I have to say—that philosophy has served me very well throughout my entire career.

Fig. 5 Joey Bretz "reading" a general chemistry textbook in 1997

I was fortunate to work for a university with progressive policies. In addition to stopping my tenure-clock for one year, I was able to also benefit from a "modified duties" policy. This policy provides for circumstances when the 6–8-week recovery period *after* delivering a baby falls during an academic semester. This policy advises modifying one's duties from a combination of teaching/research/service to just research/service. So I didn't teach that fall. I worked on designing a new course instead. It was a fabulous policy and still exists at the University of Michigan. I have recommended it to many women at many universities in the years since.

Joey was born in October—nearly 3 weeks early. Rich continued to teach as an adjunct; Suzannah went to the Child Development Center for a few mornings each week. By the time Joey (Fig. 5) was old enough to join Suzannah at the Child Development Center, Rich was itching to return to work full time and was hired in a diesel engine research lab by Ford Motor Company. Because Rich had an "8–5" job, I was the parent with the more flexible schedule. I dropped off the kids and picked them up. My hours were not as long as everyone else's in the lab, but the moment they both were fast asleep, I went back into the office and worked many long nights for a few years. Life had reached a new "normal" of sorts.

One time I returned from a conference, and a colleague asked which child had been sick for the last few days? I came to learn that because I wasn't in the office at the same hours as many of the other faculty, some had presumed I wasn't working or that I was home again with a sick child. I decided to "announce" in the next faculty meeting that my schedule may not be typical but that if colleagues needed to reach me, I returned to the building most every evening from 8 to midnight. To this day, whenever I travel professionally, I print and hang a sign on my door that says "Dr. Stacey Lowery Bretz will be out of town from mm/dd/yy to mm/dd/yy at University of XYZ or the ACS Conference or. . ." No one ever need wonder where I am again or default to the assumption that my absence must mean I am at home being a mother.

In early 1999, I was pregnant once more with what we hoped to be our third, and last, child, but once again, I miscarried. We really wanted a third child, but were learning once more we cannot dictate the timing of everything in our lives. I think this is a particularly challenging lesson for some academics to learn. We go to school. College is 4 years, graduate school will be 5, postdoc for 2, tenure clock for 5, etc. We have it all planned out, and for the most part, we hit the milestones on time and just as we had planned them. Pregnancy and childbirth, however, are oblivious to our time clocks and planned milestones. And this can be a difficult thing to accept for most people, but I think especially for academics.

Associate Professor (2000–2005)

About that same time, I had realized that despite loving my job, I craved the opportunity to mentor graduate students in my research. I would go to conferences and see what colleagues were accomplishing with graduate students. I found myself with a severe case of "research envy." I needed to find a way to "move up the food chain" from an undergraduate institution to one that had a graduate program. This would be difficult to do given the expectations for research productivity and what I had been able to accomplish with limited resources at my current institution. But, in early 2000, I was invited to apply for a job opening at Youngstown State University. They were looking for someone to grow their MS program to include the discipline of CER.

When they called, I was already 8 months pregnant with our third child. I thought I would have to wait until after she was born, but that would be at least 2–3 months as I knew I would be having my third C-section. Ultimately, my obstetrician advised me to go now, *before* the baby was born. He wanted my husband to drive me there (4 hours away) just in case there were any unexpected surprises. But we also had a 4-year-old and a 2-year-old we couldn't leave behind. The solution? We stopped in our hometown on the way and picked up my in-laws. Yes, that's right. I went on a job interview 10 days before giving birth to our third child with my husband, my two young children, and my in-laws. We stayed in one suite at the Residence Inn. (I'll pause here while you stop chuckling.)

To say that I was "huge" would be kind—it was my third pregnancy. (My first baby weighed over 9 pounds, my son was over 11 pounds and nearly 23 inches long. This baby was going to be no different.) They broke a typical 1-day interview schedule into 2 half days. After just one meeting, they made everyone walk to the chemistry department, including the deans, for fear that asking me to walk around campus that much would send me into labor (they only told me *that* more than a year later!). The job was great, and I wanted it. Just one problem: Rich was working at Ford and would have to give up a full-time job he loved to start job hunting once again in Youngstown, OH (not exactly a place of booming chemical industry). We

Fig. 6 Mikaela Bretz helping her mom with science experiments in fall 2005

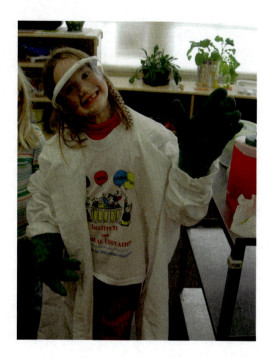

talked it through, and once again, my husband put his career needs second so mine could take another important step forward. Mikaela was born in March 2000 (Fig. 6), and we moved during the summer.

Rich once again took up part-time teaching in the chemistry department and began to volunteer with ACS Career Services. He found it very fulfilling and was becoming an expert on the "two-body" problem. One challenge we faced repeatedly was how to travel to conferences. Childcare was nonexistent at ACS meetings at that time. We tried lots of different "arrangements," but none was a permanent solution. We brought grandparents with us. We left the kids at home with their grandparents. We "divided up" the meeting—he went for the first few days, I went for the last few days, and we literally were up in the air in two different planes at the same time—hoping for one of us to make it home in time to meet the school bus. We hired an undergraduate to join us and paid her travel and a stipend to watch the kids and entertain them while we were both busy at the same time (Fig. 7). One of us went, and the other stayed home, missing the entire meeting. And, sometimes we just took the kids with us, no help whatsoever, and we juggled at the meeting in real time. My kids learned early on how to navigate the expositions at ACS meetings and how to sit quietly in the back of a research talk that Momma was giving. They've been to San Diego, San Francisco, Anaheim, Denver, Salt Lake City, Washington DC, Orlando, Boston, New York, Philadelphia, Chicago, and probably other places I'm forgetting. Did I miss out on some great talks? Yes. Did I miss opportunities to go out for drinks with colleagues and form new collaborations? Probably. But, did our family get to visit zoos and museums and try new restaurants and see how many forms of

Fig. 7 Stacey, Michelle (a college student hired to babysit), Mikaela, Suzannah, Rich, and Joey at an ACS meeting in Philadelphia in August 2004

transportation we could take on one trip? Yes! And we've made great, great memories doing so. (Now as teenagers, they beg to stay home and "take care of themselves," urging Rich and me to go by ourselves and "have fun!")

Rich was soon hired in the Provost's Office to work on initiatives for the recruitment and retention of underrepresented students. He was instrumental in the opening of the first Early College High School in Ohio on the Youngstown State Campus, an initiative funded by the Bill and Melinda Gates Foundation. We decided to add to the balancing act by bringing two basset hound puppies into our lives—Rensselaer and Cornell. We had once again established a new equilibrium.

Which is not to say there were unexpected "hiccups" in the daily routine. Children get sick and can't go to school. Or, they get pink eye and aren't permitted in school. More than once I worked from home on days I wasn't teaching. Or, I brought a child with me to sit in the front of my classroom while I taught class. They loved to bring their backpack of crayons and coloring books and sit in a desk "like the big kids." One time, Suzannah toddled up to the board to answer a question about equilibrium that had stumped the class for a few seconds!

Professor (2005–?)

By 2005, my career was progressing rapidly. I was elected by my peers to chair the Gordon Research Conference on Chemistry Education Research and Practice—the very same GRC I had attended just 12 years earlier as a bright-eyed young graduate student. (It still seems surreal to me sometimes.)

I was also offered the opportunity to move to Miami University and have PhD students in CER. There were fewer than two dozen chemistry departments in the country where I could do this. To have a job offer from the one program in Ohio was tempting. But accepting this job would once again mean asking Rich to step down from a full-time job he found exceptionally fulfilling both personally and professionally. Miami and Youngstown were 5 hours apart. Commuting was not an option. Living apart was not, either.

We both went back to Miami on the second visit trip after they made me the offer. Not only were we going to negotiate my position, but we were going to negotiate one for him, too. Once again, we took our children and that meant bringing my in-laws to watch them while we were negotiating (I combine my career and motherhood in lots of ways, but trying to negotiate space and salary while wiping a nose or changing a diaper seemed a bit much, even for me!) I put to use every strategy I had learned in the COACh workshops (Committee for Advancement of Women Chemists). They worked, and worked well. (When the meeting was over, my department chair asked me where I had learned to do that. He then wanted to know if COACh would ever let men attend their workshops!) In the end, we chose to move to Miami.

Rich was hired as a visiting assistant professor. It was full-time work, but only temporary. The job hunt for a permanent position for my husband was on once again. During the next three years, he moved into a half-time administration/half-time instructor position, but in reality, it was like he had two full-time jobs.

When we moved to Miami, our youngest was just starting kindergarten. Never once have all three children been in the same school building and on the same schedule in terms of when they get on the bus and when they get off the bus. We've "split" the day—Rich is an early riser, so he would get up and head into the office early. I'd sleep a bit later and then drop off the children at the elementary school (we transferred our children from the elementary out near our township home to the one closest to the university). He would pick them up and bring them back to the department. We convinced the department chair to move Rich's office next to mine. The kids would play on our computers, eat snacks from our always stocked mini-refrigerators, and wait for us to decide it was time to go home. Eventually, we were able to convince the school district to route one of the buses through campus. Our middle school children enjoyed the independence of walking one block to our offices from the bus stop.

I am very proud to share that Rich was hired in 2012 as an Assistant Professor of Chemistry at the Miami University Hamilton campus. Once again he is working with nontraditional students and our family is "back on the tenure-track." It's my turn to

Fig. 8 Ellen Yezierski and Stacey unexpectedly wore the same outfit to work one day

make sure if there are any sacrifices to be made in terms of schedule or family conflicts that his career takes priority now.

Moving to Miami catalyzed my research into high gear. The CER program at Miami has become one of the top in the country. I have an amazingly talented colleague named Ellen Yezierski (Fig. 8) who makes going to work each day exciting and challenging. I've learned a lot from her. I've had the privilege to mentor 4 postdocs, 19 graduate students, and dozens of undergraduate research students. These talented people have helped amass a record of 85 publications and counting, more than 150 invited seminars, 260+ conference presentations and posters, and research findings that are changing what we know about how to measure student learning.

I also know that the students whom I have mentored watch and learn from Rich and me that balancing motherhood with a career as a professor is an ongoing dynamic equilibrium. Some days are more in balance than others. Our family dynamic has changed many times over the years, with each age and school year of our children bringing new challenges. When our oldest child Suzannah started college at Cornell University and then 2 years later when Joey started at Miami University, we had to figure out how to get two children moved into two colleges when both parents needed to be at their own universities to get the school year underway—all with our younger daughter Mikaela at home with the life of a busy teenager (but one who didn't yet drive.) Just when we think we've figured it all out, we learn we have more to learn.

Having It All vs. Having Both

I started this chapter by evoking the memory of the 1970s and 1980s "superwomen" who could "have it all"—family *and* a career. While I used to think that dual reality was possible, what I've learned is it's an unrealistic standard to try to live up to. To put pressure on oneself to never fall short at work and to never fall short at home—that's a recipe for stress to be sure. You will miss an important conference or talk. You will miss the occasional spelling bee or field trip.

But, you can have *both* a family and a career. And you can find both to be incredibly fulfilling. You will choose to miss a faculty meeting so you can cuddle your sick child on the couch. You will explain to your children that you cannot stay home with them just because school was closed due to a snowstorm. The university is still open, and you must teach your class because that's how Momma and Daddy pay the bills.

In the end, it is pointless to worry about what your colleagues think of your choices. The only people you need answer to are yourself and your family. I've learned that children grow up very quickly (Fig. 9). There will be plenty of time to spend grading papers and going to conferences once they're out of the house. Then it can be all the work I want. In the meantime, I plan to continue indulging in both my family and my career. The two are inextricably linked, and both bring me great joy (Fig. 10).

Fig. 9 Suzannah, Mikaela, and Joe at Christmas 2016

Fig. 10 Rich and Stacey at her 25th high school reunion

Last But Not Least

Right before we moved in the summer of 2000, the Biennial Conference on Chemical Education was held at the University of Michigan. One night there was a social event held at the Henry Ford Museum and Greenfield Village in Dearborn. I went to the reception and took Mikaela (who was an infant) with me. My colleague, and dear friend, Marcy Towns (Fig. 11), was also at that meeting with her son, Jimmy, who was just 6 weeks older than Mikaela. At one point in the evening, we both wandered off into a remote corner of the museum, looking to find a place where we could quietly breastfeed our babies. Neither one of us expected what happened next. Slowly, one by one, women graduate students and young women faculty approached us,

Fig. 11 Stacey and Marcy Towns at a $2YC_3$ (Two-Year College Chemistry Consortium) conference banquet in 2011

often timidly asking "Can I ask you a question?" They had many questions on their minds. When did you tell your chair you were pregnant? How did you decide when to have a child—pre tenure or to wait until after? How did you avoid being in the teaching lab if there were chemicals thought to be harmful to your baby? Before long, Marcy and I realized we were surrounded by nearly 20 young women looking to us for sage advice on combining motherhood with being a professor. We certainly didn't pretend to have all the answers or even any answers that might be useful in the circumstances of any one of those young women. All we could do is tell our story, as I do here once again. In doing so, it is my hope that some of you have found a new thought or idea here that might be useful to you and your family in the search for balance and equilibrium.

Acknowledgments I want to express my deep gratitude to my parents, James and Ruth Lowery, for their love and unending encouragement to chase my dreams. I thank my children, Suzannah, Joe, and Mikaela, for teaching me about unconditional love. And I thank my husband, Richard, for making our life together more than I could ever hope for.

About the Author

Education and Professional Career

1989	BA Chemistry, Cornell University, NY
1991	MS Chemistry, Penn State University, PA
1994	PhD Chemistry Education Research, Cornell University, NY
1994–1995	Postdoctoral Fellow, University of California Berkeley, CA
1995–2000	Assistant Professor, University of Michigan-Dearborn, MI
2000–2005	Associate Professor, Youngstown State University, OH
2005–2015	Professor, Miami University, OH
2015–present	University Distinguished Professor, Miami University, OH

Honors and Awards (Selected)

2009–2014	Chair of the Board of Trustees for the American Chemical Society, Division of Chemical Education Examinations Institute
2010	Fellow, American Association for the Advancement of Science
2012	Fellow, American Chemical Society
2010–2012	Member, National Research Council Committee on Discipline Based Education Research

In 2017, two general chemistry textbooks featuring pedagogy grounded in the findings of Stacey's chemistry education research were published by W.W. Norton, Inc.:

- *Chemistry: The Science in Context* (5e) by Gilbert, Kirss, Foster, Bretz & Davies, 2017
- *Chemistry: An Atoms-Focused Approach* (2e) by Gilbert, Kirss, Foster, & Bretz, 2017

If at First You Don't Succeed, Don't Give Up on Your Dreams

Pamela Ann McElroy Brown

Photo Credit: Kevin Ragjaram

In the Beginning

I first fell in love with chemistry in high school, admiring the patterns in the periodic table which allowed prediction of the properties of the different elements, enjoying the challenge of balancing equations and stoichiometry, and recognizing the potential for solving societal problems with chemistry. Even before that, I had fallen in love with babies. My sister was born when I was 10 years old, and we shared a room. Her sweet smile and joy at discovering the world were irresistible.

My goals when I left for college were to get a degree, maybe even a master's, get married, work for a few years, have a baby, stop working, and in the 5 years before kindergarten earn a PhD. I would then get a job as a college professor when my child started school, where I could continue to conduct research and help to educate the next generation. I really wanted to be around in those important preschool years

P. A. M. Brown (✉)
New York City College of Technology – City University of New York, Brooklyn, NY, USA
e-mail: pbrown@citytech.cuny.edu

which I had seen with my sister were precious but fleeting. It did not matter to me that no woman in my family had even attended college; this was my plan! My own mother, a farm girl from Central California, had dropped out of high school after 10th grade and was married on her 17th birthday, the first day she was legally able to do so without parental permission. I was born when she was still a teen and was often mistaken for her younger sister. She felt trapped by her lack of education and domestic responsibilities and constantly urged me to get an education so that I could be economically self-sufficient. My father was 8 years older than her and earned a good living as a radio announcer. He felt, however, that since he was the sole wage earner, all childcare and housework were my mother's responsibility. Finding a husband when I was in college, who would help with the housework and childcare and was supportive of my dreams, was an important component of "the plan."

The Plan Unfolds

I began dating my future husband Harvey in college. After earning my BS in chemistry from the State University of New York at Albany, my tuition and living expenses were paid for by a Gulf Oil fellowship while I worked toward a master's degree at Massachusetts Institute of Technology (MIT). After finishing my master's, I was employed a year at American Cyanamid doing research on hydrodesulfurization catalysts. We saved enough to get married and buy our first home. I needed to relocate as my husband worked in the family demolition business and did not have that flexibility. I was able to find employment at Pall Corporation, a manufacturer of medical and industrial filters, working in technical service. This was my dream job. I would arrange for potential customers to send samples to our labs to determine which of the products would best meet their filtration needs. I would also travel to their facilities and bring equipment so that they could evaluate the filters themselves under process conditions. I would help existing customers if their specifications changed or if they encountered problems. Once I discovered I was pregnant, I decided that I would return to work after a few months of maternity leave. I enjoyed my job, had family living nearby to help out, and also felt the financial pressures of home ownership.

Motherhood

My daughter Heather was born when I was 26. She was beautiful beyond words and seemed perfect. Unfortunately, when she was a few weeks old, it was discovered that she had a congenital birth defect—a dislocated hip. She was born without a hip socket and if left untreated would walk with a limp and develop painful arthritis at an early age. The treatment was a brace that needed to be worn essentially 24 hours a day, 7 days a week for a year or more. The brace pushed the legs into the pelvis, slowly creating a hip socket in the soft, infant bones. If the brace wasn't readjusted correctly after diaper changes, etc., the treatment would be ineffective. I decided to resign from my job and stay home and take care of her. My second daughter Vanessa was born 17 months later. When the girls were 18 and 35 months old, I returned to college to begin work on a PhD. I had been awarded a teaching assistantship which paid for my tuition and provided a stipend for expenses, which I used for childcare. I took classes in the evenings and taught a 6-hour lab on Fridays. Because I already had a master's degree and had also taken graduate courses as an undergraduate, I was able to complete my required coursework in one academic year and pass the qualifying exams. I made arrangements with one of the faculty members to begin research on my dissertation the following fall, again receiving a stipend and tuition paid through a research assistantship. This is one of the many advantages of pursuing graduate work in the chemical sciences—the opportunities to earn a degree with no out-of-pocket expenses in exchange for working as a teaching or research assistant. My education was supported by the National Science Foundation and a James Lago Fellowship.

My dissertation project involved a novel process for the purification of terephthalic acid (TPA), which is used to synthesize the long-chain polyester, polyethylene terephthalate (PET). PET is used to make products ranging from plastic soda bottles to fabric. The PET produced today is much stronger because it is made from purer TPA. My work contributed to this improvement.

A little over 3 years after starting research, I successfully defended my dissertation. During this time, I became an efficiency expert—I would plan meals a week in advance, so I would only need to make one trip to the supermarket. While my fellow graduate students would socialize in the mornings, I would get right to work, arriving early and leaving late to avoid rush hour traffic. About a year before finishing my dissertation, my son William was born. Fortunately, all experimental work had been completed, and my advisor, who was very supportive, allowed me to finish up from home and continued my financial support, although at a reduced level. During naps I finished writing computer simulations and my dissertation. After earning my PhD, I was exhausted and took several months off. I then continued to work for my advisor for a couple of years on a part-time basis as a consultant—I would come in and run experiments, do literature surveys, and write reports. I also published two papers on my dissertation research in peer-reviewed journals during this time (Fig. 1).

A Career in Academia

I decided that a career in academia would allow for good work-life balance and was a better fit for my interests and the realities of the local economy. The chemical industry in the metropolitan New York area was shrinking, and positions were few and far between. The academic calendar was much more flexible than that of the

Fig. 1 Heather, Vanessa, and Will, first day of school 1993

corporate world. When my son was in kindergarten, I started a 1-day-a-week position teaching a physical chemistry laboratory at Barnard College. After a year I was offered a full-time position as a visiting assistant professor at Stevens Institute of Technology. It was a 9-month-a-year-nontenure-track-teaching appointment. I hadn't even applied for the position but had been highly recommended by my dissertation advisor when someone else left just before the start of the semester. I was thrilled! My coworkers were nice, and the students were a delight to work with. I stayed at Stevens for 4½ years but became increasingly frustrated by the lack of job security and professional growth potential—there was always uncertainty waiting for another annual reappointment and no opportunities to conduct research. My salary was also considerably less than that of the tenure-line faculty. I enjoyed developing new curricular materials and managed to present at educational conferences and publish in educational journals. One of my happiest memories was taking two of my children with me to present at a conference in Washington, DC.

Finally, I was offered a tenure-track position as an assistant professor at New York City College of Technology (then New York City Technical College), a branch of the City University of New York, in a 2-year chemical technology program. The department, Physical and Biological Sciences, included faculty in biology, chemistry, and physics. By now my children were 16, 14, and 10—they were growing up fast! Since my commute was 90 minutes each way, arranging car pools to after-school activities was a perennial challenge. Meals were planned a week in advance, and fortunately my oldest daughter enjoyed cooking and frequently prepared dinner. When she was busy, we often had "frozen food feasts" to save time but still have meals together as a family. After dinner everyone would finish their homework at the kitchen table, and I would help them out as needed.

Their academic success was a high priority. Harvey was also a big help despite his long hours. I continued to have childcare available after school and on school holidays, but my oldest daughter in particular was beginning to resent having a "stranger" in the house.

There are three major areas of responsibility for tenure-line faculty—teaching; service to the department, college, and university; and scholarship. In 2-year programs, the major focus is on teaching and service, although there is the expectation of scholarship as evidenced by publications, presentations, and other scholarly works. Faculty employed at the City University of New York, which includes community colleges, 4-year institutions, and graduate and professional programs, are employed under a collective bargaining agreement. The expectations of productivity in all three areas are the same, although the emphasis is different depending on the focus of the institution. When I began, my teaching load was 27 hours annually, typically 15 hours one semester (five sections meeting 3 hours each week) and 12 the other (four sections meeting 3 hours each week)—teaching and service were the main focus. I also advised students, helped those in my classes during office hours, was the faculty advisor to the chemistry club, wrote letters of recommendation, helped students obtain internships, and served on several campus-wide committees. I ran for College Council, the faculty governance body, and eventually chaired the committee making policies related to students. I began a research program with undergraduates, presented at regional and national conferences, and continued to develop new curricular materials and publish in educational journals. Every summer I applied for, and was awarded, grants through the American Chemical Society's Project SEED, which provided generous stipends to promising but underprivileged high school students. The undergraduates developed their own mentoring skills working with the SEED students and had fun. One project even resulted in a publication in the *Journal of Undergraduate Chemistry Research*, while one of the contributors was still in high school!

After 3 years I applied for and was awarded a promotion to associate professor, and after 5 years I had tenure at long last! My two daughters had graduated high school and went off to college, and my son was now in high school—where had the years gone?

Teaching in a 2-year program allowed me the opportunity to teach the subject that I loved and really make a difference in the lives of my students. Many of the students that I worked with went on to graduate and professional schools. Some are even adjunct chemistry faculty members on my campus—sources of inspiration to current students. Mentoring promising high school students was a delight. A moment I will never forget is when I heard a knock on my office door one summer while on the phone. Not wanting to get up, I asked, "Who is it?" My SEED student responded, "A budding scientist."

When the chemical technology program coordinator was appointed as acting dean, I became the program coordinator. In addition to teaching, my responsibilities now included scheduling chemistry courses and faculty, interviewing and recommending adjunct faculty for part-time teaching positions, handling student complaints, evaluating transfer credits, and facilitating monthly meetings with all chemistry faculty members to discuss departmental issues such as curriculum development, textbook selection, and social activities. I set up the annual meeting with our External Advisory Board, made up of alumni and other representatives of local industries, to assure that our curriculum met workforce needs. I continued to conduct research with undergraduates and involved high school SEED students during the summer.

Sixteen months after being asked to be the program coordinator, I was asked to serve as acting dean of the School of Arts and Sciences. Two years later after a national search, I was appointed dean. In that capacity I oversaw eight academic departments. Part of my responsibilities included fund raising through grant writing. I was the principal investigator on a $1 million National Science Foundation (NSF) grant to increase the number of STEM graduates and was the co-PI on several other NSF grants, totaling over another $2 million.

After 6 years as dean, I had the privilege of serving as a program director for the National Science Foundation in the Division of Undergraduate Education, in Arlington, Virginia, under a 1-year appointment. The National Science Foundation is the largest government agency responsible for supporting external research. Scientists and engineers submit proposals which are reviewed by experts in the field, to help assure that only the most meritorious are funded. In my capacity as a program director, I managed a portfolio of ~50 current awards and made funding recommendations totaling ~$13.5 million. I identified reviewers, assembled and oversaw review panels, and contributed to new solicitations. I also traveled around the country providing outreach to the larger undergraduate education community through presentations on funding opportunities and NSF initiatives and also offered grant writing workshops.

Upon my return to the college, I was promoted to associate provost. My responsibilities now include oversight of program review and reaccreditation, program and curriculum development, faculty professional development, and coordinated undergraduate education.

My oldest daughter, Heather, has chosen a life in academia and is a lecturer in Health Economics at Newcastle University, in Newcastle, England. She earned a PhD from Sheffield University in England and is married to Adam. They met in Scotland, while she was doing a postdoc. They are parents to preschoolers Sadie and Jonah. Vanessa works in human resources, is married to Paul, and lives in Manhattan, New York. William graduated from law school and works in labor law in Queens, New York, where he lives with Jane. My husband Harvey and I are still happily married, are enjoying traveling and local cultural activities, and are looking ahead to retirement in a few years.

Reflections, Advice, and Lessons Learned

While I consider raising my children to be my greatest life work, my career was always a source of great pride and joy. A job in academia allowed me to do both. My oldest daughter tells me that my insights helped her earn her PhD and obtain her position. She teaches graduate students, and her research is her most important professional priority. Vanessa fondly remembers Saturdays at science fairs, and William always enjoyed the kitchen table science experiments I tested on him before demonstrating to my classes (red cabbage juice is a great pH indicator!)

The greatest challenge that I encountered in my career was obtaining a tenure-track position once I returned to the workforce full-time. While tenure-line positions were once the norm, there is a growing shift toward part-time instructors. While some are employed full-time in industry and teach part-time for personal satisfaction or supplemental income, many patch together full-time employment by working part-time at multiple campuses. For those in this situation, there is little security, much hard work, few if any opportunities for research and scholarly growth, and typically annual salaries much lower than those of tenure-line faculty. At my campus about 50% of the classes are taught by part-time faculty members.

In my opinion the best way to prepare for a tenure-line position is to speak to as many people in academia as possible to learn about the culture and expectations. Ask about the hiring process and the characteristics of successful applicants. Peer-reviewed publications are a benchmark of scholarly accomplishments. Plan for publishing even before commencing research. Read the journals and model your writing after successful authors. Obtain teaching experience—you can start as a tutor or peer mentor as an undergraduate and serve as a teaching assistant (TA). Look for teaching opportunities in graduate school. Read journals such as the *Chronicle of Higher Education* to learn more about the issues facing higher education today. Assessment of student learning and cycles of continuous improvement are mandates of the accrediting agencies. Learn about assessment and if possible get solid experience. Read journals such as the *Journal of Chemical Education* to learn more about pedagogy, curriculum development, and effective practices in the classroom. Learn about strategies for writing proposals—in many cases grants will fund your research and other projects. In summary, academia is a challenging career, with many demands, requiring years of preparation. It is also extremely rewarding, with opportunities for work-life balance and personal satisfaction.

Reflections from Pam's Spouse
Harvey Brown

It was exciting, demanding, but never boring being the husband of Mom the chemistry professor. The additional responsibilities involved in taking care of three children, and supporting the household both economically and physically, while Pam pursued her advanced education were daunting at times. Pam's focus and endless energy made it easier to keep the end goal in sight. Having the help of nearby family made it easier to get through this challenging period.

Looking back I can't believe Pam was able to go to school full-time and raise three now very successful children. The same reflection finds it hard to

(continued)

believe that I worked 12–14 hours a day and came home to give the kids a bath. It makes me proud what we did together.

Reflections from Pam's Daughter
Vanessa Brown

There were many benefits to growing up with a chemistry professor mother. She taught us to always look for different ways to solve a problem. And she made the pursuit of knowledge fun. My advice to other mothers in the field: don't be afraid to expose your children to information and knowledge that is beyond their current age or grade level. I recall my mother teaching me the basics of calculus derivatives while I was in middle school. As I advanced in my education and began taking physics and calculus, I already had a fundamental understanding of how they worked. While I watched my classmates struggle, I was able to thrive; they were rote learning, I was comprehending (Fig. 2).

Fig. 2 Graduation celebration—from left to right: Harvey, Vanessa, Pam, William, and Heather

Acknowledgments There were many people along the way who served as my mentors. As an undergraduate at the State University of New York, my physical chemistry professor, Dr. A.J. Yencha, provided me with the opportunity to conduct research with him. He provided invaluable advice when I applied to graduate school. My PhD dissertation advisor, Dr. Alan Myerson, was extremely understanding and supportive of my family needs. He helped me to develop a research project where much of the work could be done at home using the computer. He made sure that I had access to the resources needed in the laboratory. He was a brilliant scientist combining his responsibilities as department chair with an active research program. He was patient when experiments lead to a dead end and helped me to develop more successful approaches.

Throughout my professional life, mentors continued to play an important role. As mentioned earlier, once I became the dean of Arts and Sciences, I wrote a successful National Science Foundation $1 million STEP grant to increase the number of graduates in science, technology, engineering and mathematics (STEM) at my institution. My program director, Dr. Susan Hixson, provided much-needed advice on how to successfully manage a large, multidimensional grant. Through annual meetings for principal investigators, I became part of a national community dedicated to student success, sharing information and ideas. I was asked to review proposals and met Program Director Elizabeth Dorland, who purposely arranged connections within the chemical education community. Through her intentional networking, I met Dr. Rick Moog, principal investigator (PI) on several NSF grants, and was able to introduce process-oriented guided-inquiry learning (POGIL) on my campus. I also met Dr. David Burns, another PI, and helped my campus join the Science Education for New Civic Engagements and Responsibilities (SENCER) community. Both initiatives led to innovative curriculum and improved teaching.

The support and love of my husband, Harvey, and my three children made becoming "mom the chemistry professor" possible. For that I am forever grateful.

About the Author

Education and Professional Career

1977	BS Chemistry, summa cum laude, State University of New York at Albany (now University at Albany—SUNY), Albany, NY
1979	SM Chemical Engineering Practice, Massachusetts Institute of Technology, Cambridge, MA
1989	PhD Chemical Engineering, Polytechnic University (now NYU Tandon School of Engineering), Brooklyn, NY
2012–present	Associate Provost, New York City College of Technology, City University of New York (CUNY), Brooklyn, NY
2005–2012	Dean of the School of Arts and Science, New York City College of Technology, City University of New York (CUNY), Brooklyn, NY
2011–2012	Program Director, NSF DUE
2003–present	Associate Professor: New York City College of Technology, City University of New York (CUNY), Brooklyn, NY
1998–2002	Assistant Professor: New York City College of Technology, City University of New York (CUNY), Brooklyn, NY

1993–1998	Visiting Assistant Prof., Chemical Engineering Department, Stevens Institute of Technology, Hoboken, NJ
1980–1981	Staff Scientist, PALL Corporation, Glen Cove, NY
1978–1980	Chemical Engineer, Catalyst Research Dept., American Cyanamid, Stamford, CT

Honors and Awards (Selected)

| 2013–2015 | Selected to serve on the National Research Council of the National Academy of Sciences Committee on Barriers and Opportunities in Completing Two and Four Year STEM Degrees |
| 2013–2014 | Member of the American Chemical Society Committee on Technician Affairs |

Pam was honored by the American Chemical Society (ACS) for 5 years of service to the Project SEED program for serving as a volunteer summer research mentor for economically disadvantaged but promising high school (HS) students interested in chemistry.

Invictus

Amanda Bryant-Friedrich

Photo Credit: Mike Olliver

As I sit here and look at this thing called my curriculum vitae, I see only words. Words that are supposed to be a description of my life's journey. But there are so many words that are not there, sacrifice, fear, fatigue, racism, sisterhood, joy, and guilt. That is when it becomes clear that this document is only a small part of who or what I am. These missing words describe the real journey. This is what I want my children to know about, and this is what I want the world to remember: what it takes to be a chemist and a mother not only to my biological offspring but to the children of science.

A. Bryant-Friedrich (✉)
The University of Toledo, Toledo, OH, USA
e-mail: Amanda.bryant-friedrich@utoledo.edu

© Springer International Publishing AG, part of Springer Nature 2018
K. Woznack et al. (eds.), *Mom the Chemistry Professor*,
https://doi.org/10.1007/978-3-319-78972-9_6

Motherhood?

I didn't always want to be a mother. As a girl growing up in the 1970s and 1980s in rural North Carolina, the idea of marriage and motherhood did not fascinate me. I had no dreams of white picket fences and a knight in shining armor to come to my rescue. Being raised a good Baptist girl, I knew that *first comes love, then comes marriage, then comes the baby in the baby carriage.* No husband, no children. Being raised by a father, an African-American man, who did the very best that he could for me, my mother, and my brother with the little resources that he had, is something that I am thankful for every day, but it did not make me feel like I needed someone to take care of me. I needed to learn. I wanted to be productive and help people. Life circumstances showed me that becoming a "medicine maker" to help people feel better was what I wanted to do.

A Fool's Game

The complexities of choosing this as my life's work were unknown and unimportant to me. Did I know that women were expected to be mothers? Sure. Did I know that women did different kinds of work than men? Of course. Was this important to me? No. I just wanted to be what I wanted to be. I believe that this approach to my life was built on a foundation of knowing very early on that I had to have a purpose, and I knew what that purpose was: to help people through the treatment of disease. I started to develop this purpose as a vague vision while an elementary student. It was not until I was fully engrossed in my undergraduate academic pursuits that I became aware of the fact that life was more than one's career goals and that maybe I didn't have to choose only being a top-notch chemist, but maybe I could enjoy the things that should make my life richer, like love or maybe even companionship. But where were the examples of those who had mastered this magic trick? My professors in chemistry and physics and math and almost everything else at North Carolina Central University were men. I did internships as an undergraduate at two very prestigious corporations. I did not see women who were married or raising children, or, if they were, it was kept a secret from me. Brown women were so few and far between in my academic life that their presence as role models didn't even cross my mind. But, I still pursued my professional dreams and began to believe that I could enjoy any life I wanted; I just needed to design the model because I had no template.

At North Carolina Central University, I had the opportunity to perform research very early in my training there. Dr. John Myers invited me into his laboratory, and it was there that I fell in love with synthetic organic chemistry. Making molecules that had the potential to treat disease was definitely the definition of a "medicine maker!" Federal funding from the National Institutes of Health (NIH) Minority Biomedical Research Support (MBRS) program provided me the room and mental time to focus on science and becoming a great chemist. This work, along with extensive research experience at Dow Chemical and Merck Sharp and Dohme research laboratories, opened the doors to Duke University, which provided me with the opportunity to pursue my PhD and meet the most important person who would change the direction of my life.

Duke University was a destination on my journey that slipped into my vision very early on. I was told by a wealthy (by rural NC standards) white woman, for whom my mother made clothes, that education was the only thing that no one can take away. She told me Duke was a great university, and I never forgot that. My high school counselors encouraged me to attend North Carolina Central University for my undergraduate studies instead of Duke. Thank God, but the desire to attend never went away. If I had not arrived on Duke's campus when I did, I would never have started to work for Richard Polniaszek nor would I have met Klaus Friedrich.

Dr. Polniaszek was an excellent scientific mentor. He trained me to perform synthetic organic chemistry with the expertise of the masters. He had a seriousness for the discipline that was unrivaled. He had high academic and scientific standards which made me fight for my dream in a way I never imagined I would have to.... And that is where Klaus Friedrich came into the picture.

Yes, I met this wonderful tall German Adonis as just another dude who came to the USA to fulfill his own destiny in the realm of chemistry. He worked in the laboratory of one of the other two people of African descent in our program. Little did I know that, through seeking the support of my ethnic brother, I would find the support of a human being who would become my soul mate. Klaus noticed early on in our relationship that my struggles were different than the struggles of my classmates. My academic progression was in jeopardy when we met, and he became determined to make sure that I would obtain my master's degree. This obstacle on my road to the PhD was put into place as a requirement for my continuation in the program due to my less-than-stellar academic performance in the classroom. Klaus taught me concepts in chemistry that I had only been introduced to as an undergraduate and had little time to comprehend in depth during my graduate studies. He worked with me hour after hour and day after day to make sure I was ready for my exam. It was clear to me that he wanted me to succeed. When I successfully defended my thesis, not only was I ready to move on to a new phase of my professional life, but I had found the companion that I never really looked for.

Only months after I successfully defended my thesis, Klaus returned to Germany to pursue his dream of becoming an industrial chemist, and I continued with my PhD work. It became clear quickly that the powers of the universe had a different plan for me. The pursuit of my PhD was placed on shaky ground due to forces beyond my control surrounding my research advisor, and my heart was soundly intertwined with Klaus'. Recognizing that I wanted to pursue **all** of my dreams, in 1993, I moved to Germany to begin my PhD work with Professor Richard Neidlein at University of Heidelberg, and I moved in with the love of my life. Was it difficult? Yes! The clash of cultures, language barriers, and just being took on an entirely new meaning. It is clear to me that all of this was preparation for a life, full and abundant. We married in 1994, and I received my PhD in 1997. The decision to join the academy came early in my graduate studies, as I struggled at Duke, so postdoc studies were deemed a must. I joined the laboratory of Dr. Bernd Giese at the University of Basel in 1997, and it was during this time that motherhood took on a new meaning.

Under the Bludgeonings of Chance

While attending a conference during the first year of my postdoctoral studies, I started to experience extreme pain in my abdomen. Upon my return home, I was examined by a doctor, and it was determined that I had a severely enlarged ovary that had to be removed. Cancer was mentioned as a possibility, and the strong probability of not being able to bear children became a reality. With fear and trepidation, Klaus and I discussed what this would mean for our future. As career-oriented and ambitious individuals, we were in no hurry to have children, but the prospect of not having the opportunity to have any gave us pause. Even in the face of this reality, Klaus told me, "I would rather have you than children I have not yet seen." I underwent the surgery; the ovary was removed and cancer was not found. Did we decide to stop everything and have children? No. We continued our journey together finding satisfaction in our lives and careers with family, friends, civic engagement, and community service.

After my recovery and completion of my postdoc, we decided to move to the USA. Klaus continued his career in the chemical industry, and I started the life of an academic. Yes, we had a two-body problem, and I took on the role of trailing spouse, but the goal was always for both Klaus and I to fulfill our career aspirations.

I was a visitor and then a lecturer at Wayne State University before I landed the prized tenure-track assistant professor position at Oakland University in Rochester, Michigan. What a gift! I was welcomed by my colleagues who respected me for my scientific abilities and recognized the assets that my race, ethnicity, and gender added to the faculty. I landed NSF funding and had wonderful undergraduate researchers and great colleagues. I also landed an excellent graduate student, Georges Lahoud. With a community of excellent students and colleagues, I was happy and content. I credit this feeling of satisfaction for giving me the clarity to recognize that Klaus and I were ready to begin our family. We were no longer youngsters, since I was 34 and Klaus was 41. With one ovary and a history of reproductive issues, we went to our doctor and asked that I be given fertility treatments. After the second round of treatments and an emergency flight home from a business trip for Klaus, we were pregnant.

We were delighted and afraid. After 8 years of marriage, we were now going to be parents. Our philosophy on our life as parents was the same as our life as a couple: everything had to be integrated. We wanted our child to be an integral part of our lives both professionally and personally. We wanted our parents and extended family to be a part of our child's life and wanted this baby to reap the benefits of the wonderful tribe we belonged to at work, at play, and in service. The stage for this impossible standard of integration came in a way that I had not imagined. During my fourth month of pregnancy, only days after a trip in which Klaus and I were in France for me to present my work on the international stage for the first time as an independent researcher, my doctor performed an ultrasound and determined that I had a condition that would eventually lead to premature birth if I did not immediately go home and stay in bed until my baby was fully developed. Yes, this meant 5 months of bedrest. This made my impending motherhood a departmental priority and my baby the department's baby. My colleagues stepped in to teach my courses, helped manage my lab, and divvied up my extensive service load. Were they happy with this? I don't think so, but I never heard a complaint.

During my time on bedrest, I was still intimately involved in directing the intellectual pursuits of my research group and writing grants to acquire funding. I wrote a proposal for the coveted NSF Career Award, which was funded upon my return to work. One of my colleagues and scientific collaborators from Wayne State, Dr. Christine Chow, would stay in contact with me and check on me and help me to keep me in touch with science and the academic community. She is now a huge influence in the lives of my children.

Recognizing the sacrifices that the faculty in my department made for me, I decided to return to work quickly after Klaus Bryant was born and resume my duties as quickly as possible. I did not stop my tenure clock nor did I take maternity leave additional to my time of bedrest. My mother came to help me with the care of my son for the first month of his life, my mother- and father-in-law came for the next 4 months, and my mother returned to make sure that Klaus Bryant didn't enter daycare until he was 6 months old.

The choice of daycare was critical for me to remain focused on my career while taking care of my first priority, Little Klaus. The powers of the universe introduced us to a woman who lived in our neighborhood, who was dedicated to providing high-quality daycare for the children of Detroit out of her own home. She took care of Little Klaus like he was her own. She lived a 3-minute walk from my home, so not only was she a caregiver for Klaus, but she was also a part of his community. The importance of having peace of mind about my child's well-being while I spent my days educating the children of other mothers was vital for my career success. When Klaus was old enough to understand that mama was going away and leaving him in Detroit, I decided that it was time to expand his learning environment and bring him closer to my work life. At the age of three, he started to attend Lowry Center for Early Childhood Education, a preschool on the campus of Oakland University. This move fully integrated him into my work life through our combined commute, development of friendships with families of other faculty and staff members of the university, and participation in university events. He began to see the university as a second home and my students and colleagues as a part of his extended family.

One of the hardest decisions I had to make was moving Little Klaus. Even though my career was flourishing at Oakland, I had the desire to expand my research program and my capabilities by bringing in more graduate students and having access to more on-campus collaboration. When it was time for Klaus to enter kindergarten, we decided to make the move to the University of Toledo. Even though this did not mean Klaus Sr. moving to a new job, it did mean moving to a

Fig. 1 Graduation spring 2013, with Amanda's son Klaus and Amanda's academic son, Nicholas Amato

new city. Little Klaus was not happy. He struggled academically in his new environment, and I blamed all of this on my selfish move to expand my career. Oh what a load of guilt I carried. A few years later, we found out that the move was not to blame at all. Now the guilt came from the fact that I would have known he had some learning difficulties if I were not so consumed with my own career. Oh well.... Klaus is now 15 and a freshman in high school. He is resilient and happy and a son to be proud of. No more guilt? Ha! (Fig. 1)

And Yet the Menace of the Years

When I moved to the University of Toledo, I was required to go through the tenure process even though I had successfully navigated this process before. I decided to take my time to allow myself to give the attention to my family that it needed, as well as give my research group, which moved with me, time to acclimate to their new home. This group consisted of two women, Buthina Al-Oudat and Suaad Audat, sisters, who hailed from the Kingdom of Jordan. They were both married and mothers. They watched my chaotic life wondering what it would mean to follow their professional dreams and be excellent mothers. Luckily, the experience did not lead them to believe it not to be possible, as they are now professors in their home country.

When I made this move, I was 39 years old, and the thought of our family growing was not on my mind. In March of 2009, we found out we were pregnant. What a surprise! I was going through tenure for the second time and pregnant, and I had the nerve to be 41 years old. I was overjoyed by the prospect of having a baby, but I knew I had to plan carefully to prevent a possible long stint of bedrest. I went to the doctor ready to advocate for my baby and my career. And my OB-GYN fully understood how important their integration was to me. Luckily, knowledge of my

Fig. 2 Amanda and Cornelia (Photo by Dan Miller)

previous history by my doctor facilitated the completion of a medical procedure early during the pregnancy that allowed me to continue to work normally for the duration of the pregnancy. I announced the pregnancy to my department and other colleagues, and when Cornelia was born, she took her place at my side in my office, at faculty meetings, and at university functions (Fig. 2).

Now with two children 9 years apart, 6 and 15, successful completion of a second bout with tenure behind me, and promotion to the position of Dean of the College of Graduate Studies, you would think I have these motherhood and academic things tied down. Not! With baseball, ballet, music lessons, school boards, community groups, play groups, and, oh yes, my wonderful husband, there are not enough hours in the day for me to do everything alone. I have learned to rely on many key pillars that sit on a strong foundation of love of family and self. My children are biracial, bilingual, and binational. Embracing who they are and the fact that my husband and I merged to create them gives us strength that we would not have if we ignored this important element of our togetherness. Love of family allows me to let my husband play all the roles that he is equipped to fulfill even when my maternal instincts want to take over. He can be a ballet dad, when mom is not there. I can be a baseball mom when he is unavailable. We have never let the confines of gender roles hinder our path. With fingers and thumbs both found on the hands of men and women, mom and dad, grandma, Opa, and Oma can all change diapers, feed babies, bathe babies, and style the wonderful and unruly hair of the multiethnic child. Are things always done the way I want them to be done? No. But, they don't have to be if our children are happy and well.

I am the Master of My Fate

Love of self provided me with the strength I needed to preserve when others criticized me for working too much or traveling too much and not taking care of my children. It helped me to be assured when people questioned if my children were being raised culturally aware and happy with themselves as Germans and African-Americans. I felt confident when I was asked how I would ever let my son fly across the Atlantic to see his grandparents without me or how I would ever leave Klaus to take care of Cornelia without me. I knew that person I was raised to be, confident and aware of who I am and what is valuable, was enough for me to take care of my children, my husband, and myself.

Career decisions are family decisions. Keeping our family under one roof has always been a priority for me and my husband. This has been tested over the years with changes in institutions and loss of employment, but we stayed true to our creed and made sure that family was first. My husband is now an assistant professor at an institution an hour away from our home. Family has been integrated into his career just as it has been in mine, and we are thankful for this. We have a tribe of individuals, aunties, uncles, second moms, and substitute family, across many states and countries, who are essential to the development of our children and the stability of our family. We recognized the significance of this tribe during the sixth birthday of our son when we gave him a mad scientist birthday party in which all the facilitators were "aunties and uncles" with PhDs in STEM. We have a nanny/housekeeper/personal assistant/friend who is always there for us when chaos takes over and we need stability for our children at home. For this we are thankful.

That my gender and ethnicity have never been ignored by our family is an understood reason that I particularly need to stay the course. My son refers to himself as a German-African-American and understands what that means to his own identity and those in his circle. He knows that my struggle to get to where I am has been tainted at times by those who don't believe I should sit at the table with them. He is beginning to see that sometimes he will have to bring his own chair to the table just as I have, and that is something I am really proud of. My 6-year-old daughter knows that she has to share mommy with a lot of other babies. This is of course not easy, but, when she gets to come along and sit with these wonderful students, she is happy that mommy can help them too. She then knows that she is a part of a bigger story, and for that I am truly thankful (Fig. 3).

Reflections

Returning to the USA as a highly trained and skilled chemist who happened to be female and black shaped many of my decisions around career and family. It was during this time that I learned that there was no seat at the table for me. Returning to my country of birth as one of a handful of women of color trained in physical organic chemistry, I thought those in my field would be willing to make room for me to help train a diverse and globally conscious group of young scientists. This was not the case. I was perceived as the "diversity" candidate, in the pool only to check a box. Being perceived as not being as competent as those who didn't look like me, and noticing the struggles of students who did, made the acquisition of that chair a

Fig. 3 Cornelia's debut art exhibit, Maumee Valley Country Day School, selfie exhibit, with Amanda, Cornelia, Klaus, and Papa

priority. As I have made my way through the system, I pulled my chair up to tables to influence policies that negatively impact the admission of students of color to high-demand programs and limit access to education for the socioeconomically disadvantaged and to influence decisions on the allocation of federal dollars to **all** institutions of higher learning. Why? Because at the end of the day, I want my children to grow up in a better and safer world.

A theme that runs through my life as a chemist and a mother is my need to be the person I want to be. Our definition of self is often so heavily influenced by those who do not live the experience we live or have lived. This simple fact makes us question the portrait of ourselves as scientific professionals just as often as we wish to erase our motherhood caricature. Well, using a phrase my father often used, to hell? with it! Be yourself, live your destiny, follow your path, and say to yourself every single day, **I am the captain of my soul!**

Acknowledgments It goes without saying that without the support of my husband Klaus and my children Klaus Bryant and Cornelia, this work would not have been inspired. Without the protection, participation, and support of my tribe, nothing would be achieved. Without my creator, I would not exist.

About the Author

Education and Professional Career

1990	BS Chemistry, North Carolina Central University, Durham, NC
1992	MS Chemistry, Duke University
1997	Dr. rer. nat. (PhD), Ruprecht Karls Universität, Heidelberg, Germany
1997–1999	Postdoctoral fellowship, Universität Basel, Basel, Switzerland
2000–2006	Assistant Professor, Oakland University, Rochester, Michigan
2007–2016	Associate Professor, the University of Toledo
2016–present	Dean of the College of Graduate Studies

Honors and Awards (Selected)

2015	Leadership Fellow of the American Association of Colleges of Pharmacy
2015	Alice Skeen's Outstanding Women Award from the University of Toledo University Women's Commission
2014	Recipient of the Stanley Israel Regional Award for Promoting Diversity in the Chemical Sciences from the American Chemical Society

Amanda is an active member of the American Chemical Society, holding leadership roles in the divisions of Chemical Toxicology and Medicinal Chemistry. She is also a member of the Association for Women in Science and the Society of STEM Women of Color.

My Circus: Please Note That I Have No Formal Training in Juggling

Amber Flynn Charlebois

Photo Credit: SUNY Geneseo

It was a beautiful Friday afternoon, and I had just finished tabulating the responses to the bonus from the organic chemistry exam I was grading. I was very much interested in the students' answers to the following question:

Bonus: I am due to have my third child soon; if it is a girl, what should her name be?

1. Sarah
2. Victoria
3. Elizabeth
4. Pepper

A. F. Charlebois (✉)
Department of Chemistry and Biochemistry, Nazareth College, Rochester, NY, USA

You see, my husband, Jay, somehow got into his mind that the name Pepper was appropriate for a girl. In addressing the stalemate we had reached in coming up with a good girl name, I convinced him to let me poll my organic chemistry class in an effort to gather additional data. In my class of 50 something students, only 3 chose Pepper. Additionally, many students wrote in things like "Pepper is my dog's name!" which only helped my case. Luckily these results allowed us to take Pepper out of the running. It turned out that the winning name with almost 20 votes was Victoria. My daughter's name is in fact Victoria.

It is experiences like this one that prompted me to jump at the chance to contribute to this work. I am personally deep in the trenches of trying to successfully blend two of the most amazing things, motherhood and professor-hood. It is my hope that sharing my experiences will provide not only information but also insight and hope to the reader. I have discovered that this *blend* is more of a juggling act, rather than a balancing act, as there are very often additional variables that need to be considered. Here are *my* specific roles in this thing *I* call life; in other words, here is what I have to juggle.

- I am the mother of three amazing children and aunt to my wonderful niece whom lives with us full-time; Steven is my smart and funny 14-year-old, Matthew is my resourceful and hard-working 12-year-old, and Victoria is my curious and empathetic 9-year-old, and Kyra Rae is my patient and artistic "niece-daughter" 18-year-old blessing.
- I am a lecturer of organic and biochemistry in the chemistry department, coordinator of supplemental instruction, and pre-health advisor at SUNY Geneseo (formerly an associate professor of chemistry at Fairleigh Dickinson University).
- I am a wife to my best friend and fellow chemistry geek, Jay. Yes I use the line "There is a lot of chemistry between us!" entirely too often.
- I am a sister, friend, mentor, and aunt to some of the most amazing people on this earth.
- I am an active volunteer for the American Chemical Society at the local, regional, and national levels. I also volunteer in my children's classrooms as Dr. Demo.

In that order, most of the time.

There are days in my life when I feel like "I got this!" and that I am doing an amazing job at all of the roles I have in life. There are, however, also some days when I feel like I have failed in all of these roles. Usually, I can confidently say that of my five roles, I have dominated at least three on that given day. Maybe someday, I can claim I have this professor-mother thing mastered, but today I will simply share with you some of my experiences. In this chapter I will tell my story, which I hope is helpful and insightful. My disclaimer is that many of the challenges I share here are the same challenges I continue to work on myself, *still* (Fig. 1).

Fig. 1 Amber as Dr. Demo in Steven's second-grade classroom

How It All Began

My parents were amazing people whom I admire to no end. My father was a self-employed dairy farmer, and my mother worked in retail. They spent most of their lives working very hard so that my siblings and I could have everything we needed and most of what we wanted. So from very early on in my life, I was exposed to a working mother lifestyle, and I did not know any different. Growing up, I never contemplated being a stay-at-home mom, not because my mom worked but because it was not in my personality. But all along, I knew I wanted to have children. I guess I was a bit naive, but I never spent any time or energy thinking about how I would pull it all off; I just did it.

I cannot remember a time in my childhood when college was not in my plan. My father earned his associates degree, and my mother attended college for one semester, but I knew I wanted to get a 4-year degree for sure, and I did. I received my bachelor's degree in communications (specifically photography) from Syracuse University right after high school. At some point after I worked several years in photography, I realized I wanted more of a challenge and returned to the classroom to obtain my associates degree in chemistry from Monroe Community College (MCC). Professor John Cullen my organic chemistry professor at MCC was a true inspiration. He was an amazing lecturer and easily became my first role model/mentor in chemistry and in education. I remember the day he told me that he recommended me for a scholarship/internship for the summer at Eastman Kodak Company. This summer research experience was life changing. It was during the final presentation I had to give on my summer project that I realized my true calling, to teach.

The State University of New York at Buffalo (UB), accepted me into the PhD program with my AAS degree in Chemistry combined with my BS in Communications and a few additional chemistry courses under my belt. I had a wonderful experience at UB, completed my PhD there, and continued on as a postdoc at the University of Illinois at Urbana-Champaign (UIUC), IL, under the direction of Professor Scott Silverman. Scott was not only a mentor and friend but quickly became my advocate. He was incredibly supportive in my career goals, allowing me to teach an undergraduate course each semester I was in his lab. He pushed me to be the best I could be in the lab and in the classroom and inspired me with his energy, ability, and organization. Scott played a huge role in my career and taught me what it is to advocate for someone.

Interesting side note: The spring of 2000 was a busy one for me; I defended my thesis and graduated with my PhD in the same month that my father at the age of 59 earned his bachelor's degree. It was fun to share this experience with him.

Stop Trying to Control Everything; Sometimes You Have to Just Let Go

When I was in graduate school (UB), I met Jay, my study partner/drinking buddy turned soul mate. We realized quickly that we had a good thing, and so we tied the knot. Both Jay and I were nontraditional graduate students in that we were almost 10 years older than most of the "just out of undergrad-grad" students. We had both worked in the chemical industry for several years before we decided to attend grad school. So my biological clock was ticking when we began to think about starting a family, at the same time as trying to find a postdoc appointment. I knew that I ultimately wanted to teach at a primarily undergraduate institution. I thought that would be the best place for me to have both a family and a successful and satisfying career. Therefore, as scientists, *of course*, Jay and I started calculating when to get pregnant so that the baby could be born during the summer between my postdoc (University of Illinois at Urbana-Champaign, IL) and starting my first academic position. It was perfect! We were due to have our first baby in July 2002, and I was getting my CV sent out to all these awesome undergraduate institutions, and we had timed it all perfectly. Until I miscarried.

WAIT! That was not part of the plan. NO! I had it all set; I was going to get everything I wanted, and it was going to be perfect. NOT. I was 36 years old and not pregnant. After getting over the shock of the loss, Jay and I decided to just try again. We never really talked about what it meant, and I never really thought about what it was going to be like, but we agreed to just deal with the timing issues of childbirth, whenever it happened. In other words, we were flying by the seat of our pants, but we were doing it together, so it felt comfortable.

I targeted my academic job search for areas of the country where Jay, also a PhD chemist, could find a position. Jay was interested in working in industry, so we knew his job search would have to wait until we were closer to the actual relocation date. I was interviewed at several institutions and accepted a position in New Jersey, and Jay and I started planning the next chapter of our lives together as we organized the move from Urbana-Champaign. New Jersey seemed like as good a place as any for Jay to find a chemistry job, so we were well on our way.

And guess what, we were successful at the pregnancy thing too, and we were due to have our baby right around Thanksgiving in 2002. Okay so wait; that meant that I was having my first baby smack-dab in the middle of my first semester ever of teaching. Not an ideal situation, but we could do this, right? As the due date got closer, Jay was still deep in his job search, and I was so very excited about my teaching position. In chatting with HR at my institution, we discovered that there is a requirement to work for 12 weeks in NJ before disability kicked in. My start date was September 1st. The middle of November would not be 12 weeks no matter how you counted it. I was not eligible for disability. I had NO sick time accrued to date, and Jay still had not found employment. Some members of administration and faculty offered to donate their sick time to me, for the birth of my child, but HR denied this request because pregnancy was a planned event. They did inform me that if the birth had to be cesarean, then I could use donated time. Looking back, I think I was not as stressed as I should have been.

My adorable 8 lbs 7 oz bundle of joy, Steven Flynn, arrived 3 weeks early, on November 3rd. Members of the chemistry department graciously volunteered to cover my classes until I returned. Not one of my classes had to be canceled, and I was back at work after 10 days. At that 10-day mark, I was feeling great and being in front of the classroom was fine. This was all possible because Jay had still not found his dream job, and so he was able to stay home with Steven, while I went back into the classroom. As I reflect, I think I got the better deal because Steven was colicky. I was able to kind of escape some of the extra stress by going to work. Have I mentioned yet that my husband is amazing? The whole thing worked out because I only had to be on campus for a little over 4 weeks, and then I had the entire winter break to spend with my new little Steven. Things started taking off for Jay as well with several interviews and finally a great job offer. He started his new job in January, and so with both of us gainfully employed, we could both relax and enjoy our new little life.

Funny little side note: HR in doing their job to the best of their ability actually made an inquiry to the dean when the insurance papers came in requesting me to add my son to the policy. They were concerned because I had not taken any time off of work, and yet I was adding a child to the policy.

Almost like déjà vu, we successfully calculated the timing of my next pregnancy, which also did not come to fruition. This time miscarriage was even more difficult for me. I had made it past the 3-month mark; I was actually in the 5th month. Ironically, it seemed to me that there was some force not allowing me to give birth during a school break. It took much longer to heal from this one and to decide to try again. But again, we mustered up the energy and the hope to try again, and so my second son, Matthew Michael, was born March 1, 2005 at a whopping 8 lbs 15 oz. It turned out that for this pregnancy, I was able to plan more effectively and was able to develop and teach my first online course. This allowed me to work from home most days once Matthew was born.

No, Really. Just Let Go

My point with these stories is that this idea of "control" is not possible for all things. In theory, planning your pregnancy around semesters of teaching sounds perfect, but may not always happen the way you envisioned. Don't get me wrong; it is worth a try. And I know many examples of summer babies of women faculty, but if the timing does not work perfectly, just make it happen the way it works for you. And you should know that my third child, my beautiful 9 lbs 7 oz girl, Victoria Rose, was born June 2, 2008, so I got my girl *and* I got my summer baby.

But the "control freak" in me comes out in other ways. I struggle with letting other people help me and with letting go. For example:

My husband does not fold the towels correctly. Most of the time, Jay does not contribute to the laundry process. It is because he is in charge of most of the shopping and all the cooking (plus all the outside stuff)—*I scored big time, right?*—and I do the laundry and cleaning (and much of the kid's stuff). That is just how we break it up and it works for us. He does so much around the house, and I feel a little weird complaining about it, but it drives me crazy when he folds the laundry. He folds the towels all wrong, and they don't fit in the linen closets correctly, so I have to refold them. I realize logically that this is just senselessness on my part and a waste of my time when I stop everything and refold. But when I am deep in the daily grind, I can't help myself. I need to remind myself regularly of two things:

1. I cannot possibly do everything myself.
2. When I get the help I need, I have to be open to different interpretations of success. I need to stop trying to control the folding of the towels and let it go. When I catch myself refolding, and think about it, I quickly realize it is often my overall approach to life that needs adjusting.

Trust Your Gut

There are times in your life when something deep inside your being tells you to run away fast. Those internal mechanisms are real, and they are important to heed. With that said, I confess that I did not follow my instincts this one time, and it completely changed who I am today.

It was at one of the interviews I had in New Jersey. It was a medium-sized state school, and I was interviewed to teach biochemistry in their program. At the very beginning of my individual interview session with one of the more senior members of the faculty, he introduced himself and stated that he was the biochemist and that the new hire would replace him as he was going to retire soon. He looked at my CV in front of him, looked up at me, and said, "Well, it is good that you are a woman; it would be better if you were a black woman, but it is good that you are a woman!"

Seriously, I should have run fast and far, but 2–3 months later, when I got the phone call and the offer, and no other offers had come yet, I accepted it. I had convinced myself that although I did not appreciate this mentality, this person was on the edge of retiring, and everyone else in the department seemed fine, and it was a job offer after all. So I accepted this position formally. Ironically, 2 days later, I got a phone call from a second institution asking me to be patient because there was an offer coming, and they were very excited about bringing me on board. My heart sank as I told them I had already accepted a position. So, in the end, I honored my acceptance of the original institution and started my academic career that September at the state school that was happy to have me because I was a woman, which brings me to my next cliché.

Should I Stay or Should I Go Now?

After 4 years at that institution, I tendered my resignation. At the department meeting when my resignation was announced, the comment I heard from the former chair of the department was "Well that is *great news*!" Let me interpret the meaning of "great news" in this statement. He was happy that I was leaving; he could not care less where I was going or what kind of position I was taking; he was simply happy that I would no longer be at *his* institution. The original interview comment and the final "great news" comment were the two bookends of that chapter in my life. At that time, I was so relieved to have accepted the offer from a fantastic small liberal arts college, also in New Jersey, and was most excited to be getting out of there. The challenges that came with my new position were that I had to take a substantial pay cut and I had to start the tenure process entirely over.

The decision to move was probably the most difficult one I ever had to make. I was at the end of my fourth year at an institution that had great benefits, great students, and *mostly* wonderful colleagues. All indications were that I would be getting tenure that next year. But tenure meant that I would be stuck there forever, and deep down I knew that I had to get out, and I had to get out NOW.

When I reflect back on those 4 years, it is hard for me to comprehend that it was me going through it. I was in the middle of a hostile work environment based on my gender. It was never anything sexual; it was just a constant disregard and disrespect aimed at me and the other untenured woman faculty, Anita. There were two men in the department who were the drivers behind this treatment, and the rest of the faculty in the department seemed to let just it happen. These two men (one was the chair and one was that biochemist from my interview that never actually retired) screamed at us both in meetings behind closed doors and in the hallways, while students and colleagues were present. I was even physically threatened in front of students. We were regularly referred to as "the ladies" or "those two," and these two bullies never learned our names and got us mixed up all the time (we look nothing alike, except that we were both women). There is not enough room in this chapter for me to describe the numerous different experiences and situations we were exposed to, and there is also no way that I could truly describe on paper the way these painful experiences made us feel. We had each other to both validate and corroborate our experiences and our emotions, but it didn't dull the horror. My morale was depleted completely, and self-confidence was nonexistent.

It all came to a head in the spring of 2006, when it was clear that Anita was not being renewed after her fourth year. I still have no idea how this was allowed to happen. You have to understand, she was an amazing teacher, she had publications (including a children's book about the periodic table), she had been awarded an American Chemical Society Petroleum Research Fund (PRF) Grant, and she participated in plenty of service. She was *better* than me. But looking back it is clear she was more of a threat to them; Anita was stronger than I was. So they fired her. It became evident that once her nonrenewal was announced, they were starting to target me. I dreaded department meetings and any interaction with the two men who were so dead-set against women being in their department. The other faculty in the department began "standing up for me," at this point, but honestly it was too little too late. This situation had finally started to affect my inner being, my health (my body was starting to respond to all the stress), and my family (my temper was short and I was spending many hours at home crying and sobbing). I was a mess. Even worse, I was starting to believe what they were implying and saying about me, and I was starting to really second-guess everything in my life. I had to leave; I had to get out. And so I did.

My new position was at Fairleigh Dickinson University, and by my seventh year, I successfully reached the associate professor rank with tenure, and I had no regrets whatsoever. My amazing new colleagues and I provided each other challenge, respect, accountability, and collaboration so that we could offer our students the best possible learning environment and college experience. This was an atmosphere where I was excited to bring my children when needed. This was a place where the faculty, staff, and administration all support the notion of family and all that goes along with it. This was a place where my children loved to visit and where they dreamed of attending college someday. This was how it was in my dreams.

Bottom line: The motherhood/professor-hood balance is easier to pull off when you are happy and satisfied in your day job.

Choose Your Battles Wisely

The second most difficult decision in my life took almost a year to come to terms with, but finally Anita and I decided to file a lawsuit against our former university for a hostile work environment based on gender. We felt that we finally needed to stand up for ourselves. Anita lost her job and I lost the salary and the time toward tenure. In making the decision, we especially wanted to be role models to the many young women (and men) who had watched us be treated so horribly. If we had not fought back with the lawsuit, then the students would be left thinking that it was okay for women to be treated in that disrespectful and degrading manner. When it was all over and the dust had settled, we threw a Vindication Party to celebrate our success. We invited all the many wonderful people in our lives who helped us while we were experiencing the hostility and all the people who helped us with the lawsuit process. We made sure we invited all the chemistry majors during our time there to celebrate with us. We had to show them that what they saw us experience was not acceptable, that people were held accountable, and that we had survived.

When we met with our lawyer, Sam, for the first time, he asked us what we wanted from the suit and if we wanted our jobs back, to which we replied "Absolutely NOT!" I told him that I wanted institutional change and the two bullies to be fired. He told me that a lawsuit can really only get you financial restitution, which sometimes may cause institutional change and might cause the university to modify employment terms. At that time, that was good enough for me. By no means did we get rich from winning this lawsuit, but we're content in the knowledge that we caused some institutional change. We have paved the way for any future women to work in the chemistry department and for any female students in that department as well. Most of all I think we were validated. It became public knowledge that we were treated unfairly and without the respect we deserved. This validation was instrumental in my healing process and in my moving forward.

Looking back, one of the things that helped me when I was in the middle of the hostile work environment was that I documented everything. I had a journal and kept track of (with dates) everything and every interaction (what happened and how I felt). Then, when the time was right, I worked with the dean and followed the proper channels within my institution to file a grievance against these two men. Since we were unhappy with the outcome of the grievance, we met with Sam and filed the lawsuit, which took more than 3 years from start to finish and involved many depositions and expert witness' testimonies. The process was long and sometimes difficult, but ultimately it was a cathartic and freeing experience.

Lean on Me

I have to give a shout-out to a program that helped me through this entire decade of my life. I attended an Effective Negotiation Workshop offered through the University of Oregon and ACS Committee on the Advancement of Women Chemists, or COACh. It is a grassroots organization that works to increase the number and career success of women scientists through innovative programs and strategies. More information can be found on their website http://coach.uoregon.edu/coach/.

It was attending this workshop while I was deep in the thrall of the hostile work environment that gave me the strength to keep going. I remember learning how and where to sit in a room to have the most influence and presence. The other thing that sticks in my mind clear as day is when you are preparing for a meeting, it is important to be completely prepared, of course, but to also give yourself 5 minutes just before the meeting to just BE, to mentally prepare yourself. These two very small pieces of insight gave me the strength to go back to my university and sit tall and have a strong presence in every meeting (even though on the inside I was a basket case). It is what gave me the strength to search for another job and move on with my life. Looking back, this workshop was a very instrumental part to my healing and my recovery.

A second shout-out goes to the American Chemical Society's Women Chemists Committee (WCC). Once I became an associate member of this national committee, my life was changed. I was able to network with women and men from around the nation who valued the importance of women in the chemical enterprise. This group also helped me see the important role women had in chemistry and allowed me to help them with their mission, to attract, develop, promote, and advocate for women chemists. In this group, I found a safe place for me to learn how to be a more effective leader and a better person. I found a new family.

Day-to-Day: You Win Some, You Lose Some

Being effective in all your roles on a day-to-day basis is not always possible. There are days I have had where I feel like I have failed in all of my roles simultaneously. These days are few and far between, but they do exist. If I wrote this chapter and pretended they did not exist, I would not be being truthful.

I clearly remember this one day when I was not successful in any of my roles. Victoria was almost a month old, and she had to have an MRI of her back. The only opening they had (unless we wanted to wait a month) was one Tuesday morning in July. It was the same Tuesday that I had agreed to teach some local high school students a 3-hour laboratory course about biodiesel and other alternative fuels. It was a good source of summer income and was only for 3 hours, so I had agreed to it back in May. But that meant I could not go to the MRI with my newborn daughter, and instead I would teach this group of future chemists about how to synthesize biodiesel. Of course Jay agreed to take the day off and take Victoria to her appointment in New York City without me. I was not okay with this arrangement; I really wanted to be there with my new baby. I checked and none of the other faculty could switch with me, so I was forced to come to terms with the situation. I started to accept that Jay could do this and so could I.

This dreaded Tuesday finally arrived. To top things off, Steven, my 5-year-old, woke up with a fever and could not go to daycare. Jay could not take Steven to the hospital with him as he would be busy with Victoria. So there you have it, I would tote my sick 5-year-old in with me to campus and teach this lab. During the school year when my kids are not feeling 100%, I have been known to bring them with me to campus and hire past students to babysit during my lectures/labs so that I would not be distracted from my teaching. But during the summer, there were no students around for hire. So I improvised and set my son up on the student lounge couch (just around the corner from the organic lab where I was teaching) with some snacks, some blankets, and the DVD player for the 3-hour lab. I checked on him every 15–20 minutes to make sure he was fine. The secretary and lab technician helped me by checking in on him regularly as well. And he was fine. He actually did great.

Everything was fine.

But it was not fine with me. I let my beautiful daughter down because I could not be there with her for her MRI. I let my feverish son down because he was not feeling well and I stuck him in a room for 3 hours. I sucked at teaching biodiesel, because I was not totally present with them as I worried about my daughter and even more immediately my son, as I ducked out of the room frequently.

But in reality, everything was fine. Jay did great and Victoria's MRI came back totally normal. Steven's fever went away and went back to daycare the next day. And finally, my student evaluations did not reflect my internal turmoil but rather indicated that most students thoroughly enjoyed the lab and learned a lot from the experience. Everything was fine.

Everything really was fine.

Finally, Admitting I Cannot Do It All and Asking for Help

However, my feelings of inadequacy on this day were overwhelming. Actually, these feelings occurred whenever any of my children were sick, or there was a snow day, or if school was closed. What the heck was I doing, thinking I could really pull this all off? Something needed to change. I had been toying with an idea my friend and mentor Kelly had shared with me as we discussed work-life balance. She told me that if I could afford to pay someone to help me in my life that would also make me more effective in the rest of my life, then I should go for it. And I agree. It was time to get some help. It was shortly after this Tuesday experience that we decided to get a nanny.

There is stuff that goes with the nanny thing. For example, I had to deal with some guilt about having a nanny, and I am sure some people judged me for it, but it was what I needed at that time to maintain my sanity. My first nanny was perfect, and our world was a better place with Nina in it. She became part of the family, and my kids still talk about her and beg for her to come and visit. A year and a half later, my second nanny was not so perfect, but it was still better than no nanny. Ultimately, having a nanny that was just OK made it easier to decide I could do it all on my own again. All said and done, I had a nanny for almost 2 years, and it saved my life. I returned to babysitter/daycare mode once Steven was in school, which just so happened to be the fall I was going up for tenure. I had successfully used the extra time in my daily schedule during the 2-year nanny phase to get myself completely ready for my tenure submission. It was perfect timing.

Fig. 2 The whole gang (left to right) Kyra, Matty, Amber, Victoria, Jay, and Steven) sporting their freshly minted tie-dyes in 2015

My "Still" Expanding Family

In April of 2014, I seized the opportunity to take in my then 14-year-old niece, Kyra. We welcomed the opportunity to have Kyra live with us so that she would have access to a supportive and structured home that valued hard work and education. Jay and the kids were always completely supportive, and the new family dynamic was exciting! It must have been particularly interesting and challenging for Kyra's laid-back and passive personality to join our driven, focused, and constantly moving lifestyle. She still teases me today that she hates it when I say "*You* are in charge of *your* own destiny!" (Fig. 2).

She did, in fact, take charge of her own destiny and has graduated from high school (*cum laude*) and will be attending college this fall. The idea of her going to college was on the table from day 1, and we never veered from that path. She shared with a close friend that she believes that both graduating high school and attending college are things that would not have happened if she did not move in with us. As I reflect on this aspect of my life, I realize that we never went through the decision of whether or not to have Kyra move in; we just did it. I am so glad we did; she has added a new dimension to our family and our lives, and she is well on her way to being a productive and impactful member of society.

Should I Stay or Should I Go Now?: Take Two

During this entire time, Jay had been working in industry (pharma) and has never really been satisfied with his career. He had also been teaching several night labs at FDU as an adjunct and very much enjoyed it and the little bit of play money that came with it. So through the years he had consistently been perusing the job postings just in case his dream job appeared. Well one day in the fall of 2014, he asked if he should/could apply to a position in upstate NY at SUNY Geneseo. I said sure, considering that was about 30 minutes from my hometown and 20 minutes from my family. About a month later, he asked if he should/could go on the interview, to which I replied, sure! I figured we would simply cross that bridge when we came to it. Well when he got the offer to be instrument support specialist at Geneseo, we had arrived at the bridge, and we had a decision to make.

Recall that Jay served as the trailing spouse right after our postdocs. So was it my turn to fill that role? In addition, we would be able to slow our lives down quite a bit, as the pace in upstate NY is much slower than in North Jersey. It was an incredibly difficult but very easy decision to make. We moved to NY and we have not looked back. In terms of my job search, during the weeks he was "deciding" whether to accept the position or not, serendipitously, and an organic chemistry lecture position at SUNY Geneseo was posted. I applied and was hired. We were so incredibly lucky to both have great offers from the same institution. The kids are so much happier, the hubby is happier, and I am learning how to slow down a little bit, kind of anyway.

I should note, however, that being a lecturer is very different (I knew it would be, but I did not expect it to be so different). It is all the same work, but, somehow, it is different. It may be subconscious, but colleagues treat you differently. In effort to more fully satisfy my career aspirations, I have taken on other leadership roles on campus. I am a coordinator of the supplemental instructor program, and starting this fall I will serve as the pre-health advisor too. I have mostly stopped my undergraduate research projects, and I will receive one course release/semester to allow more time for these other roles. I am ok with letting the research piece go, partly because my bottom line is that I love to mentor and advise students. I have that in these additional roles, and I love that. I am establishing myself as a productive and hopefully invaluable contributor to campus, so that every 3 years they decide to keep me on the payroll. I do have a better overall life in terms of my juggling act; however, in my new upstate NY life, I had to be strategic in maintaining my career satisfaction, and I can say that I am getting there.

Two Generations of Role Models

My children are with me on campus all the time. Anytime they can come with me they do. They know many of my colleagues by name, they know where the cool trees to climb are located, and they for sure know where the dining hall is. They know the

Fig. 3 Victoria, Steven, and Matthew in England, 2014

code to get in my office, and they know where the markers and pens are in my desk. They know where all the outlets are in my office for charging their phones, iPads, and DVD players. Most importantly, my children know many of the students. I think it is amazing that my students interact with my children in a positive way and become role models for my kids. Several students have donated their old children's books or decks of Pokémon cards to my children. I just think that interaction is priceless, as my kids may now aspire to be like my students (the good ones anyway, wink wink).

Then it goes the other way too. Most of my students each semester have met/seen my kids at some point. I try very hard to be a positive role model for my students when they see me interact with my children. I love that the young women (and men too) see me balancing my family with my career. I am not saying it is easy or perfect; I am just saying it is doable and it is great fun!

As a matter of fact, during the January break in 2014, our whole family was able to travel to the FDU Wroxton campus for a weeklong excursion. I was serving as the advisor/mentor/chaperone for the nine FDU chemistry and biochemistry students who were traveling to take a two-credit liberal arts course as part of the chemistry living learning community experience. The students were amazing with my children, and my kids were in awe of the students. I consider myself incredibly lucky to have had this experience with my students and my family (Fig. 3).

I do have to say, though, that there is a high-energy barrier for me to be a professor and mother at the same time. When I have my children with me around my students, I feel torn. I feel like it is hard to be firm, nurturing, supportive, and instructive to both my children and my students at the same time. It almost feels like I have a double personality. My voice changes and it feels like my personality changes too as I go back and forth between the two types of interactions. That being said, without a doubt, I will continue these interactions because I love that my students see me as a mother and, even more so, I love that my children see me as a professor.

Reflections from Amber's Children
Steven Charlebois
My mother is pretty cool sometimes but also can be super not cool. She works a ton, so us kids usually have to do a lot of chores, while she is gone. I don't mind it though, as because she works so much that we get to travel to different places with her on work trips. I have even gone to San Diego with her and traveled internationally when we went to England in 2014. But because she is a teacher, I have an advantage over the other kids at my school because she helps me with my homework and I know how and what a teacher looks for to grade a project. I think it is cool how mom juggles all her work and family and still takes out time to let me hit an extra PokèStop on the way home from the grocery store. It also helps being able to walk around the campus she works on and get an in-depth view of how college life works. This extra information is super useful in class, and the advice she gives me about how to get teachers to like me as well as how to get good grades helps a lot (Fig. 4).

Victoria Charlebois
My mom is the best mom I could ever have; she is nice, caring, and fun, and she perseveres. She is the best mom because she plans days she can come into my classroom and do fun experiments like mini volcanos, tie-dyes, magnets, dry ice, and chromatography butterflies. She also can be pretty boring from time to time like sometimes she can't come to tuck me in at night because she's too busy grading or writing an exam, and she makes us do chores, and she makes me go to bed at 8:00 that is pretty early in my mind. I love helping my mom with writing down exam scores; sometimes when a person gets a good score, I get to write a smiley face on their paper. And that, my friends, is why my mom is so awesome. I could say she is the most "awesomest" mom in the history of moms (Fig. 5).

Matthew Charlebois
In my opinion having a chemistry professor mom is pretty cool. I get to go on business trips like San Francisco, Washington DC, and England. We also get to go to the campus and eat at the dining hall and play Pokèmon GO. I also get to meet some of her students, and some of these students are really cool. Some bad parts are she is not always home, so she makes us do lots of chores (Fig. 6).

Kyra Flynn
Being an honorary child of the household, I haven't dealt with having a chemistry professor parent as long as the other kids have. But I have dealt with it long enough to know that it is a game of pros and cons. For instance, a pro is that she gives me a leg up in my chemistry class by helping me with homework. A con is she has trouble juggling her many responsibilities

(continued)

sometimes and overbooks. There have been a few times that her work has gotten in the way of my plans, not that I can blame her with the busy schedule she has. Another pro is she knows how college works. Just the other day, she was able to help me with scheduling and registration as I make my own transition into college. I have also been able to view campus life through her job. It is very exciting to already have an idea about how college works before going into it. I definitely feel a bit more comfortable than others knowing I can ask her questions about college. While it is a curse and a blessing, I would definitely say it leans more toward the blessing side of the spectrum (Fig. 7).

Fig. 4 Steven and Amber with Milli Mole at the ACS Meeting in San Diego in 2016

Fig. 5 Victoria and Amber at the ACS NERM 2016 Awards Dinner

Fig. 6 Matthew and Amber in San Francisco in 2017

Fig. 7 Kyra and Aunt Amber at the ACS meeting in Philly at the Chemluminaries in 2016

Reflections

I enjoy my job so much that, honestly, most days it does not seem like work. I feel like somehow I am getting away with something and that one day someone will find me out and make me do real work. I feel blessed that I have found a job where I feel happy, safe, challenged, and satisfied. Mix that in with my fun and crazy little family, and you have my life. Don't get me wrong; there are some days that life is not all calm and peaceful, but the majority of my days are a crazy-busy circus that I absolutely live for.

Below are the rules (guidelines) that I had made every effort to live by in effort to achieve my personal balance (Fig. 8).

- Find a partner (if that is in your cards) who wants the same things out of life that you do, so that you can save the world together.
- Learn how to say NO! In this insanely busy thing we call life, there are things that need to be done, but you do not have to be that person every time.
- Ask for help; it takes a village, or in my case a full circus staff, to pull this mom the chem prof thing off; do not be afraid to ask for and accept help.
- Make sure you have mentors/role models in all areas of your life. Identify people who you are like-minded and you aspire to be like. Then utilize their friendship and guidance whenever you need it.
- Stay true to yourself. Always follow your heart. But remember to challenge yourself and step out of your comfort zone once in a while. This is how you will learn and grow.
- Slow down and enjoy your life. (My sister, Patty, has to remind me of this one all the time.)
- Don't worry about the towels.

Fig. 8 The Charlebois family Matthew, Victoria, Amber, Kyra, Jay, and Steven, plus pets, Georgia (dog) and Colonel Saunders (chicken), at Kyra's graduation in 2017

My Circus: Please Note That I Have No Formal Training in Juggling

Acknowledgments I am indebted to numerous people who have pushed, challenged, inspired, supported, and believed in me through the years, ultimately getting me where I am today, starting with my parents, who believed in me and taught me to follow my values and to work hard. Instrumental professors Cullen (MCC) and Morris (SUNY Brockport) introduced me to what is looks like to be an amazing teacher at the undergraduate level. UIUC's Scott Silverman for being a true mentor and champion as my postdoc advisor. Troy Wood (UB) for believing in me and inviting me to be his research collaborator on a very cool project. Brian Mauro (FDU) for providing many opportunities to step out of my comfort zone and tackle new projects and programs. Wendy Pogozelski, Celia Easton, and Ken Kallio for recognizing and acknowledging that I am more than a lecturer and that I have much more to offer SUNY Geneseo. In addition, I am incredibly blessed with a number of women whom I consider to be my mentors, confidants, and friends. Then there is my family. Nothing I do (have done) would be possible without their support, encouragement, and love. My children still love me (and even give me hugs) despite the fact that I cannot always be there for them at every stop. My husband Jay has supported me in all that I do. He must think I am truly a crazy person for getting involved in so many things and throwing so many balls into the air at one time. But as crazy as he thinks I am, he is always my grounding and stabilizing force; he is the person in my juggling act who helps to catch some of the balls I miss and more importantly helps me pick up the pieces when a ball crashes to the ground. It is only with all of this amazing support that I am able to continue my circus performance, my juggling act.

Editor's Note

Between this chapter's submission and the publication of this book, Dr. Amber Charlebois has changed institutions. Dr. Charlebois is now back on the tenure-track as an assistant professor of chemistry at Nazareth College.

About the Author

Education and Professional Career

1988	BS Photography, Syracuse University, NY
1995	AAS Chemistry, Monroe Community College, NY
2001	PhD Organic Chemistry, University at Buffalo, NY
2002	Research and Teaching Postdoctoral Associate, University of Illinois at Urbana-Champaign, IL
2002–2006	Assistant Professor, William Paterson University of New Jersey, NJ
2006–2011	Assistant Professor, Fairleigh Dickinson University, NJ
2011–2015	Associate Professor, Fairleigh Dickinson University, NJ
2015–2018	Lecturer, Coordinator of Supplemental Instruction, and Pre-Health Advisor, SUNY, Geneseo, NY
2018–present	Assistant Professor, Nazareth College, Rochester, NY

Honors and Awards (Selected)

2015 Fellow, American Chemical Society
2010 E. Emmet Reid Award for Outstanding Teaching of Chemistry at Small Colleges from the Middle Atlantic Regional Meeting of the American Chemical Society in DE
2010 Becton College Teacher of the Year Award
2009 Innovative Challenge Award winner for business "Dr. Demo: On Site Science"

Besides her ongoing commitment to the American Chemical Society (ACS) locally and nationally, Amber aims to inspire future science students and chemistry majors with Dr. Demo, on-site science demonstrations for children ages 3–10.

Planned Serendipity

Renée Cole

Photo credit: University of Iowa

"I'm a bitch, I'm a lover, I'm a child, I'm a mother..." I could add "I'm a professor" to the chorus of Meredith Brooks' song. The different aspects of my life each inform the other and make me who I am, and I really wouldn't have it any other way. At different times in my life, the balance among these roles has shifted, but overall I can say that I have been blessed with a fantastic partner, delightful daughter, and an engaging and successful career.

R. Cole (✉)
Department of Chemistry, University of Iowa, Iowa City, IA, USA
e-mail: renee-cole@uiowa.edu

© Springer International Publishing AG, part of Springer Nature 2018
K. Woznack et al. (eds.), *Mom the Chemistry Professor*,
https://doi.org/10.1007/978-3-319-78972-9_8

I've often been asked to describe my career path and how I came to be a successful woman in science. During a leadership workshop, it occurred to me that the best answer was "planned serendipity." Many of the opportunities in my life often seemed to me to be the result of good fortune and circumstance but, upon further reflection, are linked to the choices I made that put me in the right place with the right connections or skill set to be offered those opportunities. I think this is also true of my journey as a mother and professor.

Starting Out

I've enjoyed math and science as long as I can remember. I took every math class I could in high school and almost every science class. The one class I avoided was advanced biology ... you were required to dissect a fetal pig by yourself, and I didn't think that was something I could stomach. I never have liked gooey, squishy things or anything resembling gore. (Fortunately my husband has always been willing to take care of those tasks that I couldn't handle.) I did have math and science teachers who encouraged me to excel and who supported extracurricular activities in those disciplines. Science fairs and math competitions were something to look forward to, and I usually did well in those competitions. Choosing to participate in these events opened many doors to me, including travel and scholarship opportunities.

My goal from a very young age was to get a college degree, so choosing to attend college was an easy decision. My goal through most of high school was making sure I had an academic record that would help me get scholarships to fund my college degree. I had the great fortune to attend Hendrix College, where I still have close ties and many fond memories. I started college as a math major, although I didn't really know what I wanted to do other than something in math or science. By the end of my freshman year, I decided I much preferred chemistry to calculus and changed my major to chemistry. Hendrix had (and still has) a great chemistry department where I was encouraged to get involved in research. They also make it a point to take students who are engaged in research to the spring American Chemical Society (ACS) meeting, so I was able to attend meetings in Atlanta and San Francisco as an undergraduate. These experiences helped me see the value of being engaged in the community and had a significant impact on how I mentor undergraduate and graduate students in my role as a faculty member.

Of course, another significant event in my time at Hendrix was meeting my now husband. Our first date was a blind date set up by my friends as part of a social event sponsored by the dorm where I lived. The date went well, and we quickly became a couple. That was in January of my sophomore year of college. Greg transferred to Oklahoma State in August, so we had a long distance relationship for the rest of my college career. We were engaged in the spring of my junior year and got married the weekend after I graduated from college. It was important to me to have a spouse who was truly a partner, and my "plan" was fulfilled better than I could have hoped.

Unlike many of my peers and the students and postdoctoral fellows I speak to, the decision about the timing of getting married and starting a family was not one that I really struggled with. I grew up in a family and culture where marrying young and starting a family were pretty much the norm. Many of my peers and colleagues have commented that I got married really young, but it wasn't something that I really thought about at the time. I actually got married at an older age than most of the women in my family, so I didn't think getting married at 22 seemed particularly unusual.

My role models growing up were my mother and grandmother, both of whom worked outside the home and changed careers multiple times. Balancing education and finding a career path you enjoy were natural activities to me. My grandmother went back to school to become a nurse when she was 50, and my mother returned to school to complete a master's degree in nutrition while I was finishing up my PhD.

Next Steps

I didn't really know what I wanted to do when I started college, but I found activities that were related to teaching were things that I really enjoyed. I started tutoring students in math and science when I was in middle school, and this continued after I started college. In addition to tutoring, I worked as a laboratory and stockroom assistant. These activities combined my love of science and the joy in helping someone understand a concept or process. By my senior year of college, I knew I wanted to be a college professor, and a PhD was a required step to achieve that dream. While at Hendrix, I had the opportunity to participate in two Research Experiences for Undergraduates programs—one at the University of Arizona and one at the University of Oklahoma. My experiences at Arizona and Oklahoma gave me a good idea of what to expect in graduate school, and I never considered doing anything else after I graduated. I chose to attend the University of Oklahoma (OU) and continue my work with Roger Frech, working toward a PhD in physical chemistry.

Married graduate students were not uncommon at OU, and there were a mix of single and married graduate students in Roger's group. Being married meant that I tried not to work too much at night or on weekends, which meant being more focused while I was in the lab. I still participated in study groups while I was completing coursework, and regularly spent some time in the lab on Saturdays. Most of our friends were other couples from the chemistry department, and we would regularly get together on the weekends to play games or just hang out. Greg often describes himself as a chemistry groupie because he has spent so much time at chemistry-related social events.

In addition to a PhD, I also knew that I wanted a family, and my husband and I decided that the end of graduate school while I was writing up my dissertation would be good timing. For most of my friends, it had taken 6 months to a year for them to get pregnant, but it didn't take nearly that long for us. The result was that my daughter was born about 18 months before I was ready to defend my PhD. Particularly at the time, and regularly even now, when people find out I had a baby during graduate school, the reaction is "I can't imagine having a baby while working on my PhD." My usual response is that I can't imagine having a baby while trying to get tenure or waiting until afterward (although I know plenty of women who chose both of those routes). The follow-up message is that there is no perfect time to have a baby in terms of your career. When you and your partner decide you are ready to have a child, you figure out how to make it work. In my case, I was fortunate to have an extremely supportive husband and an understanding graduate advisor.

My research involved solid-state spectroscopy, so there were no particular hazards that I had to avoid to continue doing my research while pregnant. I actually don't know what the university's maternity policy was at the time because I never asked. I was lucky to have a relatively easy pregnancy and delivery, so I was able to resume my work in the lab after a couple of weeks, although with an altered schedule. My husband worked a fairly early shift and was usually home by 2:30 pm. This allowed me to go into the lab in the afternoons for group meetings and to do any experimental work. I could do much of the analysis and writing up of results at home, so the schedule worked for what I needed to do. Once my daughter was old enough for daycare, we found a licensed daycare home with a wonderful caregiver. I would go into the lab in the late morning, and Greg was home early in the afternoon, so this also gave our daughter quality time with each parent. Because I was working as a research associate when my daughter was a baby, it also meant that it was relatively easy to stay home when she was sick or to take off part of a day for well-baby checkups.

After I finished my PhD (Fig. 1), I had to choose between a faculty position at a very small college (I would have been one of two chemistry faculty) and a postdoctoral position. Although I enjoyed my dissertation work in solid-state spectroscopy, I found that I was very engaged by my work in updating the undergraduate physical chemistry laboratory and I wanted to do more work in this area. I was fortunate to discover chemistry education research (thank you to Palmer Graves and the rest of the Abraham group), but I did not have much experience in this field. This led me to accept the postdoctoral fellowship in chemistry education at the University of Wisconsin, and we moved to Madison, WI, when my daughter was 18 months old. Childcare in Madison was significantly more expensive than it had been in Norman, particularly for a child under the age of 2, so Greg stayed home with her until she turned 2. Having a husband who was willing to follow me to different positions and to take on the role of running the household definitely made it easier for me to pursue my career goals. Choosing the postdoctoral position was also a very good choice in this regard. I had good mentors who introduced me to the chemistry education research community and encouraged me to get involved with activities on a national level.

Fig. 1 Renée celebrating with Greg and Mackenzie after graduating with her PhD (May 1998)

Growing Up

When it came time to look for faculty positions, I was most interested in finding a position at a primarily undergraduate institution, preferably in an area of the country that was closer to our families (who were mostly in Arkansas, Oklahoma, and Texas). At this point, my primary focus was on teaching, although I still wanted to stay engaged in some research. I was offered a position at Central Missouri State University, which later became the University of Central Missouri (UCM), and we moved to Warrensburg, MO in January 2000.

Greg went back to school to finish his degree after I started my faculty position, but he still took care of a lot of the household and childcare responsibilities. Being able to support him in completing a college education was a sweet reward after the years he spent working factory jobs so we could have good insurance while I was going to school. He graduated with a bachelor's degree with a major in secondary education and social studies in 2003, and a few weeks later, we bought our first house.

He did a lot of substitute teaching the first year after he graduated, which led to a full-time position teaching seventh grade. It was shortly after he started teaching full-time that I commented one night that life seemed a lot more complicated in trying to get everything done. Greg commented that it was because I no longer had a personal assistant. Shortly afterward we hired someone to come in and clean every other week. That was one of the best gifts we gave ourselves. This meant we didn't have to spend every weekend cleaning the house (and provided an incentive to keep everything picked up).

The chemistry department at UCM was very family friendly, which made the transition to professor and mother much easier. Several of my colleagues had children about the same age as my daughter, and no one seemed to mind if I brought her up to my office. It wasn't long before I had a small love seat and drawer with books and toys in my office so that she had a place when she came to the office with me. She grew up playing with molecular modeling kits like they were Tinkertoys. When she took organic chemistry in college, she commented that it brought back flashbacks to her childhood and almost seemed a little odd to being using them for their intended purpose. Liquid nitrogen ice cream was a treat, but not necessarily a novelty for her because it was a regular part of departmental picnics and gatherings. I guess that's one advantage of growing up in a chemistry department. She participated in hands-on science activities when we would do demo shows, workshops, or activities with different school groups. She was usually with me when I was preparing for Science Olympiad events or classroom demonstrations, so she was engaged in science from a very young age.

I think being a mother also encouraged me to be more engaged with science education activities in the local schools. I was one of the sponsors of the UCM student ACS chapter, and we regularly did chemistry demonstration shows and hands-on science activities. I also became involved with mentoring middle school students for some of the chemistry-related Science Olympiad events. Once my daughter entered middle school, she joined the team, and I became even more involved. However, not all the students knew me outside of being Mackenzie's mom. One afternoon, when I showed up a little before the end of practice, a few students were working on an event focused on science terminology. They were struggling with one of the terms, and I corrected their definition and made some suggestions for how to remember it. One of the students looked at me and said, "What do you know? You're not a scientist!" The students who knew who I was and what I did for a living just looked at him with open mouths and then started laughing. I don't think he ever quite lived it down for the rest of the time he was on the team.

I was atypical in my department in terms of my level of engagement at the national level and in how much I traveled. I have attended the spring ACS meeting almost every year since I was a junior undergraduate, only missing a few years during graduate school. As I became involved in national chemistry education initiatives, I also traveled to conduct workshops or for project meetings. Balancing my travel schedule with parenting created its own challenges. When my daughter was small, she refused to talk to me on the phone and didn't seem to really mind too much that I was gone, although I can't say the same for my husband. As she got older, I think she missed me more when I was gone. This was particularly true as she took higher-level math and science classes and wanted help with homework. There have been many times when I've gone back to my hotel room for homework sessions. As technology has advanced, Skype and FaceTime make this easier, although it can still be challenging.

For many years, I used a planner at work, and we had a family calendar at home. This worked fine until the year I ended up committing to present at a national meeting and then discovered it was the same weekend as my daughter's first dance recital. I missed the recital but was fortunate to have good friends who could take care of fixing her hair and makeup—my husband wasn't quite up to that task, although he did learn to put her hair up in a ponytail. After that, I kept a single calendar, although conflicts still arose when some work commitments were scheduled further in advance than school activities. These days, electronic calendars that can be shared among all members of the family make it much easier to avoid conflicts, although the timing of when events are announced can still cause problems.

Moving On

As I moved through the ranks at UCM, my research program in chemistry education grew to the point where I had to make a decision to scale back or move to a new position with more support for research. I loved teaching and my students, but the teaching and service load at UCM were more than I could sustain along with my research agenda. I began applying for positions at research universities when my daughter was in middle school, but nothing worked out for a couple of years. When she was a freshman in high school, I was offered a position in the chemistry department at the University of Iowa. This was a fantastic opportunity for me, but it did give rise to a number of discussions about the impact it would have on my daughter who had spent most of her life in Warrensburg and was not happy to leave her friends. As it turned it out, it was a great move for the entire family. The schools in Iowa City were much better at meeting my daughter's academic needs while still providing opportunities for her to engage in choir and show choir (Fig. 2). We made regular trips to Missouri the first year after we moved, so she could see her friends, which also helped smooth the transition. She has now left the nest to pursue her undergraduate degree at Wellesley College, working on a double major in chemistry and Russian. I'm very proud of the young woman she has grown into.

The move was good for my career as well. I have graduated two PhD students and have an energetic and talented group of graduate students to work with. It's exciting to move forward mentoring graduate and undergraduate students and helping them achieve their dreams.

Fig. 2 Greg, Mackenzie, and Renée hanging out at a show choir competition. Long days, but we enjoy supporting Mackenzie's activities (2014)

Reflections

I can't say that I really had role models early in my career for how to balance being a mother and professor. During both my graduate and undergraduate studies, I never had a female professor for math or science courses. I never really thought about it until one day when a visiting scientist commented that I always said "he" when I referred to instructors and mentors I had had to that point. When I started my faculty position, I was the only woman in the department. There were a couple of women in biology, but they had chosen to not have children. I did find good friends and colleagues in other departments, which helped with some feelings of isolation. I also had good friends and mentors in the chemistry community. My interactions with other women faculty with children at American Chemical Society meetings and the Biennial Conference on Chemical Education gave me the opportunity to talk to other women who had similar challenges and rewards. Some of these women have become very good friends through the years.

I think my position as a professor and a mother also made me more understanding of nontraditional students who were trying to balance being a mother and student. Particularly at UCM, I had many students who were single mothers returning to school to earn a degree to make a better life for themselves and their children. I had to bring my daughter to work with me many times when school was out, so I didn't object to students bringing children to class on those occasions as long as they were well-behaved and didn't distract the other students. I also counted it as an excused absence when students missed class to stay home with a sick child.

A concern of many faculty, particularly women with children, is finding the right work-life balance (Fig. 3). Sometimes I feel like I do a good job of finding balance, and other times I feel like I'm on a merry-go-round hanging on for dear life. Some days it all works well, and other days I wonder how it will all work out. Lynn Zettler, who was presenting a workshop for the ACS Women Chemists Committee, recently introduced me to a new model. Instead of the model of finding balance, she suggested it was better to think of tending a garden. The different aspects of your life require different degrees of attention at different times, but it's important to make sure everything important in your life gets enough attention and to watch out to make sure the "weeds" don't choke out things you care about. I think this model is useful in deciding what to commit to and what to say no to. Hopefully it will provide a framework that will help me in the future, particularly when I'm feeling overwhelmed.

There have been many times when I've wondered if I was present enough for my daughter, but an event that occurred during my daughter's senior year of high school assured me that I was a good role model if nothing else. Mackenzie was taking a course at the university that required her to leave the high school a few minutes before the end of class. Another student was in a similar situation but was afraid to discuss it with the teacher. Even though Mackenzie had never had any interactions

Fig. 3 Mackenzie and Renée hanging out with the dinosaurs at the Natural History Museum in New York (2005)

with the teacher, she went to talk to her on behalf of her friend. When I asked her why she wasn't as intimidated by this teacher as her friends, she replied that she guessed she had grown up watching me and other smart, intense women, so this teacher didn't phase her much.

As I reflect on the past and think about the future, I have no regrets about the path I created. I have an incredibly supportive spouse, a daughter who has grown into a bright, independent young woman, and a career that continues to develop. All in all, the rewards far outweighed the challenges, and I'm looking forward to see where the future leads.

Reflections from Renée's Daughter
Mackenzie Cole (age 20)

One of my clearest early memories is of sprawling on the peculiar red cement floor in the back of her office on a snow day in second grade, listening to her hold office hours. It would be years before I had even a rudimentary grasp of the meanings of the terms "thermodynamics" and "electrochemistry," but I was already sure that I wanted to be Dr. Cole when I grew up, just like her. I was convinced she knew everything, and as a child possessed of curiosity spanning light-years, I was determined to have that wealth of knowledge for my own. She gave it to me freely—she answered whatever wild questions I asked and taught me the things I did not know to ask for. What's more, when what she knew proved to be less than inexhaustible, she found the answers that I sought regardless of how strange or trivial.

I think I learned a great deal of my spatial reasoning skills from playing with molecular modeling kits in the back of her office; complicated math concepts were much less daunting because I knew mom could explain if I didn't get it immediately.

A large part of my ability not to be intimidated is because I grew up around brilliant, confident, incredibly intense women—if I'd let that make me nervous, I don't think I'd have survived until middle school. This gets more useful every year; it's much easier to communicate with authority figures if you're not scared of them. Obviously it's not unmitigated sweetness and light; she's gone regularly—sometimes only a weekend, sometimes 5 or 6 days, but rarely longer than that, except in summers—but I've never felt that she was neglecting me for her work. It was sometimes frustrating to have her gone (for one thing, neither my dad nor I can do anything with hair) but never a matter for resentment. It does help that she's careful not to miss the events that are truly important to me. Even after I realized that my mother's profession wasn't exactly the norm; I've mostly regarded it as an advantage.

Acknowledgments I have to thank my husband, Gregory Cole, for all his support through this crazy journey. I couldn't have done it without him. I also thank the graduate students at the University of Iowa who read through drafts of the chapter and reassured me this was on the right track and something they wanted to read more of.

About the Author

Education and Professional Career

1992	BA Chemistry, Hendrix College, AR
1995	MS Physical Chemistry, University of Oklahoma, OK
1998	PhD Physical Chemistry, University of Oklahoma, OK
1998–1999	Postdoctoral Research Associate, University of Wisconsin-Madison, WI
2000–2003	Assistant Professor, Central Missouri State University, MO
2003–2008	Associate Professor, Central Missouri State University, MO
2008–2011	Professor, University of Central Missouri, MO
2011–present	Associate Professor, University of Iowa, IA

Honors and Awards (Selected)

2015	American Chemical Society Western Connecticut Section Visiting Scientist
2015	Fellow, American Chemical Society
2014	Iowa Women of Innovation Award for Academic Innovation and Leadership
2010	UCM College of Science & Technology Award for Excellence in Teaching
2009	UCM Fall 2009 Convocation Speaker
2009	Missouri Governor's Award for Excellence in Education

Renée is active in chemical education research, focusing on issues related to how students learn chemistry and how that guides the design of instructional materials and teaching strategies. She is actively involved in the POGIL (Process-Oriented Guided Inquiry Learning) project at a national and international level. She also serves as an associate editor for the *Journal of Chemical Education*.

Readymade Family

Mary Ann Crawford

A chance meeting in the college eatery called the Gizmo during my senior year changed my life forever. Twenty-eight years ago, I met Kel, who had three children ages 4, 6, and 7. We both wanted to finish our undergraduate degrees and go to graduate school. I would graduate from college in just a few weeks, and Kel still had 2 more years to go. Of course my friends thought that I was insane to even consider dating a woman with three kids. I loved children! I had nine nieces and nephews, and I had been an aunt since I was 9 years old. I always told my mom that I never wanted to physically have kids (it hurt), but I would love to adopt some of my own. My other reason, besides being terrified of childbirth, was that I believe there are too many kids already in the world who need to be loved.

M. A. Crawford (✉)
Department of Chemistry, Knox College, Galesburg, IL, USA
e-mail: mcrawfor@knox.edu

Graduate School

Before I met Kel, I was considering a graduate school in upstate New York. After meeting Kel and the kids, I picked a graduate school not too far away. I bonded with the kids the summer before I started graduate school. I stayed home with them while Kel worked. I did the majority of the cooking. We always ate all our meals together and did things together as a family while the kids were young. We would take walks after dinner, play some games, go to movies, or watch a movie on video.

My first year of graduate school was very lonely and isolating. The environment at the graduate school turned out to be very hostile toward me. I was the only African-American in the entire chemistry program, and I found out I was the department's attempt to diversify its graduate program. Most of my classmates had attended large schools and had not interacted with a person of color before. I found myself in a program with no support emotionally and very little support academically. I was interested in the area of physical chemistry, but only one professor was interested in working with me as a graduate student. For the first time in my academic life, I felt alienated. The academic environment only made me miss the kids and Kel more. It was difficult to see the children every other weekend, and financially maintaining two households was tough on our family. I applied and was fortunate to win a Graduate Education for Minorities (GEM) fellowship that year. As part of the fellowship, I spent the summer interning at Union Carbide, which brought some financial relief, but I also found out the GEM fellowship was portable, and since Kel had another year of undergraduate work, I decided to take a year off and apply to other GEM graduate programs. I applied for an opening in the office of admissions so that I could support our family, while I waited for Kel to finish her undergraduate degree. We then embarked on our graduate work together. She studied social psychology, and I studied analytical/physical chemistry.

We chose to go to Purdue University, but we did not realize how difficult it might be to find a place for an interracial same-sex couple with three kids in the early 1990s. We definitely experienced housing discrimination. I remember one manager pointed to the bedrooms and made it clear that I could not sleep in the same room with my partner nor could the kids sleep in a mixed gender situation. In other words, there were no accommodations available for us. We had all but given up, but fortunately, we managed to get on the waiting list of a place that had a three-bedroom town house with no inappropriate questions asked. I called daily for weeks until something opened up. The town house would be our home until we moved out of Indiana. The kids loved it. There were other families with children their age, and it had a swimming pool, a game room, and plenty of empty parking spaces for one of their favorite activities, rollerblade hockey.

Readymade Family 129

Parenting

Kel and I had very different parenting styles. I was more of the disciplinarian, stern but fair. Kel was the softy. She would try to be stern but could always be made to smile with the quick wit of one of our children. My parenting skills were mostly influenced by my mother, father, maternal grandfather, and paternal grandparents. I tried to incorporate the best of all their styles and do the things that I wished had been done for me. For example, unlike my home, all the kids, no matter their gender, had weekly chores. Our daughter has just as much social freedom as our two sons. We were very lenient in letting them spend time with their friends as they got older. However, I did not care what the times dictated in the 1990s, I was never going for the mixed gender sleepovers.

Managing graduate school and being a parent presented challenges. I cooked and was home for dinner almost every night. We often had to take the kids to basketball practice, tae kwon do, or soccer practice. In order to move to Indiana, my partner had to petition the state of Illinois to move the children out of state, but they still needed to have visitation with their father. As part of this agreement, we had agreed to drive the kids to their aunt and uncle's house every other weekend to visit their father.

Financially, we struggled. We took out student loans and there were times when I sold plasma. We took on extra jobs when they were available. I tutored high school students in chemistry and taught review sessions for other departments and worked several summer science programs designed to encourage underrepresented students like myself to pursue a career in STEM. Kel taught in the summer whenever she could. I always remembered something my mother told me about money and bills. She said "Mary if you pay your bills, someone will always be willing to help you eat." She was so right.

In spite of the financial struggles, we found ways to spend time with our kids. We could go to the Purdue women's basketball games for $10 for our family of five; we would often go to the dollar show, rent movies, go camping in the nearby state parks, and get to a few White Sox games in Chicago. It took us longer than most people to finish our degrees, but your kids only have one childhood, and we always tried to put their needs ahead of our own.

Finding the Right Mentor

Academically, I struggled to find the right mentor. My father very suddenly died in the middle of the selection process for first year students like myself to find a graduate school thesis advisor, and I did not realize at the time how his death truly impacted me. I knew I had found a home at Purdue because upon my return from my father's funeral, my classmates showered me with care. The students in the program were amazing people. They filled my desk with sympathy cards; they took class notes for me and even dropped by my office to take me out to lunch.

In the end, it took me three attempts to find the right thesis advisor. My first advisor never really gave me a project, and I spent my first summer fixing equipment, setting up a BET isotherm apparatus, and setting up Schlenk lines for other people in the lab. The last straw for me was when he did not attend the meeting to assign my grade for my second year seminar and we got into an argument over my grade in his class. I lost it completely, so there was really no going back to the group. He changed my grade in the course, but I felt the trust that I should have toward my mentor was not there anymore, and so I moved on to another group.

Calling my second advisor misogynistic would be an understatement. I remember when he had to hire a new secretary and he said, "Well, I know who I am going to hire but the only question is whether or not she can type." Of course, being gay, I was mostly treated like one of the guys, and there were no other women in my group at the time. I passed my oral exam and was preparing for the final hurdle in the program, which was to develop an original research proposal and propose the topic of my dissertation. My second advisor had started dating a graduate student who had taken one of his courses. He was out of town with her at a conference, while I worked on the proposals, but we communicated on the phone. He instructed me to delete several sections of my dissertation proposal during this process. On the day of my oral exam, I passed the original research proposal, but I was told that I did not have enough in my thesis proposal for a PhD. I was devastated. I spoke to a mentor of mine in the graduate school that night, and he arranged for me to meet with someone he thought would be willing to take me in his group so that I could continue on the PhD track. Afterward, I went into talk to the ombudsman in the chemistry department and showed him the drafts where I was told to exclude portions. At this time it was revealed to me that the day before I presented, my advisor had also spoken to ombudsman about my getting a masters.

It was pretty obvious to everyone that my advisor had decided that he wanted his new bride (he married her the day after my oral exam) to continue on my project and he wanted me to leave the chemistry program with a terminal masters.

My department overall was very supportive of my decision to leave this person's group. However, working with my new advisor meant leaving my family for a while. I had to move to Michigan to work in the labs at Ford Motor Company. It was one of the toughest decisions I have ever had to make. In the end, I had to think about what was in the best interest for me and for the financial future of my family. If I did not make the move to Michigan, I was not going to get my PhD. This was not an option for me because my career goals did not line up with a terminal masters. I wanted to work in industry as a PhD scientist or teach at an undergraduate college. I wanted my family to see me succeed and be proud of me.

My new advisor was aware of my family situation and he was very supportive. Every other weekend I was allowed to leave work at noon on Friday and return around noon on Monday, which allowed me to visit my family all day Saturday and Sunday. Despite missing Kel and the kids, the research was fun, challenging, and interesting. I was able to learn very quickly and get a significant amount of research done to propose my new dissertation topic and continue working toward my PhD in chemistry.

I looked forward to going to the lab every day, and I learned a lot in the 2 years I spent at Ford Motor Company. My research involved looking at reactions of hydroxyl radicals with cyclopentylperoxy and acetylperoxy, respectively, using the techniques of time-resolved UV spectroscopy, transient IR, and FTIR smog chamber. One of the most important lessons I learned was that science is not done in a vacuum. At the scientific research lab, organic chemists, physical chemists, physicists, and engineers worked together to solve problems. You need to be able to communicate effectively in order to get things done.

A Career in Academia

Toward the end of my time at Ford, I was offered a Howard Hughes Medical Institute (HHMI) teaching fellowship by my alma mater, Knox College. At first, I declined the offer, but then my undergraduate mentor, Dr. Robert Kooser, called and said he really needed me to come. Now this man had supported me since I was a prospective student. When I met him on my visit to Knox, he offered me a research position in his lab. When I was in graduate school the first time, he gave me a copy of a book used in one of my classes because the book was out of print and one of my classmates stole the copy out of the library and my professor had refused to replace it. When I moved to Michigan, he gave me the money to get established in an apartment, and when I told him I had no idea when I would be able to pay him back, he said that it was okay because he had money and I didn't. So if he needed me at Knox, then I would have to talk to Kel and make it happen. Instead of renting an apartment, I rented the extra bedroom upstairs in Dr. Kooser's house. I was only supposed to be at Knox for 1 year.

The job at Knox was a lot closer to Kel and the kids. I was still working on writing a paper and writing up my dissertation when I accepted the job at Knox. Of course, now I always encourage my students to never leave graduate school without first completing their dissertation because teaching full-time and writing was not a lot of fun. During winter break, I went home to Indiana and worked on a paper with my Ford colleagues. I loved teaching general chemistry my first term at Knox. The students were so smart and enthusiastic most days. Over break, Dr. Kooser called and asked if I might be interested in staying on at Knox. I told him we could talk about it around Christmas because for the first time I was going to Kel's family for Christmas. You see, for years, I was not allowed to go to my in-law's home because my father- in-law, Kel's stepdad, objected. So as a compromise, Kel and I would do Christmas in the morning with the kids and then drive to Illinois. I would stay at Dr. Kooser's home in Galesburg, while she visited her family and the kids saw their dad's side of the family in the Quad City area. Dr. Kooser, his son, his partner, and I would exchange gifts, eat, and talk most of the night. This year was different because my brother-in-law and his family were hosting Christmas at their home in Dunlap and I was invited.

Well not having spent Christmas with Dr. Kooser, I wanted to stop by and spend time with him. So Kel and I were going to stay in my room for a night or so. There was a party at a colleague's house that evening. Dr. Kooser was in the kitchen watching a football game and he was on the phone. I was in the bathroom brushing my teeth. As Dr. Kooser was talking to Kel and walking up the stairs, he fell and we now know he suffered a major stroke and he passed away a few weeks later with me holding his hand.

At first the thought of staying at Knox without him was too painful to consider, and I went on the job market. It was on my final interview that I realized how much I loved teaching, so I decided to accept the job offer for the tenure-track position at Knox, and the college also offered my partner a position in the departments of psychology and gender and women studies.

I would struggle the first year in that job. The students I had in my first physical chemistry class were not the easiest crowd to get along with, and I think we were all struggling with the sudden death of Dr. Kooser. He was the person I would always confide in and seek advice and guidance. I felt so alone. I was the first African-American tenure line professor in the sciences and for many of my students the first African-American role model they had encountered. A few students made my life miserable, but the vast majority of my students were amazing! I had to learn to focus on the positive, learn from the negatives, and grow from the constructive criticisms that would make me a better teacher in the end. I was never going to be able to change my race or gender, so I had to stand up for myself when students disrespected me in ways that I thought took away from the educational experience of other students. For example, there was a group of students who talked whenever I was lecturing and writing on the board. I sent them an email requesting they stop talking because they were disturbing other students. When the behavior continued, I kicked them out of class. I had one student get angry and disturb my class because she wasn't understanding something I was saying and I told her to wait that I would help her in a minute and she continued to be disruptive. I dismissed her from class and she later came to my office and asked me "Were her questions too difficult for me?" I almost fell off my chair. I respectfully told her when she was the only one that did not understand what was occurring in the lecture out of the entire class perhaps she was the one that needed to reevaluate what she thought she knew. I am sure my white colleagues have never had such a question posed to them.

Throughout my career I have worked with some amazing students in the classroom and in my research lab. My teaching philosophy is to make students think critically and not be afraid to challenge themselves. Learning is a lifelong process and science is forever changing and discovering new things. I received tenure in 2004. I became the first African-American woman in the college's 175-year history to reach the rank of full professor in 2011. I served as my department's chair from 2012 to 2016. I currently serve as chair of the Gay, Transgender and Allies subdivision of the Professional Relations Division of the American Chemical Society.

As of July 2017, I will have been at Knox for 20 years. My research interest in kinetics of atmospheric reactions, solution kinetics, and teaching physical chemistry

Readymade Family

has been the focus of my career. I also became interested in computational chemistry. I have had several students work in my lab over the years, and many of them are now doctors, industrial chemists, psychologists, and PhD chemists. My students are my legacy, and I think of them as also a part of my family. Like Dr. Kooser, I support them even after they graduate because I believe being their professor is a lifelong job.

I was so proud when one of my research students sent me a picture of himself at his dissertation defense with a caption that proclaimed that "my little Chihuahua was done!" Little Chihuahua was a term of endearment bestowed upon him in my lab because he was always a nervous wreck. I had the privilege of watching him grow and mature into a wonderful scientist and young man. I was present at his wedding, and his parents thanked me for mentoring him. I remember attending one of his oral presentations at an ACS meeting and having his graduate thesis advisor invite me to lunch with his research group afterward. On the walk over to where we were dining, his thesis advisor commented on what a wonderful job I had done. But my most joyous moment was getting a picture of a sonogram and announcing he was going to be a dad. Of course, this means I get to be a grandma! This is why I teach at a small liberal arts college. I love getting students ready for the next step and seeing them flourish.

Our children are now in their 30s and our first grandchild is 4 years old. Being a mom in graduate school and in the early part of my academic career was not easy. I considered leaving Knox at one point because my youngest son was not happy in Galesburg. He left all of his friends in Indiana when we moved here, and it was very tough for him to adjust to life in Galesburg. We ended up letting him drop out of school, and he started taking classes at Knox. He eventually graduated with a degree in history from Knox. He has a job at an Illinois historical site. He also just bought a home right around the corner from us for himself and his young son. I was so happy when he asked me to help him through the process of finding and buying his home. Our older son also works and lives in town, and he was married in September of 2017 to his girlfriend of 5 years, and he also became a step dad of three daughters. Our daughter lives in a town two and a half hours away where she works and spends time with her wonderful dog named Taco (Figs. 1, 2 and 3).

Advice

As of September 2017, Kel and I divorced after a 28-year relationship. It was an amicable breakup, but it was painful nonetheless. It was important to both Kel and I that our relationship with our children remain strong. We agreed to continue to parent and grandparent together. I think my youngest son said it best when he told me, "the only thing that would change about our relationship is my address."

I find myself single for the first time since I was 22 years old. I am still a mother to my three children and a grandmother to my grandson. I have gained three

Fig. 1 Morgan, Ross, and Tara from Mary's 50th birthday party

Fig. 2 Mary's grandson Carter (age 4) and Gee (he calls Mary "Gee")

granddaughters courtesy of my newly married son and daughter-in-law. I was very happy to host their rehearsal dinner in my new home.

The advice I would give to anyone is to find a balance between parenting, marriage, career, and self-care. I think it is safe to say that my priorities were probably backward between my career and my marriage. I have to own how my actions contributed to the end of that relationship. Sometimes loving someone is not enough and you grow apart before you know it. Sometimes letting go is necessary so the person you love can be happy.

Professionally, it was important to seek colleagues that would encourage you to publish, not give up on the grant writing when you get rejected (and you will get rejected) and collaborate with others so that you can hold each other accountable for getting stuff done.

Fig. 3 Ross and Brandy's Wedding, September 2017

I am a first-generation African-American female, and a lot of people would say the odds are stacked against me. I would say that no matter who you are, there will be obstacles that you will have to face, but the secret to your success is how you handle your failures. Every obstacle that I have endured has taught me something about myself. It is important to build a support system for yourself and to have mentors that believe in you and push you to reach your fullest potential. I have been fortunate to have great mentors in my life, along with the support of my family and close friends.

There is no greater gift than a person letting you help them raise a child. There is a lot of responsibility that comes with such a gift, and you should never make a decision to become a parent if you are not willing to be selfless for a while. I remember taking my daughter to work when she was in her teenage years, and she was so excited that one of her high school friends was pregnant. I asked my daughter how much money she made a week at her job? She said 100 dollars. I said imagine having to give 90 dollars of your money to someone else because being a parent means you are responsible for someone else's well-being. Becoming a parent and a grandparent is the best thing that has ever happened to me. Spend time with your kids when they are younger, and they will want to spend time with you when they are adults.

My daughter lives the farthest away, but she still calls me and invites me to her house for visits. I had a pizza party with my younger three grandchildren, followed by sleepover with my son and grandson the first evening in my new empty home. My son and his new bride have already extended an invitation to their home for Thanksgiving this year. My life has changed a lot in the last few months, but my children are still a big part of my life. I am grateful for every chance I get to spend with them.

Acknowledgments This chapter would not be possible without my family. So I would like to thank and dedicate this chapter to my ex-wife, Kel, and our children, Tara, Ross, and Morgan. Our family will always be an important part of my life. Kel, thank you for allowing me to become a part of our readymade family. Kids, I love you and thank you for opening your hearts to me. I am honored to be your mom.

About the Author

Education and Professional Career

1989	BA Chemistry, Knox College, Galesburg, IL
1999	PhD Analytical Chemistry, Purdue University, West Lafayette, IN
1997–1999	Instructor of Chemistry, Knox College, Galesburg, IL
1999–2004	Assistant Professor, Knox College, Galesburg, IL
2004–2011	Associate Professor, Knox College, Galesburg, IL
2012–2016	Chair of Chemistry, Knox College, Galesburg, IL
2011–present	Professor, Knox College, Galesburg, IL

Honors and Awards (Selected)

2014	Faculty Exceptional Achievement Award, Knox College
2008	Trio Achievement Program Distinguished Educator Award, Knox College
2006	Howard Hughes Medical Institute Faculty Fellowship, Knox College
1997–1998	Howard Hughes Medical Institute Teaching Fellowship, Knox College

Mary has published in the Journal of Physical Chemistry, the International Journal of Chemical Kinetics, and the Journal of Chemical Education. She has received funding from NSF and is currently finishing her term as the Gay, Transgender and Allies subdivision of the Professional Relations Division of the American Chemical Society (ACS). Mary is also the Chair-Elect of the entire Division of Professional Relations. In addition to her commitment to ACS nationally, Mary works with local elementary schools and her local section of the National Association for the Advancement of Colored People (NAACP) to encourage and inspire future science students.

Mother and Community College Professor

Elizabeth Dorland

When do you decide you want to be a chemist or to teach chemistry or to be a professor? I wasn't sure what I wanted to do even after arriving in graduate school. Then suddenly I did. But that's the middle of the story, so let's go back to the beginning.

In the Beginning: Family Values and Role Models

As a kid I imagined being married and having a family in the 1950s Midwest norm of around two boys and two girls. Very few women in our small town worked outside the home. Most married soon after high school and had families immediately. By the

---

Elizabeth Dorland retired from Washington University in 2014.

E. Dorland (✉)
Washington University, St. Louis, MO, USA
e-mail: Ldorland@mac.com

time I got out of college and into grad school in the early 1970s, I had settled on two kids as a good number. In that later era of women's liberation, I also thought that it would be a good idea to be at least 25 when I married. This was in stark contrast to the norms in my small hometown.

Both of my grandfathers were Nebraska farmers. In the 1950s I lived on a farm until I was 8 years old. The closest I came to doing chemistry back then was making mud cookies to bake on warm rocks in the sun. Or maybe the time my sister and I tried to walk (then run) through a cloud of anhydrous ammonia leaking from a fertilizer tank to get to our house from where the school bus stopped at the bottom of our lane. That didn't go well, so we took a long cut around through the cornfields.

I lived in the same town as both sets of grandparents. Neither my mom nor my grandmothers worked outside the home when I was young. However, both of my grandmothers had been teachers before they were married (in their day, married ladies had to quit), and both put a high value on education. My parents both attended college but were married at age 20 and didn't complete a degree at that time. Still, I always knew I would go to college. It was just assumed. My mother went back to college in her 30s and became a teacher too. It seems to be in the blood.

School Days in the 1950–1960s in the Midwest

There were no chemistry kits in my childhood, and I set off no explosions. In my small town of 1500 residents, there was just one room of students for each class from the 1st to the 8th grades. All of my teachers were female until I reached freshman general science. I don't recall much about my science education in the early grades, but I definitely remember 7th grade. Miss Bowers filled a large beaker with table sugar. Then she poured concentrated sulfuric acid into the beaker. Readers who are chemists know what happened next. A tall, fat black column of porous carbon grew out of the beaker and towered over her head. We were amazed. That may be the only thing I remember from the whole year.

I had an excellent (male) science teacher when I was a freshman in high school. He had been a farmer for many years but had gone back to college to become a teacher. I thought that chemistry in particular was interesting. Although we learned about atoms in general science, what the juniors were doing in their chemistry class looked a lot more interesting. Strange and horrible smells came out of the room. They heated mercury and got some red powder. I couldn't wait until my turn came in a couple of years.

But fate intervened, and my mom, a single mother with three kids who had gone back to college to become a teacher, married a fellow student who also had three kids. Our blended family moved to an even smaller town in Kansas with a population of 250. The high school had physics, but no chemistry class. None.

The physics teacher (also the high school principal and our typing and book-keeping instructor) was excellent, and I did well in spite of his doubts that I would survive as the only female in the class. I always say that I chose chemistry because I

Mother and Community College Professor

had missed out on it in high school. When it came time to choose a college major, I read the descriptions in the catalog. Chemical Engineering! Plastics! Why not?

At this point I had no role models whatsoever for a college-level academic career and no idea at all about what I wanted to be when I grew up.

College and Grad School: What's the Plan? MS vs PhD

At Kansas State I did well in general chemistry lecture. I didn't really like my lab classes, but being the only female freshman major in chemical engineering wasn't much fun either. So when the chemists offered me a summer job if I changed majors, I bought in. During my college years, I worked for one summer in a chemistry lab at Kansas State and then two summers in the research labs at Kodak Park in Rochester, NY. Working in a lab still didn't charm me as a lifelong goal, but other options were not obvious.

Travel was not something my family did. I had seen the ocean only once, on a Christmas trip with a classmate during my sophomore year at college. I barely knew what a bagel was. Visiting New York City on that trip and twice later when I was working at Kodak convinced me that I would love to live in the big city. So I applied to chemistry graduate programs at Columbia, University of Chicago, and as an afterthought, to UC Berkeley.

In fact, going to graduate school was partly a way to avoid taking a job as a technician in a chemistry laboratory. I had no college role models in my family other than teachers. I didn't even know what graduate school was until I met my grad student TAs in freshman chemistry. In my hometown as a kid, I remember a gossip about the son of a prominent family who was so irresponsible that he was still a student in his late 20s. I later realized that he was a graduate student! That was not a recognizable beast in a small town in the 1960s. Being a TA would pay my way, so off I went, but not to Columbia.

In the end the attraction of California, where I had never been, was irresistible. I was 21 years old. That turned out to be a good choice for many reasons. My relatives were worried about me driving alone, and by coincidence, the son of a lifelong friend of my great aunt in Nebraska was also going to Berkeley in Chemistry. I asked him to ride along. Guess what? He did, and we got married in Golden Gate Park a year later in front of our grad school friends. We celebrate our 47th anniversary this year. But that is getting ahead of the story again.

Teach College Chemistry? Me?

During my first year at Berkeley, I was assigned to be a TA for freshman chemistry. As I recall, the course structure included a 1-hour recitation session and two 3-hour labs per week. My first obstacle was my shyness and my fear of speaking in front of any group. My parents and grandmothers were high school teachers, but I had no intention of following in their tracks. I never intended to become a teacher of any sort.

However, explaining the chemistry concepts in recitation seemed to come naturally to me. The faculty member assigned to my section complimented me on my teaching. Students liked me. Still, giving lectures in front of an audience (the only way science was taught back then) was not something I could picture myself doing. In fact, I was so terrified to give my first presentation in front of my new research group that I cried during a practice run with my future spouse. Amazingly, my advisor and lab mates complimented me on the clarity of my talk. And it was fun.

In grad school my students, peers, and professors found my presentations and explanations to be very clear. I enjoyed explaining scientific concepts very much. Lab work did not interest me as strongly. I continued as a TA in general chemistry at Berkeley and later in introductory organic chemistry. During the following months, I continued to find lab work rather tedious. I started hearing about the community college system in California and the possibility of teaching college chemistry with a master's degree. By the end of my first year, I had decided. I got my MS in organic chemistry the next year and started looking for a part-time job to test the waters.

My by-then-spouse Robert Blankenship continued with his PhD at Berkeley. We were married in Golden Gate Park in 1971. The wedding lunch was a pot-luck picnic provided by our grad school friends (Fig. 1).

Community Colleges Rock!

In 1972, I got my first position. I was very lucky. The vice president of the college liked my ideas on science education reform, but the chemistry department was less certain. They wanted to hire a male high school teacher who had been teaching evenings for them, but I got the job. It was half-time with full benefits, teaching two sections of introductory chemistry with labs at Diablo Valley College. I was their first-ever daytime adjunct. The all-male faculty and I got on very well, and as a whole, they were very talented educators.

Before the 1970s, part-time faculty members were a rare breed. The vast majority of positions in colleges and universities were full-time and tenure track. The status of adjunct faculty changed drastically as colleges all over the country started hiring part-timers to save money. No more benefits. No more pay at the same rate as full-time faculty. Luckily, my spouse had access to benefits via his postdoc position and later from his professorships. At that point I had *no* idea how many years it ultimately would be until I had my own tenured position!

I loved teaching in community college. I have never regretted my decision to leave laboratory research. But if I had known how long the pathway to a tenured, full-time and permanent job would be, I might have hesitated. I'm glad I didn't. After finishing grad school, I taught part-time at two colleges in the San Francisco Bay Area and as an associate instructor back at Berkeley. Later on I taught at two

Fig. 1 Liz and Bob just before their wedding in Golden Gate Park in 1971

different community colleges in the Seattle area and at the University of Washington. Later still, we lived in Amherst, MA, and I taught at a nearby small college in Springfield. Ultimately, we ended up in Arizona. But I'm getting ahead of the story again. Perhaps you are wondering when the mother part will begin. We will get there eventually. As you will soon see, travel (both with and without your children) is one of the best perks of being a college faculty member.

Travel as Education and Life

When my spouse finished his PhD, we decided to teach overseas for 2 or 3 years. At that time he was thinking of looking for a small college teaching job when we returned, and I wanted a full-time community college position. We decided on the American University of Beirut in 1975 and had a great 5-week trip hopping from Japan to Hong Kong to Thailand, India, Nepal, and Afghanistan. Unfortunately, by the time we reached Beirut in September, the Lebanese Civil War had begun. We stayed there for 6 weeks in a great apartment with a sea view but were never able to begin teaching. Hearing gunfire every night and watching documents being burned on the roof of the nearby British Embassy were the last straws. Faculty members and their families were evacuated eventually to Greece to wait until after Christmas to begin the school year. We spent an enjoyable 2 months in Athens and traveling the countryside, but by Christmas the fighting in Lebanon had not abated.

We returned to the USA. I took a sabbatical replacement position with my former employer, Diablo Valley College, and my spouse became a temporary postdoctoral fellow for his Berkeley PhD advisor. At the end of the spring, he was offered a postdoc in biochemistry at the University of Washington in Seattle. I auditioned for a part-time position at Seattle Central Community College and landed the job. The division chair who hired me was a chemist and innovative educator. He was fantastic to work with. During the 3 years we were in Seattle, I also taught at Shoreline Community College and in the chem department at the University of Washington.

In 1977 we began our tradition of combining academic conference trips with vacation time. This activity has continued for almost 40 years! Our first trip was to England. Bob's postdoc colleague and his wife had become our best friends in Seattle. She and I drove all over southern England on the wrong side of the road, while the guys attended the International Congress on Photosynthesis. I'm sure we had the better time! After the conference Bob and I traveled around England and Scotland. Since 1972, we have visited over 30 countries around the world. At least half of those trips included the kids and still do. These are among my favorite memories.

Kids on Hold?

But back to our narrative, when we left Seattle in 1979, I was in my late 20s, and we discussed when to start a family. I was very worried about never getting a permanent job and wanted to wait until I had a permanent position before having kids. However, the most favorable job offer my spouse got was nowhere near any community colleges. His least favored offer was. The middle course for him was near at least some small colleges, so we both compromised. After I got a permanent position, then we would discuss children. I should mention that beginning our relationship in graduate school meant that we both had the same demanding schedules. Neither of us was much of a cook yet, so we shared all of our household tasks, including cooking and cleaning. We always have. We each have our strengths and preferences, but we always negotiate. Work out a fair system early on and be flexible.

I did get a non-tenure-track full-time position teaching laboratories and supervising the stockroom at American International College, but it was not what I wanted to do permanently. But before I knew it, I was well over 30, and another compromise seemed wise. By the time our daughter was born in 1982, we had been married for 12 years. I do not regret waiting—it was the right choice for us. But there are many options depending on your situation.

Fig. 2 Liz and Larissa knead bread in Amherst, MA, around 1983

There are always trade-offs. During our daughter's first 3 years, I taught half-time at AIC. My spouse spent his first sabbatical at his home campus during the fall semester of 1982, after our daughter Larissa was born. He took care of her in his office 2–3 days a week, while I was working. The second semester she went to daycare at Mt. Holyoke College (midway to my job in Springfield) on the days that I worked. I was not on the tenure track during her early years, and that made time management much simpler.

We continued our travel tradition. The same conference that we attended in England was in Belgium in 1983. Our daughter learned to walk as a 1-year-old in a hotel room in Germany.

Our coping strategy was in place and I was teaching, but I was still very concerned that I would never be able to achieve my goal of a full-time tenured position at a community college (Fig. 2).

Back to Community College: Evening Teaching with Baby 1 + Baby 2

In the end, I worked half-time as an adjunct from when my daughter was born until my son was 3 years old and she was 7. I worried all those years that I would never find a full-time tenure-track job, but working part-time turned out to be a luxury I could afford. If I were doing it all over and a position came up sooner, I know I would take it just because there are no guarantees. But I was very lucky.

Moving to Phoenix in 1985 was also lucky, because there were far more community college opportunities in the western USA than in western Massachusetts. When our daughter was 3 years old, my spouse accepted a position at Arizona State University. He had discovered that he really liked having graduate students and a full research program. I was also somewhat optimistic about my prospects because the Phoenix area is home to the largest community college system in the country.

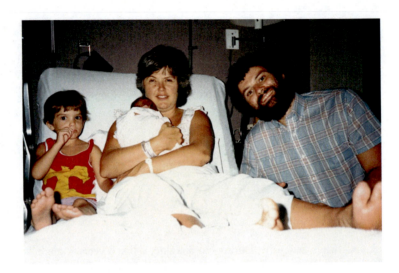

Fig. 3 Larissa (age 4) with baby Sam, mom Liz, and dad Bob at the hospital in Arizona

Fig. 4 Sam, Liz, and Larissa in their favorite activity—reading books

I immediately started teaching general chemistry lecture and lab at Glendale Community College about 45-minute drive from our home. I had child duty during the days, and I left for the GCC campus as soon as her daddy got home. In 1986, our son was born in late August. That fall semester was the only time I didn't teach during the 14 years since I had finished grad school (Fig. 3).

I went back to teaching evenings, eventually at Mesa Community College just a few blocks from our home. When my son was 3 years old, I was offered a 1-year-only sabbatical replacement position at MCC. My son was old enough for their fantastic Children's Center, and my daughter started kindergarten (Fig. 4).

When I first arrived in Phoenix, I was told that *no* full-time community college chemistry faculty had been hired anywhere in the district in the previous 10 years. But a few years later, a number of faculty members began to reach retirement age. There was hope!

Full-Time with Tenure After 20 Years?

The first year I interviewed at Glendale Community College for a full-time tenure track position, I was not hired. The following year I was. In 1990 I got a tenure-track position at Glendale, and my son was able to continue at the MCC Children's Center. I dropped him off on my way to GCC. I transferred back to Mesa after 3 years in Glendale to be closer to home. Two years after that, I received tenure in my dream job.

Have you done the math? A lot of time and luck was involved. I had a great deal of experience by this point and great references. But I almost lost the position to a very experienced professor from a mid-western university. In the end, I was told his wife refused to move to Phoenix. Thank you very much! I was 42 years old. Still 20 or 30 years of career left to enjoy. And more travel... (Figs. 5 and 6).

Fig. 5 Fiji in 1992, Larissa celebrated her birthday on the island of Fiji and then again on the plane after crossing the international dateline!

Fig. 6 Sam holds an antique Fijian War Club (1992)

Conferences and Professional Activities as a Community College Prof

One of the most valuable perks of being a professor is the opportunity to become a part of an international community of like-minded scholars. Getting help from and forming relationships with other faculty are important both for job satisfaction and career advancement. Teaching daytime classes during most of the years I was an adjunct allowed me to do this, but my disciplinary connections (in those pre-internet days) were mainly local. Once I had a full-time job, I wanted what my spouse had since graduate school—trips to professional conferences and a network of colleagues around the world. Luckily the community college district that hired me provided monetary support and encouraged these activities.

It wasn't long before my chemical education network began to expand and to inform my teaching. One of my first conferences was the 1992 Gordon Research Conference on Science and Education, the first-ever education GRC. Chemists at that meeting subsequently organized the first Chemical Education GRC in 1994. It is still going and now is known as Chemical Education Research and Practice. On the bus from the airport to the BCCE (Biennial Conference on Chemical Education) at UC Davis in 1992, I shared a seat with Kurt Sears, who was then a chemistry program officer (rotator) at the National Science Foundation in the Division of Undergraduate Education. By the time we reached the campus, he had already invited me to be on an NSF review panel. Later that week I met Susan Hixson,

DUE Chemistry Program Director. Susan also invited me to review, saying that they needed reviewers with a community college perspective. My network was beginning to form. These conferences and panels and the friends and colleagues I met there laid the groundwork for some of the most rewarding activities of my entire career.

The Macintosh SE-30 computer in my office was already connected to email in 1990 and to campus network, but not much more. But in early 1993, I joined the recently established chem-l listserv, and my love affair with online connections began in earnest. That day my Chem 101 students were extracting red cabbage juice to form an indicator solution that exhibits a rainbow of colors depending on pH. As I wandered back to my office to fetch something, I was wondering what kind of molecules were involved. I decided to ask online and sent a quick email to the list. When I checked my email at the end of the lab, I was amazed to find not just one but three detailed responses to my query. Remember, these were the days before Google and web browsers (let alone Facebook), so this was about as amazing as it could be. Chemist friends from chemed-l became in-person friends when we had our first face-to-face meet-up at the next BCCE at Bucknell University in 1994.

From my new contacts, I learned about the availability of visual chemistry tutorials and software programs that illustrated reaction mechanisms step by step from CD-ROMs. I became a pesky early adopter who was always pressing my school and department to buy these cool toys for us to use in lecture, lab, and in the campus computer lab. With the advent of the World Wide Web, the chime plugin allowed students to view and rotate molecular models on-screen by using files from the protein database. I could create my own (horribly ugly) web pages and share my learning adventures with other chemists by passing along the links on chemed-l. Perfect synergy.

More Travel: Sabbatical in Australia + NSF

Travel is one of my very favorite things, and being a professor provides many chances for adventure. Do not pass them up! In 1999 we were able to go on sabbatical for 6 months in Australia. One of my new chemistry contacts helped me to arrange a position at the University of New South Wales, while my spouse did some research with a colleague at the University of Sydney. Our kids were in what the Aussies call Year 7 and Year 11 and had a great experience attending a local high school. They learned that our Math is their Maths and that our Sports are their Sport. Surfing was one of their choices for sport, and I think they were crazy to pass that one up. But they did have a fantastic time. We negotiated with their school principals ahead of time and ultimately their credits transferred back with only a few glitches.

In the spring of 2003, I got a phone call I'll never forget. A community college-based chemistry program officer at NSF invited me to interview for the rotator position he was about to vacate in Washington, DC. At my interview talk, I spoke about my various classroom and lab teaching experiments, including the student video project that was going on in my qualitative analysis lab. I was hired. My son was a junior in high school, so I was reluctant to be gone from home. However, he was very supportive, and I promised to stay only 1 year and to be back for his senior year. My husband was department chair in his university chemistry department at that time and could not take leave, but he also agreed that I should not miss this opportunity. In the end, my college age daughter moved with me and took some classes. I went home or my family visited us about once a month. Many of the other DUE rotators and program officers are still friends.

I gave many talks on NSF programs and returned home with a much wider perspective on educational issues. I was sorry that I couldn't stay a second year and use the vast amount I had learned, but it was a good compromise. I was appointed to several NSF Project Advisory Boards, and I continue to review frequently for NSF. Take advantage of this kind of opportunity and of all of the sabbaticals that are available once you are full-time. They create fantastic family stories and are opportunities that should not be missed.

Another GRC: The Gordon Research Conference on Visualization in Science and Education

The longest trip I took during my NSF experience was to the 2004 ICCE (International Conference on Chemical Education) in Istanbul. My talk about NSF programs inspired an audience member to tell me that his country needed an NSF. I innocently asked where he was from and he said—Iraq. Amazing. I didn't know it at the time, but contacts I made at that meeting would lead to some great adventures over the next 10 years. Peter Mahaffey, a chemistry professor at Kings College in Alberta, Canada, was one of the keynote speakers. Peter was also elected to be the chair of the Gordon Research Conference on Visualization in Science and Education in 2005 at Oxford in the UK. I had always wanted to go, but a flight to England was beyond my travel budget. When I mentioned this, he said: "I think I can help." And he did, with both travel and registration support. When I returned to the GRC conference at Oxford (England) in 2007, I was elected as vice chair for 2009 and chair for 2011. As far as I know, I am the only community college faculty member ever to be a GRC Conference chair. I highly recommend this community and conference to anyone who teaches science (Fig. 7).

My science and education conference and NSF activities ultimately created just what I wanted—a fantastic international network of colleagues, friends, and collaborators who work in science education and visualization, and more recently in educational technology, online communication, and virtual worlds. Our kids both took their first trips to Europe at age 1. They remain inveterate and enthusiastic travelers (Fig. 8).

Fig. 7 Liz and her family visiting Africa in 2007. Left to right: Sam, Bob, Liz, and Larissa

Reflections from Liz's Spouse and Children
Bob, Sam, and Larissa Dorland

As far as career influencing my family, our kids are both proud of having "academics" as parents. The extensive travel opportunities tied to academic conferences, and the extra time my summers off provided were both very good for all of us. To this day, with the kids in their 30s, we still love traveling as a family and try to schedule at least one trip per year to somewhere new and exotic. We plan to continue that tradition even after the kids have their own families.

I asked the kids about the best part of having professors for parents. My daughter says that she particularly loves the travel opportunities provided by family vacations before or after academic conferences in exotic places. Some of her favorites include Australia, New Zealand, Fiji, Africa, the UK, China, Prague, and Rome. Says my son, the family wit: "The best part was being able to build dinosaurs out of their molecular models."

(continued)

Spouse Bob says, "It has been enjoyable for me to read Liz's story of her professional life and how it has intertwined with our personal and family life for now over 47 years. I have a new appreciation for the difficult balancing act that faces many women who desire to have both a career and a family. Unfortunately, the sacrifices usually fall disproportionately on them. I am very happy that in the end she was able to have both a satisfying professional life doing something that she loved and also have a great family life with a husband and two wonderful children who all love her very much."

Advice and Lessons Learned

I think that making family choices is a very individual decision. Also, the times (and the rules) have changed for the better in many places. Partly my thinking is influenced by my experiences and partly by observing grad students in my husband's and other professor's groups over the years. When I was a Berkeley grad student, none of the female students had kids either before or during their graduate years. I didn't see how they could. Most were unmarried. The "conventional wisdom" among male faculty (and probably some students as well) was that females were

Fig. 8 Liz and her children, Larissa and Sam, on a trip to China in 2010

just going to drop out to have families either during or after obtaining their PhD. Only five of us were female in a class of seventy beginning PhD students. More than one of the professors questioned why I was there.

My best advice? Try to be very, very patient if humanly possible. Also, try to work days rather than evenings, and do get acquainted with the full-time faculty in your own and other departments. Ask for their advice on teaching, and attend department meetings if allowed. Volunteer for tutoring or department outreach activities. If you are feeling exploited, it will show. Faculty members do not like the adjunct system any better than adjuncts do, but due to funding issues, the system has irreversibly changed.

Sharing childcare and household duties is something that is simply nonnegotiable for two professionals. We still share the cooking duties. After our daughter was born, we hired a college student to clean the house once a week. When we eventually could afford it, we hired regular cleaners. We never stopped. Do it!

I was very worried that having kids too soon would interfere with my ability to have a fulfilling career. In those days, there were few role models who had successfully combined both. In the end, I taught as an adjunct in community colleges and universities for 10 years before deciding that it was time to have kids, or it might be too late. After getting my MS in 1971, I taught for 4 years in California, 3 years in Seattle, and 3 years in Massachusetts. My daughter was born in 1982 and my son in 1986. Between the two, we moved in 1985 from the east coast to Arizona. I taught part-time and as a full-time OYO (1 year only) in the Maricopa Community College District in the Phoenix area for an additional 5 years before being hired full-time at Glendale Community College in 1990.

If someone had told me in 1972 that I would not have a permanent full-time position until 1990, I'm sure I would have freaked out. But in the end, it worked out just fine.

The fact is full-time community college teaching jobs were just as hard to come by in the 1970s and 1980s as university positions. Things finally opened up, at least in Phoenix, when the initial hires from the 1960s began to retire. In the end, I'm not sure that I would have found a full-time permanent position any sooner whether or not we had kids. Being a 2-body problem family perhaps had an impact, but even then there were no guarantees. But I love teaching and could never see myself doing anything else once I figured that out early on in grad school.

Choose an institution in a setting that provides the necessary support and resources, and do choose your spouse carefully. I'm only partly kidding! We have always shared household duties and chores, including cooking and childcare. Negotiate a schedule, continuously revise, and let it evolve. One thing our kids missed was having grandparents close by. If your or your spouse's parents are true gems and live in a great place, consider the benefits of proximity if you can.

I believe whether kids and grad school (or a new professorship) can be balanced comes down to how organized and efficient the individual is. But then again, finding success as a professor at a major research university requires those same skills whether you are male or female. Balancing a family with a community college teaching career certainly is easier, but that's not why I chose it. It's more like "it chose me," and I loved it (Figs. 9 and 10).

Fig. 9 Larissa, Bob, Sam, and Liz, Thanksgiving 2009 in St. Louis, MO

Fig. 10 Family in Hawaii (1991)

Mother and Community College Professor

Acknowledgments I was very fortunate to work with the many, many colleagues and friends who influenced and supported me over the last 40-plus years. Thanking all of them individually would take up way too much space, and thanking only a few is impossible to contemplate!

I am particularly grateful for the opportunity to attend a wide variety of conferences within the chemical education research and visualization in science and education communities. These interactions provided me with important insights into the nature and process of learning.

After my retirement from teaching in 2006, a flexible schedule at Washington University allowed me to continue reviewing for NSF and to attend and present at conferences with multiple perspectives on educational reform, collaborative online communities, and the effective use of technology.

But certainly, I want to thank my family above all. Here we are in Hawaii in 1991. Fun times!

About the Author

Education and Professional Career

1970	BS Chemistry, Kansas State University
1972	MS Chemistry, UC Berkeley
1972–2006	Teaching introductory, general, and organic chemistry in two liberal arts colleges, three state universities, and seven community colleges, including 21 years in the Maricopa Community College District in Phoenix, AZ
2003–2004	Program Director, DUE, National Science Foundation in Washington, DC
2006–2009	Education Specialist, Washington University in St. Louis
2009–2015	Communications Director, PARC (Photosynthetic Antenna Research Center)
2007–2012	Education in Virtual Worlds

https://www.hastac.org/blogs/ldorland/2014/09/08/connectedcourses-my-brain-edu

Honors and Awards (Selected)

2014	Invited Speaker, American Chemical Society National Conference Symposium honoring Thomas Greenbowe for winning the George C. Pimentel Award in Chemical Education
2008–2016	Reviewer for Macarthur Foundation Digital Media and Learning Competition

| 2011 | Chair of the Gordon Research Conference on Visualization in Science and Education |
| 2003–2004 | National Science Foundation Division of Undergraduate Education, Chemistry Program Director |

Liz strongly preferred community college teaching. Their unique environments and diverse student populations have provided a broad view of the nature of chemical education and its evolution over the years.

From Premed to US Professor of the Year: My Personal Journey

Amina Khalifa El-Ashmawy

Photo Credit: Nick Young/Collin College

> Mom was always a renaissance woman. Her singing voice was pitch-perfect, and she could play the piano. She could paint and draw, and her artwork was framed throughout the house and featured on t-shirts. She would drag my sister and me into the kitchen to teach us her cooking skills. Her closet was full of the most fashionable shoes, clothing, colorful scarfs, and jewelry. We had a costume drawer filled with costumes she had hand-sewn herself. Despite much resistance from us, she was everyone in the house's barber. She fixed leaky sinks and changed tires. There was nothing my mom couldn't do, and if she didn't know how, she would teach herself, try it out on us, then make us learn how to do it, too.—Mariam El-Ashmawy

A. K. El-Ashmawy (✉)
Chemistry Department, Collin College, McKinney, TX, USA
e-mail: ael-ashmawy@collin.edu

It's interesting to hear your children's perception of you. The above depiction seems perfect. However, my journey included many challenges and has taken many turns, as the case for just about everyone. Consider a decision I made in my undergraduate career. I was unsure of what road to take. Medicine was in my family, so I went for premed. You see both my mother and father were MD-PhDs. The expectations were high for my siblings and me from when we were very young. Because achieving academically was the standard, I knew I was going to graduate school since I was in junior high school. But to which graduate school and, more importantly, what field would I pursue? Now, let's start from the beginning.

My Childhood: First Forecast of Academia

I was born in Alexandria, Egypt. My mother and father both had successful careers. My mother, Aida Mohamed Geumei, was a strong, brilliant woman. As a youngster, I remember her telling us the story of when she was pregnant with my sister and then, 4 years later, my twin brother and me. My grandmother helped out as mom juggled her full-time career along with having a home, family, and young children. Mom was very particular about how we were raised and, despite her high level of achievement in her career, invested a lot of time and attention to her three young children. The focus mom had and how she praised and boasted our good performance in school let us know from a very young age that education was very important. She was the disciplinarian, and my father, Ahmed Hamed Khalifa, was the more lenient parent. They made a great team (Fig. 1).

At the age of 8, my family emigrated from Egypt to the USA, so my mother could pursue an opportunity in research. I had just finished the third grade when we moved, and I only spoke Arabic and French (Fig. 2). I started school in fourth grade, had to learn English, and assimilate into the culture. My mother would come home after a long day at work and cook dinner. At the end of the evening, we would go to bed,

Fig. 1 Aida M. Geumei (left) and Ahmed H. Khalifa (right)

Fig. 2 Amina in Egypt months before emigrating to the USA, 1968

and mom would start her evening—sitting in her special chair with the wooden board across her lap piled with papers, reading journals, and writing. I was so curious about what she did after we were in bed every night. Sometimes I would sneak out and watch her work. She received countless awards and high-level recognition. I was so proud of my mother. On top of doing research, seeing patients, making rounds at the hospital, and holding a faculty position at UT Southwestern Medical School, she was a generous hostess to friends and guests on the weekends. She was a force and my most influential role model.

Looking back, I remember my instinct to explain things to Anwar, my twin brother, and our friends as we played. Mom used to call me "Mademoiselle Deroose" ("Miss Lessons"). Although she was teasing me, I felt her sense of pride. Being an academic, she recognized my knack for teaching at such a young age.

I had several teachers in school that stood out. My sixth grade teacher, Mrs. Edith Pulley, was strict yet nurturing and encouraged me to succeed. In junior high, I remember Ms. Dwyer, my eighth grade science teacher, who took me to an academic meeting one evening with her, which I don't remember, and bought me dinner at Jay's Cafeteria. In high school, Mrs. Martha McSweeney was my first chemistry teacher. We related to each other on a level that was different from other student-teacher relationships. She selected me to represent the school for the district book review. I, along with other students from the school district, gave input on which science textbooks the district should adopt. I ended up babysitting her son. We are still in contact today. I was truly fortunate to have such teachers and role models all along who invested in me. Every experience since my childhood influenced and shaped where I ultimately ended up in my career and the type of person and parent I became.

My College Years: Making Critical Decisions

My general chemistry instructor at Kilgore Junior College in east Texas, Mrs. Anita Neeley, was a beautiful, warm, sharp biochemist. I remember sitting in her class mesmerized by her spirit and sayings. She made the subject relatable and fun. The Kilgore College chemistry faculty, Anita, Dr. Pete Rodriguez, and Dr. Dave Bugay, mentored me in the 2 years I was at Kilgore College and beyond.

When I transferred to Texas A&M University, I had to declare a major. I consulted with my father, who advised me to pursue chemistry, as it is the most versatile science. And so my major was. I finished my Bachelor of Arts in chemistry, didn't get into medical school, and decided to pursue a doctoral degree in chemistry.

Interestingly, in graduate school I found my interactions and time as a teaching assistant enriching and fulfilling more so than research. Knowing my background to this point, I'm sure you can understand why. After finishing all the coursework and setting my sight on the qualifying exams, I was advised to stop with just a master's degree. Maybe it was because traditional bench research didn't excite me so much. But, the advice made me doubt whether I was cut out for a PhD. About the same time, meeting Ahmed, my husband, awakened my desire to have a family. Continuing with a PhD would take another couple of years. With somewhat of a heavy heart, I decided to heed the advice of stopping with just a master's degree. Making the decision lifted a heavy weight off my shoulders as I was wrestling with prioritizing having a family and finishing a PhD. Deep inside, there was a faint sense of disappointment for not achieving my original goal. We'll come back to this a little later.

Ahmed and I got married in the spring of 1986. There were just a few experiments I still needed to run after we came back from our honeymoon. I finished my experiments as I started writing my thesis. Once my adviser reviewed my final draft, I moved up to the Dallas area to live with Ahmed. I hired someone to type my thesis, while I drew the figures. Ahmed and I drove down to College Station together in early fall, and I defended my thesis. Shortly thereafter, we were expecting our first child!

Employment and Involvement: A Powerful Combination

We had our first daughter, Mariam, in July 1987. By the fall, I was ready to get back to work. Through a conversation with Ahmed and some research, my decision was crystal clear. I knew I wanted to teach at a community college in particular. I applied for an adjunct position at Dallas County Community College in October 1987. In mid-January 1988 I received a phone call from Collin County Community College District (Collin College) asking if I could come in for an interview. Collin was founded in 1985, so I hadn't heard of it. Mary C. Fields, one of the original 13 faculty the college hired and the sole chemistry professor, interviewed me for a lab section. The assignment was a good fit for me given that Mariam was only 6 months old. I taught every subsequent semester till spring 1989. Ahmed worked for a private company that exported agricultural equipment to the Middle East. His work hours were flexible and accommodating of my part-time employment. He took care of Mariam while I was at work.

Personally, I found I had a lot of time on my hands. I didn't realize it then, but I have a higher capacity for multitasking and juggling many irons in the fire than many. I volunteered to teach at the Sister Clara Muhammad School in South Dallas. It was affiliated with the mosque where Ahmed and I were most active at the time. I taught 4th–8th grade math and science to underrepresented Muslim students. I really enjoyed this time. I took Mariam to school with me. The teachers, all our friends, liked taking care of her while I was in class. It was a win-win-win situation. That third win was for Mariam. The experience exposed her to diversity and taught her to appreciate people irrespective of their skin color. In the meantime, we were expecting again.

Mariam was a very curious toddler (Fig. 3). One day she was in the bathroom making a soapy mess washing her hands. She asked me why the nail polish wasn't washing off. You can imagine how excited and proud I was of her question. I enthusiastically told her about the differing polarities of water and nail polish. Though simplistic, the answer was enough to satisfy Mariam's curiosity about the nail polish

Fig. 3 Mariam, age 2 and Laila, age 5 months

Fig. 4 Mariam, 2.5 years old and Laila, 11 months old playing together in their room

not washing off. "Even as a child whose questions could have easily been dismissed, my mother always took the time to explain everything to me in accurate detail."

In spring 1989, our second daughter, Laila, arrived (Fig. 4). Ahmed's company was struggling in 1990 due to the unrest in the region and, eventually, the first Gulf War. Equipment orders decreased to a point where the company could not sustain his salary. Luckily, Collin opened a full-time chemistry position in January 1991. Ahmed and I both knew this job would be perfect for our family. Mariam was not yet 4, and Laila was 2 years old. I wore a white linen suit, a red and white blouse, and red and white shoes to my interview. My starting salary was $25,800. I was so excited and anxious at the same time. Mary took me under her wing. She was a great mentor. Ahmed bought me a mini refrigerator and a small coffee maker for my office. I took great care in how I decorated the office. I wanted it to be professional yet warm and welcoming for students and colleagues alike. Laila recalls, "The inside of Mom's office was home to our childhood art. She had silly putty, costume nerd glasses, a tie-dyed lab coat, a white board and markers. She had several sets of molecule building kits, or Legos for chemistry students." I bought a halogen floor lamp because I preferred its warmness to the cold feel of the fluorescent lights. After rearranging the furniture so I am facing the door, I ordered a small chalkboard and a bulletin board via the Mathematics and Natural Sciences division secretary (Fig. 5).

As the first year progressed, I attended Faculty Senate meetings and got to know my colleagues from different disciplines. I became more familiar with campus politics, which was crucial. In summer 1992, I attended my first national meeting for chemical education, the 12th Biennial Conference in Chemical Education in Davis, CA—thanks to the travel funding provided by the college. The conference was overwhelming and inspiring. Mary introduced me to many colleagues from all over the USA and invited me to be her guest at an American Chemical Society Examinations Institute dinner. Mary had served on the 1993 General Chemistry Exam Committee. I met Dwaine and Lucy Eubanks, Directors of the ACS Examinations Institute, who later became my mentors. Dwaine talked

Fig. 5 Portrait taken by Collin photography students, 1992 "Environmental" photo for photography class. Editor's note: This beautiful portrait was "staged" for a photo shoot and does not represent real experimental conditions or appropriate usage of personal protective equipment (PPE)

to me a bit that evening then asked me if I would be interested to serve on the next General Chemistry Exam Committee. It was the first of seven committees on which I've served. Based on Dwaine and Lucy's advice in the early 2000s, I applied for and was selected as a field tester of the ACS general chemistry textbook that was in the works. This was a huge opportunity for my learning and pedagogical development.

Mary Fields left Collin in 1993 to pursue her dream of becoming a medical doctor. I was left as the sole chemist at the college for two long years. I felt abandoned and lost. My experience and network through the exam committee helped me immensely not only in my teaching but also in handling my responsibilities as the sole chemist and "chair" of chemistry. I was so happy when we hired an organic chemist, Dr. C. Frederick Jury, in 1995 to compliment my physical chemistry background. I had a partner again. What a relief! We worked well together and got a lot accomplished as the department grew.

The stress from work faded when I went home to Ahmed and the girls. The girls' academics were my department. It was so gratifying to talk to them about their day at school and what they learned, ask them questions, and tell them stories. Having a routine made things much easier for both the girls and us. Meetings with teachers were important to me, as the girls quickly learned. They also learned they were responsible for their actions and work. One day when I was at work, Mariam asked Ahmed for help with her math homework. He took a look at the problem and gave her an answer. She was surprised because she expected to get the same guidance through the problem like she did with me. The next day, she came home really upset with her father. She got that problem wrong. He told her he had made up a random number. This was a lifelong lesson for both Mariam and Laila, teaching them to always depend on themselves.

Laila started half-day kindergarten in 1994. Ahmed took her with him to run errands in the morning before he dropped her off at school for the afternoon. I picked her up after school. Ahmed and Laila developed a special bond that year, which is still very strong today. Ahmed was coaching two boys soccer teams. Mariam and Laila played soccer with others coaching them. Ahmed and I took turns taking them to practices and games. Mariam's interests shifted more to academics. Math Club took up a lot of time in middle school. In high school, she joined the color guard. Laila started soccer at 4 years old, played competitively through high school, and was on the high school soccer team. She was active in Latin Club in high school. Thinking back, it is quite interesting that both girls had an academic interest, which was my influence, and an athletic interest, which was Ahmed's influence, to varying degrees.

Trouble in Paradise: Dealing with Adversity

My dean raved about my classroom skills each of the first 3 years she visited my class. Just after the third yearly class visit, she met with me for my multi-year contract application appraisal, Collin's version of tenure. I was floored to see that she had rated me poorly and had negative comments in each of the categories. Confused and frustrated, I consulted both Collin and Division of Chemical Education colleagues for guidance. I was advised to write a rebuttal for each negative point or comment, providing my side of the story. Tom Greenbowe also suggested surveying my students and sent me sample survey questions to gather their feedback. My rebuttal was included for faculty reviewers to read in my multi-year contract application packet. After the long-awaited outcome, my multi-year contract was approved. During that time, Ahmed and the girls provided me with perspective. I don't know how I would have endured that time without their support and love and the support of my colleagues and network.

The next two or so years were rough. I was a hair away from resigning a couple of times. I had to be very careful, emailing my dean after each interaction or communication to document everything while blind-copying human resources to protect myself. I felt sick at my stomach driving to work each day. My mother advised me to ask my dean for her feedback on anything I wanted to do. Although I already knew what I needed to do in those cases, having my dean's opinion made her invested in what I planned and more likely to support me as I moved forward. This worked brilliantly. Going home to my family's daily routine was the strongest motivation to keep going. Mariam was in first grade and Laila went to preschool. Our time together became that much more precious day-by-day. Looking back, this was a time of significant growth for me. At work, I took great comfort from being in the classroom with students, knowing that I was making a positive impact. This focus helped me maintain perspective at work.

Excellence and My PhD Endeavor: Implications on Family

I was Collin's Outstanding Professor in 2004 and 2014. Let's backtrack a bit. By the late 1990s, Mariam and Laila loved coming with me to work on seldom occasions. According to Laila, it was a special treat for her to be with me for office hours.

> Mom's office was full of knowledge, technology, and fun. I had to be on my best behavior. Mom doesn't just put on her teacher hat and become a professor. She is a professor all the time. She explains things to her students just like she explains them to me. It's as if she's teaching them how to cook, or pointing out a sock they left in the corner of their room.

I gained a reputation of excellence at Collin for my national involvement in ACS and from being able to relate to students, colleagues, and staff. In 2001, my colleagues recruited me to run for Faculty Senate vice president/president-elect the next year. I never thought of myself as political. Ahmed reminded me of an elementary school talent show incident as proof of my political ability. One particular talent show act, which involved several students with their parents, made fun of and depicted Arabs very negatively. I wrote a letter to the school principal asking for the act to either be modified or stopped. Accordingly, the principal decided to stop the act from being in the show. It was not popular, but I took a stand to show my girls to be proud of whom they are, irrespective of the outcome. The children in school labeled them for several years to follow as the ones whose mother stopped the act at the talent show. I think this was a good outcome for the girls. Although their peers mocked them, they knew they couldn't mess with Mariam and Laila when it came down to it.

I digress... getting back to my endeavors, I ended up the Faculty Senate president in 2002–2003. It was an interesting year. I learned what a vote of no confidence meant and the power it carried. I learned that shared governance was a gained privilege not a right; that having a PhD offered financial, academic, and political advantage.

In early 2003, my dean (a different one from the one mentioned previously) shared with me that the University of North Texas has a chemistry education doctoral degree. My buried aspiration for a PhD resurfaced; I contacted Diana Mason for details. Ahmed, the girls, and I discussed the possibility and what it would take from all of us. Knowing how much I wanted to get my doctorate, Ahmed and the girls were very supportive. I also discussed it with my parents and sister Soumaya, who were likewise. My parents paid my tuition throughout my degree. I started my studies in January 2004, attending as a full-time student while maintaining my full-time job. Luckily, Collin had a course-banking program where faculty can work overloads and bank the credit for later. I had been banking courses for several semesters prior, which lightened my course load and made it manageable. Ahmed was in charge of the home front. Sometimes I would get home from night classes to find the girls asleep. The more stressful times coincided with exams. Mariam graduated high school in 2005 while I was in the middle of my doctoral work (Figs. 6, 7 and 8).

Fig. 6 Before Mariam's high school graduation, 2005

Fig. 7 Just after Amina's hooding, December 15, 2006. Left to right: Laila, Ahmed, Amina, and Mariam

It was a tough road, a challenge for all, but we managed to get through it. Mariam and Laila were most proud when they came to my hooding in December 2006.

I served again in 2006–2007 as the Faculty Council (name changed) president. I learned from my experiences that my analytical skills coupled with my directness were intimidating to some.

Fig. 8 At Amina's PhD reception, 2006, with Laila (left) and Mariam (right)

Administration or Faculty?

In early 2006, Collin created two new deans positions. From my leadership experience at the college, I contemplated changing career paths and heading into administration. I consulted my sister, who specialized in human resources for a large corporation in Atlanta. At the time, I was writing my dissertation. I was not selected for one position and was a finalist for the other. I felt I did well in my interview, until I met with the vice president. In an awkward meeting, I felt he was, in not so many words, asking me to assume the responsibility or blame for decisions he was making. Needless to say, I was not selected for that dean position. Very disappointed, I conferred with my sister and CHED mentors and colleagues, several of whom were my references. Ahmed told me it was for the best, so I could spend more time with Laila as she decided on and prepared for college. Things have a way of working out in ways we may not see at the time. The consolation was that I was still in the classroom where I could innovate and try different pedagogies to help the next generation of scientists and citizens.

In 2011, Arlene Russell, University of California, Los Angeles, contacted me asking if I would be interested to be a collaborator on an NSF-funded project in cognitive science. The study focused on learning technology and its impact on efficacy and durability of conceptual knowledge in chemistry. It involved collecting data in my general chemistry classes in collaboration with UCLA and University of Pennsylvania. The research design and study were fascinating. I learned a lot, and I made some really great connections. Mariam and Laila were proud to see me collaborating on such a scale. Laila spent some time in Sonora, Mexico, collecting data for her last undergraduate research project. Both girls were active in undergraduate research and presented at national meetings (Figs. 9, 10 and 11). In fact, Mariam was accepted into a combined MD-PhD program and received a Howard Hughes Medical Institute fellowship.

Fig. 9 Laila presenting undergraduate research poster at the 2009 Geological Society of America Meeting, Portland, OR

Fig. 10 (**a, b**) Mariam's White Coat reception, University of Texas Southwestern Medical School, September 2009

Fig. 11 Laila's undergraduate graduation, Arizona State University, 2011. Left to right: Mariam, Soumaya (Amina's sister), Ahmed, Laila, and Amina

Highest Recognition

The year 2015 was notable for our family. First, I was named a Minnie Stevens Piper Professor in May (Fig. 12), followed a week later by Mariam defending her dissertation on lung cancer research. In August, Laila started graduate school seeking dual masters in energy sector economics. Ahmed had triple coronary bypass surgery in early November, followed 2 weeks later by me being named the Carnegie/CASE US Professor of the Year in Washington, DC. What a roller coaster ride for our family, both personally and professionally! (Fig. 13)

Where Are They Now?

Mariam completed her bachelor's degree from the Barrett Honors College at Arizona State University, in psychology with minor in biology. She was accepted in the University of Texas Southwestern Medical School's Medical Scientist Training Program. Two years after defending her dissertation, she graduated from medical school and joined New York University's Clinical Investigator track of internal

Fig. 12 Piper Professor reception at Collin College, May 2015. Left to right: Laila, Ahmed, and Mariam

medicine residency. She is on track to become a compassionate, astute oncologist (Figs. 14 and 15).

> **Reflections from Amina's Daughter**
> **Mariam El-Ashmawy**
>   As a nerdy, reserved girl with glasses and good grades, I didn't have many friends. But the week when my mom planned to come to the elementary school to do chemistry demonstrations, everyone would talk to me and get excited. Of course, she came to school with her tie-dyed lab coat, full headscarf and eyeliner, safety goggles, and bag of tricks and would wow the entire school including the teachers with her easy-to-understand explanations, fun experiments, and hands-on activities. Those science demonstrations made me the most popular kid in school for a whole week!

Laila's bachelor's degree was in civil and environmental engineering, also from the Barrett Honors College at Arizona State University. Upon graduation, she worked for Schlumberger as a field engineer, servicing offshore oil rigs for 4 years (Fig. 16). At the downturn of oil in 2015, Laila was laid off and went back to graduate school. In December 2016, she completed dual master's degrees at

Fig. 13 Amina on stage receiving the US Professor of the Year award, Ronald Reagan Building, Washington, DC, November 19, 2015, with Sue Cunningham, CASE President and CEO (left), and Paul Le Mahieu, Carnegie Foundation for the Advancement of Teaching, Senior Vice President (right)

Colorado School of Mines and the French Institute for Petroleum (IFP) in Mineral and Energy Economics and Petroleum Economics, respectively (Fig. 17).

Just after graduation, she started working as an analyst for the International Energy Agency (IEA) in Paris, France. She still plays soccer, now for the IEA co-ed team.

> **Reflections from Amina's Daughter**
> **Laila El-Ashmawy**
> There are countless things my mom has done throughout my life to empower and encourage me, challenge and inspire me. Lately I've been thinking about the example of when she made a couple phone calls to the school administration, and saw to it that I was in advanced math upon entering middle school. Mom didn't rely on a school-administered placement test to

(continued)

Fig. 14 Just after Mariam defended her dissertation, University of Texas Southwestern Medical School, 2015. Left to right, Mariam, Amina, and Dr. Ahmed Khalifa (Amina's father)

Fig. 15 Mariam with her dual hoods, PhD (blue) and MD (green), at commencement, 2017

> indicate my abilities in math. "You are good at math," she would tell me, despite my perceived inability to do simple division and multiplication. "You are good at math!" Mom would say, ignoring all the signals that I wasn't. I was too young to see beyond missing recess or the results from a placement test. Mom always believed in me and pushed me when I was too young and didn't know how to believe in myself.

It was fascinating to observe the girls' development in intellect and personality over the years. Mariam resembles me in her perfectionism in her studies and work while being more like Ahmed in her personality. Laila was more laid back in her studies like Ahmed but is more like me in her personality and intensity at work. They now are able to identify how their childhood experiences affected different aspects of who they have become today. Both achieved academically and are solidly founded professionally. However, neither has yet to start a family.

After being laid off in 1990, Ahmed pursued his passion for soccer. He started his own competitive soccer club, the Hawks Soccer Club, in 1991. In 2003, he merged

Fig. 16 Laila, with red snapper, on Gulf of Mexico oil rig, 2012

Fig. 17 Laila and Mariam at Laila's Masters graduation, Colorado School of Mines, December 2016

the Hawks with the Dallas Texans Soccer Club and has been the Director of Coaching since. After his 2015 bypass surgery, he is doing very well. He has been and will remain the love of my life (Figs. 18, 19, 20, 21 and 22).

Advice Based on Lessons Learned:
1. Mindset: As you study chemistry, realize that science has been male-dominated through the ages. This could manifest in different ways in the twenty-first century. Don't doubt yourself! Instead, consider where others are in their thinking, whether knowingly or unknowingly. A centuries-old engrained mindset is difficult to shake completely. Don't let someone else's mindset be a deterrent from your goals or a great opportunity.
2. Choices: If an opportunity peaks your interest, consider it seriously. Do your homework. Make a list of pros and cons, including the personal impact each option may have. The choice will most likely become obvious to you. Now, make the decision and move on. Don't look back.
3. Network: Get to know your colleagues, not just in the chemistry department or your area. Forge acquaintances and friendships with as many outside the department as you can, especially those outside your institution. The network you

Fig. 18 Mariam (left) and Laila (right) at the Great Pyramids, Egypt, 2006

Fig. 19 Amina with Ahmed on the beach in San Diego, 2007

develop will be invaluable to you at the most unexpected times. Knowing your institution's administrators is good. Lastly, staff (secretary, administrative assistant, even physical plant) is extremely valuable. Treat them with respect, gratitude, and kindness. It will make your job a lot easier day-to-day.

4. Accomplishments: Keep track of what you've done and accomplished. Celebrate your victories, no matter how small. Most importantly, be kind to yourself and don't beat yourself up over things. (I've learned that the hard way and still have to work at it.)
5. Passion: Enjoy what you do. Pursue whatever it is that excites you and motivates you to go to work enthusiastically every day.

Fig. 20 Amina grading papers on the beach during family vacation, San Diego, 2007

Fig. 21 Family vacation in New Orleans, 2012, left to right: Amina, Ahmed, Laila, and Mariam

Fig. 22 On family vacation, Alexandria, Egypt, 2014, left to right: Laila, Ahmed, Mariam, and Amina

Acknowledgments My deepest love and appreciation go to my late parents, Dr. Aida M. Geumei and Dr. Ahmed H. Khalifa. Without their nurturing, many decisions that led us to the USA, solid foundation they gave me, support, and untiring love, I have no idea where I might be today. To my husband, Ahmed, thank you for being my sounding board, rock, shoulder to cry on, and, most of all, the love of my life. To my two daughters, Mariam and Laila, you are my closest friends and my pride and joy in life. To my sister, Soumaya, thank you for being my strong role model on many levels and genuine friend. To my twin brother, Anwar, what a duo we made. You are one of a kind. To Dwaine and Lucy Eubanks, and Arlene Russell, I extend to you my admiration and sincere gratitude for believing in me and supporting me professionally over the decades. To Diana Mason, thank you for mentoring and helping me finally achieve my original goal. To Cheryl Frech, my roommate at meetings and friend, thank you for your warmth and spirit of exploration. My heartfelt thanks go to others mentioned in this chapter and many others who were not.

About the Author

Education and Professional Career

1980	AS, Kilgore College
1982	BA Chemistry, Texas A&M University
1986	MS Chemistry, Texas A&M University
1987–1989	Math and Science Teacher, Grades 4–8, Sister Clara Muhammad School
1988–1991	Adjunct Faculty, Collin College
1991–present	Professor of Chemistry, Collin College
2006	PhD Chemistry, minor in Higher Education, the University of North Texas

Honors and Awards (Selected)

2015	Carnegie/Case US Professor of the Year
May 2015	Minnie Steven Piper Professorship
2013	American Chemical Society Fellow
2005 and 2014	Collin College Outstanding Professor award
2014–2015	Collin Scholars Innovate Fellow
2004	Collin College Academy of Collegiate Excellence Fellow
1999	Outstanding Faculty in Math and Science Award

Selected Professional Activities

Amina has served the American Chemical Society in numerous ways. Amina has served on the ACS Undergraduate Program Advisory Board (2015–present), the Task Force on Revision of Two-Year College Guidelines (2013–2015), the Two-Year College Advisory Board chair (2012–2014), the Task Force on Two-Year College Activities (2009–2012), the Committee on Public Relations and Communications (2012–present), and Two-Year College Chemistry Consortium 204th Conference Chair (2014). Additionally, Amina has served as the ACS Division of Chemical Education 21st BCCE Technical Program Chair (2010), on the Biennial Conference Committee (2001–2014; chair 2007–2009), the Long Range Planning Committee (2003–2008), the Regional Meetings Committee (2004–2014), the Task Force on Division Travel Reimbursement (2012), and the ACS Division of Chemical Education Exams Institute Board of Trustees (2017–2019), and she was a member of seven different General Chemistry Exam Committees (1995–2013), having chaired two committees.

Minha Vida e Minha Carreira e Minha Família (My Life and My Career and My Family)

Isabel C. Escobar

Photo Credit: University of Kentucky

I. C. Escobar (✉)
Department of Chemical & Materials Engineering, University of Kentucky, Lexington, KY, USA
e-mail: Isabel.Escobar@uky.edu

© Springer International Publishing AG, part of Springer Nature 2018
K. Woznack et al. (eds.), *Mom the Chemistry Professor*,
https://doi.org/10.1007/978-3-319-78972-9_12

A 1980s cholera outbreak in Latin America, an inspirational university professor and one not-so-inspirational professor, and an Environmental Protection Agency STAR Fellowship were all pivotal experiences in my evolution as an environmental engineer, at the University of Kentucky where I am a full professor in the Department of Chemical and Materials Engineering.

Early Years

I vividly recall the impact of a 1980s cholera epidemic. As a little girl in Brazil, I tracked the movement of the deadly waterborne disease. It began along Latin America's West Coast—where a ship unlawfully discharged its sewage—and advanced to where I lived, Rio de Janeiro, as the Amazon River became polluted. Brazilians were told not to drink the tap water without boiling it first and to avoid using the water to wash vegetables without a saline solution that I learned to make as a child when I was younger than 10 years old. I remember the songs that taught us how to mix sugar and salt in boiled water to survive. It became a powerful memory of how an environmentally harmful action can spread broadly and affect so many people far from the initial damage.

Starting College: What Is Environmental Engineering?

With that memory in mind—as well as the continual news about environmental problems—I entered the University of Central Florida (UCF) where I received my BS (1995), MS (1996), and PhD (2000) in Environmental Engineering. I had little idea of what environmental engineering was, and I knew it as "sanitation" engineering, which has a connotation of something really not clean. This is truly the last thing that an 18-year-old me wanted to be called; after all, I was still a teenager. Therefore, I initially enrolled in electrical engineering. I was excited on the first day of my first course in electrical engineering; unfortunately, I left the first class meeting feeling ashamed. The professor introduced himself and the syllabus; he discussed goals and expectations and the fact that we were going to be graded on a Gaussian Curve, which meant he already knew how many A, B... grades he was going to give. The problem was that he then proceeded to tell us that there used to be excellent mnemonics to deal with all of the equations of the course, but unfortunately we could no longer use them because "there are women in the class now."

At this point, I have to say that when I look back at that experience, I fume and know exactly how to respond and which office to go to right after to complain. Unfortunately, an 18-year-old me just sunk down in her chair and wanted to disappear. I actually remember apologizing to others in the class because I was impeding their learning. Luckily, at the same time, I took an honors ecology class taught by a professor who sealed my passion for water-related research through field

Minha Vida e Minha Carreira e Minha Família (My Life and My... 179

trips and discussions of such issues as water systems, acid rain, and global climate change. Although I had always enjoyed math and the sciences, the class revealed how engineering could be applied to address environmental problems, and I was hooked on the environment. I switched majors and got my BS degree in Environmental Engineering.

Here Comes Graduate School

Around the time I was finishing my BS, UCF hired a new assistant professor, Andrew Randall, who had that excitement to change the world along with the knowledge to make a good attempt at it. I did undergraduate research under his guidance and continued on to a MS degree with him on storm water treatment. At this point, anyone wanting an academic position would be applying to other universities, but I had no academic desires. I wanted to be a consultant or work at a research laboratory. In addition to that, as a Latina, I could not really conceive of the idea of being 23 years old and living away from my parents; plus, I had a serious boyfriend, so I stayed at UCF to get my PhD.

This is a good point to go into a sidenote and talk about my family and their influence, especially my grandmother and my father. My grandmother married my grandfather at the age of 16, while he was 27 years older than she was. From the early days of their marriage, he would tell her that she would become a young widow (she became a widow at the age of 45), so she had two choices, make something of herself and care for herself and her two daughters or marry another man in the hopes he would take care of them. She chose the former, so in the early 1930s, she would leave their young daughter with him, take a train to go to the big city, and go to school. She eventually became principal at the school where he was a math teacher. I was named after her and had her influence in my life for nearly 18 years, so I grew up thinking that the natural progression was K-PhD. She would always tell me not to rely on anyone but myself, that education was mine and could never be taken away, and that an intelligent and strong woman could do anything. I am already teaching these values to my daughter.

Now, to my dad. My dad grew up on a farm in a family with 14 children, and he did not know how to read or write. As a young teenager, an air show came to town, so he took all his money to ride on a plane and was hooked. He saved all his money and soon after left to go to the big city to become a pilot. He slept on park benches and had to steal food at one point because he did not have money, but through it all remained single-minded to become a pilot. He flew an air taxi to the Amazon rain forest mines for many years and even had an airplane crash once back in the early years of air travel. He retired as an airline Boeing 767 international pilot. Aside from being the role model of "You can achieve anything you really want in life," he always told me that my education was his top priority, so he paid out-of-state tuition for me during my entire undergraduate education since I was an international student. These days, as he plays with my daughter, he tells me that what I give her must be more than what he gave me because she is the future. Then, there was my mom. My mother and I have talked every day for as long as I can remember, and every day at 7:15 pm, she calls for anywhere from 5 minutes to 30 minutes to chat and process the day. My mom is my number one cheerleader, regardless of what I do. My very close, very into each other's lives and businesses, Latina family is the cornerstone of my life, who I am and all I can do.

Now, a second sidenote is on my serious boyfriend. He was incredibly jealous of me, so me being a graduate student in engineering sharing an office with male graduate students was a serious problem to him. He did not trust me and always thought I cheated on him, which I never did. He asked me to marry him because he was jealous of this, and I said yes because I just wanted to avoid an argument and keep life without changes. So, as I started my doctoral research, I found myself engaged to be married to someone I did not love, but I was too afraid of being alone or of changes in my life.

Back to my doctoral student life, I applied for and received the US EPA STAR Fellowship, which proved to be transformational. It opened doors I didn't know existed. I was the first student at UCF to receive the award, which my advisor strongly urged me to apply for so that I could conduct the research I wanted to pursue on the availability of nutrients for bacteria to grow in distribution systems even in the absence of necessary bacteria. With my STAR Fellowship, I had funding to present my work at both national and international conferences, where I interacted with leading experts in my field. In that way, I developed my scientific skills and formed a strong network. It was a powerful award that enabled me to research new lines of inquiry that led to writing some ten papers as a PhD candidate and separated me from so many equally stellar applicants.

Starting My Faculty Life

I still did not want to be an academic, and I loved research, so my advisor, who wanted an academic child, suggested that I should teach a course for the department. I was beyond terrified on the first day of classes and over-prepared my notes and my answers to every possible question students would ask. I expected bored students challenging my knowledge every step of the way, but instead I had a class full of excited students participating and paying attention to what I had to teach them. Needless to say, I immediately started applying to academic positions and received several interviews that turned into offers. I would always ask to meet with women in the faculty to learn from their experiences at the university. I saw everything, from no women in engineering and having to meet with female staff, to young women faculty crying because they felt used by senior faculty, to strong women speaking their minds, with the latter happening at the University of Toledo. Thus, I accepted a faculty position at the University of Toledo to start in July of 2000.

Back to my personal life, I finally married my fiancé in 1999, and on my wedding day, after all the wedding party walked through and the doors were closed in

Minha Vida e Minha Carreira e Minha Família (My Life and My... 181

preparation for my dad to walk me down the aisle, my dad leaned toward me and said "When you are done with him, call me and I will help you through it." In June of 2000, I went to a conference in Paris, and we moved to Toledo about 48 hours after my return. From the start, things were rocky since the couple dynamics changed overnight. I went from being a graduate student on a fellowship and him being a stock broker to me making a higher salary than he did. Add to that the pressure of a tenure-track position. Sometimes, life before tenure felt like hazing and work around the clock was expected.

Life changed, and epic arguments started and culminated on September 11. In 2001, I experienced a difficult professional year. I was used to being at the top of my career; after all, a doctoral student close to graduation with eight academic offers on the table feels pretty darn full of herself. Overnight, I was at the bottom as a starting assistant professor, and my most difficult critics were my husband and me. When I would share that I was afraid and that a couple of proposals were not funded, so I would think my ideas were bad, he would tell me that "Every field has a loser and you might be the loser of your field." That slowly chipped away my confidence to where I really felt like the loser of my field. On September 11, I was in Washington, DC, attending a panel and ended up trapped there with airports closed. One panel member was a professor at Purdue, and his spouse drove a minivan to pick him up, so they offered all those from Ohio and Michigan rides. I was overjoyed, but when I shared this with my husband, he screamed at me "no other man is going to save you, so stay there until I can help you." I rebelled and took the ride, which led to months of arguments at home. After that trying time, the loneliness made me want a child, but that was just not a good idea with a rocky marriage, so I got a puppy. I would get home every day, hug my puppy, Butters, and cry from all of the pressures at work.

By 2002–2003, grants and publications started to come, teaching evaluations became better and better, and my confidence started to return. His career, on the other hand, was crumbling; by 2004, he had worked for three brokerage firms and ended up unemployed. By early 2005, he got a new job at a bank and went on a 2-week training program out of town. Those were the two best weeks of my life, and I found myself faced with the knowledge that my marriage was over because I was happier alone than with my husband, so I asked for a divorce. The 2005–2006 academic year was incredibly strange with on one side, my paperwork moving up from committee to committee through the tenure process, while on the other side, my paperwork was moving from filing to scheduling court dates to lawyers and finally to the judge. I became a tenured associate professor and a divorced woman around the same time.

Short Life as an Associate Professor

My years as an associate professor were nearly all as a single woman, and my motivations in life were my dogs and making it the rank of full professor. Between the fall of 2006 and the fall of 2009, I published 17 papers, was awarded several competitive grants, and decided that it was time to become a full professor. On the little personal life I had, I made friends that became more than family, and the majority of those were women, women who went with me to court dates for my divorce, hugged me when I cried, and stood by me every second of the way and women who became my sisters. I also became a runner and learned who I was since I had never lived alone in my life, and I decided that I would be the mother of dogs instead of children.

Of course, that's when I met someone. I met Marty on a flight from Detroit to Denver on February 27, 2008. We were both going to Denver, but I was going for to do the final presentation on a grant to the Department of Interior, while Marty was going skiing. He started talking to me thanks to a running magazine on my lap, and through a funny exchange of short questions and answers, we learned he graduated with a BS in Chemical Engineering from the department where I taught and that we lived less than 10 miles away from each other and knew many of each other's co-workers.

Here Comes the Promotion to Full Professor

In May of 2010, I became a full professor, the Director of the university women's center, and associate dean for research development and outreach for the College of Engineering. I became the first minority in the administration at the College of Engineering, and it was strange to be the only woman and minority at the college senior leadership meetings. Then, I started receiving awards for my work, and Marty kept dusting off his suit to attend events that honored me. I remember one of the award events, dozens of people were nominated for the Toledo 20 Under 40 Award, but awardees were not announced ahead of time; I felt like a long shot, but heck, it was a free dinner. When they started announcing my name, I think Marty was happier than I was and was too excited to even take pictures. He is comfortable with who he is and always praised my achievements to others, and I cannot think of any arguments.

In 2012, he asked me to marry him and by the end of the year, we were married. We were pretty set in making all decisions together, so he was in charge of the invitations and photographers, while I was in charge of the location and flowers. Once businesses got over the fact that "no, my wife is really letting me make this decision alone," he kept being called ma'am. Around then, a strange thing happened, and I found myself really wanting a child. We were both 39 years old and established in our lives and careers, which is a very different place to be when being newlyweds discussing children. We decided to adopt an American child through a difficult and emotional process. After a 2-year wait, we were told a birth mom had chosen us, and since we were open to anything from a closed adoption to an open adoption, we met the birth mom. Since life is complicated, around that same time, my long-term career mentor and number one advocate recruited me to move to the University of Kentucky (UK) as a senior hire.

Life Changes: Becoming a Mother and Moving to Another Institution

I fell in love with UK from the second I arrived for my first visit, and while I had interviewed at other universities over the years, only UK felt right. Marty and I made the decision of moving together after much talking, and his employer decided to keep him working remotely, which has been a huge success. In this time of changes and decisions, the birth mom had her child on a Monday, and I was the first person to hold him. I felt like a mom and Marty was a dad, and we could not stop the tears of joy. We spent Tuesday at the hospital with our son, but on Wednesday, the adoption agency called us and said that she had changed her mind and would keep the little boy rather than place him with us. Death is the only word I can use to explain the experience, and I submerged myself into my work, as we kept on discussing if moving to UK was the right thing for us. Things became worse when I found Butters (remember my September 11 puppy?) dead at home. My world was crushed, so we decided to move to UK for a fresh emotional start at life.

As fate loves to play tricks, a month after we decided to move, we received a phone call that we had been selected by a birth mom, but the baby was already 5 weeks old; everything was set and we would have her within 1–2 weeks. We divided and conquered the task of preparing the house for her during a 2-week period that was made of purchase lists and numerous trips to baby stores. We got our daughter on May 5, 2015, also known as the happiest day of our lives. Now, I had a new position at UK that I had already accepted to start on August 1, 2015, and a new baby. I chose to take maternity leave on during my last 3 months at the University of Toledo because I both wanted to get acquainted with my daughter and send the message to others (women and men) that the leave is important. Thus, on May 4, I was at work, performing my duties as a professor and associate dean; on May 5, we got our daughter; and on May 6, I became a stay-at-home mom. I was lucky to have an excellent research group that kept the research going with updates via email, phone, and Skype and who prepared my laboratory for the move. I brought three members of my group to Kentucky with me. My maternity leave ended on my last week in Toledo, so all I did was finish packing, miss my daughter, and move!

Between my start date of August 1 and the adoption finalization of November 23, I drove 291 miles south on Mondays to Lexington and another 291 miles north to Toledo on Thursdays since Tessa, our daughter, was not allowed to move during the adoption finalization. I made it through the stress of the time thanks to Marty, who was a single dad for half of every week, to my mentor and his family who let me live in their house part of the week, and to excellent graduate students that moved with me to Toledo and landed at the university running. They set up my lab and took leadership of their projects to become the mature people that I trust blindly.

My days in Toledo were precious since I only had my family from Thursday evening through Monday morning, so trips and work were difficult but plentiful. A highlight was giving a TEDx talk on worldwide water issues, something I am happy Tessa will get to see her mom do someday thanks to their archives. I also fulfilled the dream of becoming a respected "cover girl" by appearing on the cover of the Association for Women In Science (AWIS) Summer 2016 Magazine that reached over 20,000 worldwide, which I have saved for Tessa to see her mom's work.

Now, I live in a very strange world; my colleagues with children around the age of mine (she will be 2 in March of 2017) tend to be assistant professors growing their careers and stressed out with the pre-tenure world. Many colleagues that are both around my age and my professional level of full professor tend to be dealing with teenagers becoming empty-nesters. I am changing diapers with a solid established career.

Tessa has attended conferences, and at the age of 18 months, she went to Ireland to attend a scientific conference that I co-chaired. Tessa got to dance with me in front of world-renowned colleagues at the opening reception and got to be chased by some of those too. Marty spent a lot of time with the spouses of several of the other attendees, who were mostly older women, so Tessa had many honorary grandmothers. Marty was the only male spouse and Tessa was the youngest child. What I saw was that Tessa became the collective child of the group and that having her there won me the support of so many for being the mommy who can run an international conference and be a mommy. Even more interesting were the graduate students attending the conference, especially the women. They felt safe acting like "girls" by playing with a baby during the poster session. I think that by being a senior professor of chemical engineer chairing a respected international conference series,

Fig. 1 Tessa and Isabel dancing at the opening ceremony of the Engineering Conferences International (ECI) Advanced Conferences on Membranes VII, which Isabel co-chaired in Ireland, September 2016

with large grants to support attendees, showed to them that yes, they can do this and that we support each other (Figs. 1 and 2).

I am now looking forward to taking Tessa to attend the International Conference on Membranes (ICOM) in July–August of 2017, in San Francisco, since I am president of the host society, the North American Membrane Society (NAMS). I am only the third woman to be president, so I cannot wait to have Tessa sit in the back of a room with possibly over 1000 people (and likely 2/3–3/4 male) listening to me speak for NAMS.

> **Reflections from Isabel's Spouse**
> **Martin Geithmann**
>
> When Isabel and I met on that airplane, there was definitely a good connection made; the poor teacher who was sitting between us actually asked "you aren't going to start dating now, are you?" What Isabel didn't mention was how fast she ran off the plane and down the jetway to the taxi stand. It took a little persuasion over email during the next several days to get a date! It makes for a good story but also speaks to how worried she was about getting into another relationship and how hard it was to learn to trust again after going through a tough divorce. The emotional scars of that event took a lot of time and love to heal, but eventually they did.
>
> It can be a bit awkward being at conferences as the "plus one," but it helps that I have a technical background as well; I actually like going to the poster sessions! Watching Isabel give a talk or lead a conference makes me feel proud, not threatened. I want our daughter to grow up knowing that she can reach for any opportunity, and handle any challenge, just like her mom!

Reflections

Looking back, and really while I have been a full professor for nearly 7 years and feel I am nowhere close to the end of my career, I feel I made several mistakes with not thinking through my personal life. At the start, I was so focused on getting tenure that I neglected my personal happiness. I allowed a fear of being alone to rule my life, and I chose a roommate instead of a partner to share my life with me. I did not mention before that I did a lot of therapy after my divorce, so that I would feel comfortable and confident with who I was. I think that was the only reason I was able to identify Marty as the loving and supportive partner he turned out to be. I learned to like him, respect him, and trust him in addition to loving him. I learned that I could learn to change diapers by just doing it twice at the age of 40. So often I travel for my career with my heart in pieces because of leaving Tessa and Marty behind, but I

Fig. 2 Tessa (wearing her suit), Marty (Martin Geithmann), and Isabel at the ECI banquette, where Isabel was recognized as conference co-chair

know that I have now become a role model not only for Tessa to go after her dreams but also for women looking ahead at their futures in academic scientific fields. It sounds like I am pretty full of myself by saying I am a role model, but after seeing how students change once they see being a mom while discussing the future of research with renowned people, I know they are watching.

Lessons Learned

Thus, my words of encouragement are to find support, which comes from friends, partners, mentors, and so many others in your path. A successful senior woman, who is the VP of a large publically traded international chemical engineering firm and who was the first woman in their senior leadership team, once told me that she was told early on that mothers did not make to their high-level positions, so now that she is there, she makes sure to support every woman in her chosen path. I take that as my calling, to support every woman that comes behind me (Fig. 3).

Fig. 3 Tessa and Isabel (Marty took the picture) at the Women's March, and yes, Tessa also joined Isabel at the Scientist March

Acknowledgment and Dedication Too many thanks go out toward my family. My late grandmother and aunt, my parents, and my husband encouraged me with love and unconditional support to keep pushing forward, especially when it became too hard, and my Tessa. My closest friends (Maria, Cyndee, Anna, and Wendell) never judged me, taught me to be kind to myself, and always "had my back" as I had theirs. Lastly, my past and present research group since I have blessed to work with people I respect and care for as my own children.

About the Author

Education and Professional Career

1995	BS Environmental Engineering, University of Central Florida, FL
1996	MS Environmental Engineering, University of Central Florida, FL
2000	PhD Environmental Engineering, University of Central Florida, FL
2000–2006	Assistant Professor, Chemical and Environmental Engineering Department, The University of Toledo, OH
2006–2010	Associate Professor, Chemical and Environmental Engineering Department, The University of Toledo, OH

2009–present	Associate Editor of *Environmental Progress and Sustainable Energy* Journal, a quarterly publication of the American Institute of Chemical Engineers
2010–2015	Professor, Chemical and Environmental Engineering Department, The University of Toledo, OH
2010–2011	Acting Director, Catharine S. Eberly Center for Women, The University of Toledo, OH
2010–2014	Interim Assistant Dean for Research Development and Outreach for the College of Engineering, The University of Toledo, OH
2014–2015	Associate Dean for Research Development and Outreach for the College of Engineering, The University of Toledo, OH
2015–present	Professor, Chemical and Materials Engineering Department, University of Kentucky, KY

Honors and Awards (Selected)

2015	TEDx: https://www.youtube.com/watch?v=-wbHD77kMWE
2015	The University of Toledo Edith Rathbun Outreach and Engagement Excellence Award
2011	American Institute of Chemical Engineers (AIChE) Separations Division FRI/John G. Kunesh Award
2011	The University of Toledo Culture Ambassadors Diversity Award
2009	Recipient of the Toledo 20 Under 40 Leadership Awards Recognition Program
2009	The University of Toledo College of Engineering Outstanding Teacher Award
2009	Recipient of YWCA's Milestones: A Tribute to Women Award for Education
2009	Ohio Academic Leadership Association (OALA) Fellow (8/2008–5/2009)
1997–2000	Environmental Protection Agency (EPA) Science to Achieve Results (STAR) Fellowship

Isabel has edited two books, *Sustainable Water for the Future—Water Recycling versus Desalination* and *Modern Applications in Membrane Science and Technology*. During the 2014 water crisis in Lake Erie, Isabel Escobar participated in the community outreach in addressing and responding to the issues and has made numerous media appearances, including the Wall Street Journal, NPR, and Al Jazeera America. Escobar is the 2016–2017 President of the North American Membrane Society. She is active member in many associations, including North American Membrane Society, Association for Women in Science, and American Institute of Chemical Engineers.

Chemistry in the Family

Cheryl B. Frech

Photo Credit: Dan Smith

Growing Up in a Chemistry Family

As a girl with no brothers growing up in the 1960s, I got to participate in both traditional boy and girl chores and activities: mow the lawn and weed the garden, clean and cook the fish from a fishing expedition, and play with Barbies® and with Lego®. My father worked as a synthetic chemist and frequently came home smelling

C. B. Frech (✉)
Department of Chemistry, University of Central Oklahoma, Edmond, OK, USA
e-mail: cfrech@uco.edu

like organic chemicals or, more frighteningly now, the diborane gas that was piped into one of his labs for research on rocket fuel formulations. He sometimes took me to the research facility with him on weekends, and I would sit in his office playing with the giant desk-sized calculator. I delighted in one display of brightly colored inorganic compounds and another of the more pastel hues of the displayed rare earth compounds. By the time I was in high school, I was often one of just a few girls in my chemistry, physics, and advanced math classes. Yet not one person told me that the study of science was not for girls. So when I enrolled in college, I selected biochemistry as my major at Oklahoma State University.

College and Graduate School Years

A fairly typical undergraduate, I took my classes, studied, and had fun. My 17-hour first semester included 1 hour of marching band, which led in the second semester to a stint in the wrestling and basketball pep band and participation in concert band. In my sophomore year, I joined a sorority and continued with concert band. While no one in my science classes discouraged me from pursuing my major, I was more actively discouraged in the sorority. When it came time for officers to be selected, I was slated only for positions that required a lot of drudgework: service projects chairman and reference chairman. When I inquired about seeking more of a leadership position, I was told that would not be possible since I was "never around." Around where, I thought? The television room, where girls congregated in the afternoon to watch soap operas and smoke? That was when I was in the laboratory for quantitative analysis and organic chemistry. I had missed a couple of days during initiation week, but that was because the concert band toured the state visiting high schools. Despite this less-than-optimal situation, I stayed in the sorority and graduated in 4 years with a respectable GPA and headed to graduate school in chemistry at the University of Oklahoma (OU).

When I arrived at OU in 1981, there were no women on the faculty, but just the year before, the Department of Chemistry had begun an initiative to recruit outstanding women for the graduate program. Waiting for me were a small group of women who were 1 year into the PhD program and who would become my peer mentors and friends. While our undergraduate and high school friends had gotten married and started families, or gone off to work in banks and offices in their women's suits, shoulder pads, and silk scarves arranged like neckties, we were wearing lab coats and immersed in stopped-flow apparatus, rotovaps, and nuclear magnetic resonance. We were required to spend at least 1 year as a teaching assistant. I found myself at 21 years old (and looking younger) standing in front of my first general chemistry laboratory section and looking completely indistinguishable from the students. In order to maintain order and to be recognized as the instructor, I realized I would have to dress the part and be a little bit strict, at least at the beginning of the semester. That was the origin of two habits that I have maintained throughout

Chemistry in the Family

my career: always dress like you are the instructor (because you are) and establish some rules for your classroom (you can always ease up later in the term).

Throughout graduate school, I excelled at teaching and at presenting the seminars that constitute part of your degree requirements. Members of my graduate committee would say, "You explained that concept very well. Have you considered a teaching career?" I would always answer no, since I fully expected to work for a chemical or petroleum company, which is where my father worked and where my graduate school colleagues were finding jobs upon graduation. I met my husband in graduate school. Roger was already on the OU chemistry faculty, had been married previously, and had children who were in their teens at the time. We were married shortly before I completed my doctorate. I was able to complete one postdoctoral fellowship at OU before we moved to Mainz, Germany, where we both worked at a Max Planck Institute: he on sabbatical as an advanced researcher and I completed a second postdoc.

Starting My Own Chemistry Family

By the time we were in Germany, my biological clock had triggered, which in turn initiated many conversations about the possibility of us having children. Roger had entered the university system as part of the post-Sputnik science boom generation. Almost exclusively men, they toiled away in the lab, wrote papers and grants, taught their classes, and traveled to conferences and meetings. Most of them had wives at home who managed the day-to-day tasks of the household and essentially raised the children. We knew if we had children, the dynamic had to be different, mostly because I could not be that kind of wife.

When we returned from Germany, I was hired for 2 years as a visiting assistant professor at OU, teaching very large lecture sections of general chemistry. I didn't know there were better teaching situations than 250 students in a lecture room, but I loved teaching and started to work to bring more active learning into the classroom. In 1991, a permanent tenure-track position was advertised at a regional university in the Oklahoma City metropolitan area, and I was encouraged to apply. Central State University, soon to become the University of Central Oklahoma (UCO), was hiring someone to teach general chemistry and serve as the general chemistry coordinator: it was if the job description had been written for me. The only downside was the 37-mile one-way commute from our home in Norman. I sought advice and someone asked, "If there was no commute, would you take the job?" I said yes, and he replied, "then take it anyway." I have now been at UCO for 23 years, commuting daily from the southern edge of the Oklahoma City metropolitan area to the northern edge. When I first started commuting, I listened to audiobooks on cassette tape to cope with the 45- to 60-minute drive. The technology has changed from tapes to CDs to downloads on my smartphone, but I am hopelessly addicted to listening to books and am never without one in the car. People inquire about the stress of the commute. The time and distance intervals of the commute have been an excellent buffer between university and home concerns.

A few women taught in the UCO Chemistry Department before me, but none were in tenure-track positions. The year that I was hired, another woman who already had a young child was also hired. Some faculty in the department had older children and some had grandchildren. In my second year at UCO, I became pregnant. This was a pregnancy with major complications, and I missed several weeks of the fall semester after emergency surgery. My department chair treated this absence as he would have any other illness. As my due date neared, I planned to take off the entire spring semester and return to teaching the following fall. The state of Oklahoma offers no parental leave, so I was required to take an absence without pay. (This will no doubt be factored into my retirement calculations at some point, so there will be a financial repercussion.) Our daughter, Alison, was born in February 1993.

When I returned to work at UCO in the fall of 1993, we hired an in-home caregiver. Wanda was an older woman with a lot of experience in childcare who delighted in caring for Alison. I was still nursing Alison that semester and was able to nurse her in the morning before I left, in the evening when I returned, and before she went to sleep. Neither of my children would ever take a bottle, so Wanda taught them to use a sippy cup from an early age. By the spring semester, I had weaned Alison so that I could resume travel since I had recently been elected chair of the local section of the American Chemical Society (ACS) and needed to attend a training conference.

My only sister is 8 years younger than me. We were never close until we both became adults. I vowed that if I had more than one child, they would be close in age so they would have the potential to be closer friends and siblings. And so I became pregnant again when Alison was just a year old, and our second daughter, Emily, was born in April 1995. After her birth, we needed a little longer time as a family, so I took off another semester and returned to teaching in the spring 1996 semester. We again hired Wanda to care for two small children 4 days a week. On the single day that I was not teaching, I dropped the girls off at my parents' house in Oklahoma City on my way to Edmond, so that the girls could get to know their grandparents.

Because my work and home life are geographically separated, my children were not as present in the UCO Chemistry Department as other children might have been. And as such, I was not perceived as a parent perhaps as much as others who had to bring a sick child to work with them or who had to leave suddenly to pick up a child from school or daycare. Of course, this would not have been possible without Roger in Norman to respond to emergencies and to fill in when Wanda was unavailable. I am grateful that he was at a point in his career and his life where he could be the responsive and available parent when I was often an hour away.

Mom and Chemistry Professor

The University of Central Oklahoma is a primarily undergraduate institution (PUI), and we do not have a graduate program in chemistry. When I was hired in 1991, the university was not far removed from its normal school/teacher's college roots. UCO faculty have always had a heavy teaching load, usually four courses per semester. Although the department has good instrumentation in chemistry, we have limited laboratory space for research. It's only been in the past year that we have carved out shared space for faculty to conduct research outside of teaching laboratories. In order to progress through the promotion process, I have participated in scholarly activity such as writing and reviewing chemistry teaching materials, writing invited book reviews, and serving as an associate editor for a chemistry journal. Being able to focus on these sorts of scholarly activities has been far less stressful for me than for my peers at research universities.

As the girls grew and started school, Wanda moved on to other babies, and we hired college students to pick up our girls after school and stay until Roger or I got home. I never felt terribly guilty about not being around my children constantly. I completely subscribe to the "it takes a village" approach of raising children. Each of the women who cared for our children brought something to our daughters' lives that we could not. Wanda had an excellent singing voice and had sung extensively to the girls. Two of the college students, Auva and Tonya, were excellent at crafts and taught the girls how to make simple decorations for their rooms. Michaela was an engineering major who studied calculus on the couch in the afternoons: Emily was in awe. Today there is tangible evidence of the multiplication of these gifts: both Alison and Emily have worked in childcare and served as caregivers for other families.

At UCO I always made a point to tell students in my classes that I am a mother and to let them know that I have done all the things that their mothers might have done: gone to dance recitals, attended softball and soccer games, and served as a band parent. Our university has a large population of commuter students, many of whom attend school while raising their own children. Knowing that their instructor is also a parent makes it easier for a student-parent to let me know that they have missed class because of a doctor's appointment for their sick child: otherwise I might never find out the reason for the absence.

When I arrived home in the evening, the four of us would prepare a meal and eat together. Over the years the girls heard many dinner table conversations about our students, departments, and chemistry, and perhaps it prepared them well to face their own education head on. Even though we are both educators, unlike some of our daughters' friends' parents, we did not intervene in our daughters' public school education. If either girl had a teacher they did not particularly like, we told them to stick it out, rather than go to the school and demand a change. The girls learned early on to take responsibility for their own education issues. Emily was only in sixth grade when she realized that she was playing on a competitive soccer team while also enrolled in a rather lame physical education class at school. Figuring she had enough exercise playing soccer, she went to the office and arranged a schedule

194 C. B. Frech

change and became an office aide for the rest of the year. Surprisingly, the school never contacted us for our approval.

When the girls were 5 and 7, both Roger and I applied for and were granted sabbaticals, and our family moved to Scotland for a year. Roger worked on various research projects at the University of St. Andrews, and I taught two chemistry tutorials and worked on several writing projects. We lived in a small fishing village, and I was able to work mostly from home and be there for the girls before and after school in a way that I could not in Oklahoma. This was a delightful respite and a cherished year for all of us.

Mom, Professor, and Department Chair

Shortly after we returned from Scotland, I was elected Chair of the UCO Chemistry Department. The previous chair was male and single and lived only a few miles from the university. Again, I told my colleagues that I could perform this job, but I would do it differently than my predecessor. In fact, the previous two department chairs had spent many nights and weekends handling work in their offices, but I knew I would not be able to do this and be present for my daughters, who were 9 and 11 at the time. My approach was to organize the work and take what needed to be done home to complete after the girls had gone to bed. I also strove to delegate tasks to get more people involved and to mentor new faculty in projects that benefitted the department. I ended up completing two 4-year terms as department chair. During this time our department grew in number of faculty members, credit hour production, and chemistry majors, we completed several assessment reviews, and we modified our curriculum: a successful run, at least by my account.

In the department I heard minimal grumbling about my commitment to UCO in my role as a multitasking mother, professor, and chair. However, I served under one administrator who did question my availability. When my daughters were in elementary and middle school, my schedule was such that I could drive them to school and then drive the 37 miles to the university, arriving between 8:45 and 9:00 before my classes began at 10:00 am on Monday, Wednesday, and Friday. I was never late to class. I was not, however, around at 7:30 am when this person sometimes wanted to speak to me. During an evaluation conversation, he mentioned that I was not always available when he called my office. I responded that I was always available by mobile phone and that I was on campus on a regular schedule. He continued to push, blustering about how I might be needed to respond to a vague "incident" that might occur someday. I kept asking, "What, in particular, have I missed up until this point?" He had no answer, and I kept taking my daughters to school until they were able to drive themselves.

Role Models

In addition to my career at UCO, I have been professionally active in the American Chemical Society (ACS), at the local, regional, and national level. I have met and worked with a number of outstanding chemists through the ACS. One of them is Helen Free, now in her 90s, who, among her many accomplishments, was president of the ACS and has been recognized for inventing diabetes test strips. I served on an ACS committee with Helen when she was well into her 80s and she still attends national meetings. Her husband had children from a previous marriage, and they had six additional children together. I once asked her how she had seemingly "done it all." She replied that it was perfectly acceptable to pay for someone to do the things that you do not have the time to do or that you do not enjoy, such as cleaning house or cooking. What a relief! I immediately hired someone for biweekly housecleaning.

I was born in 1959 and I did not know very many women of my mother's age who worked. My own mother taught school for a few years before I was born. The only professional woman I knew growing up was Anne Plyer, the mother of one of my friends and a chemist at the same company where my father worked. She was an outstanding role model because not only was she raising two accomplished daughters and working as a chemist, she was a kind, caring, and engaged woman in her church and community and a fabulous cook.

Reflections

My two children were born while I was a tenure-track faculty member at a regional university. While I cannot say for sure that having a family influenced my career, I am certain that my career has influenced my family. My two daughters have only and always known their mother to work. Both of them, now college graduates, are self-sufficient and successful young women.

Many of today's young women are more aware of their options than I was at their age. They have heard about the fertility struggles of women who postponed child-bearing into their 30s and beyond. My only pieces of advice would be: always do your best (then you have no regrets) and don't worry about what other people think of your choices (they are too busy worrying about their own business).

Being a professor and a mother is a challenge that will be unique for each woman's family, career, and institution(s). There is no one "right" way to be successful. Strive to let go of being perfect, or at least perfect in all aspects of your life at the same time. You will miss a few of your children's events, and you probably will not be as fit, relaxed, or current with your hobbies as you might have otherwise been. Even during the busiest times, cultivate your hobbies and friends for less busy days. Don't forget your spouse/partner: they are an integral part of the equation, and they will still be around when the children make their way into the world (Fig. 1).

Fig. 1 Cheryl Frech, left, with Roger, Emily, and Alison Frech

Reflections from Cheryl's Daughter
My Mom and Science
Alison Frech (age 25)

Thanks to my mom, I wasn't privy to the fact that women hadn't always been encouraged to pursue a career in academia until I was nearly in middle school. I had long assumed that every mother was as delighted and involved with her career as my mother. This was enhanced by the fact that my parents conducted an annual chemistry demonstration at my elementary school, which elevated me to a celebrity status among my peers during the week they were scheduled to perform their science show. Together my mom and dad would dazzle us with beakers spurting copious amounts of foam or concoct liquid nitrogen ice cream. My mom was no assistant; the two of them ran the demonstration garbed in crisp lab coats, working as a team to interest the students before them in chemistry. This is illustrative of how I viewed their careers. Although I later learned about their separate focuses within chemistry, the perception that my mom was equally as capable as my dad has remained constant. Having such a strong role model in my life has allowed me to grow up sure of the fact that I am able to pursue and excel any career I desire, for my success relies on my ability, not my gender (Fig. 2).

Fig. 2 Alison, Cheryl, and Emily Frech on a return visit to Scotland in 2011

Acknowledgments Everyone's journey is unique, but the people you meet and with whom you share the journey are what make life special. First, I thank my UCO and ACS colleague, Terrill D. Smith, for modeling volunteer service at all levels. I am deeply grateful for the support and encouragement of my entire family on my chemistry journey. My father, Roger Baldwin, showed me the beauty of chemistry and never thought there was anything I couldn't do. My mother, Evalyn, and my sister, Lynne, have always been strong supporters. My daughters, Alison and Emily, have been inspirational by just being themselves. And finally, my husband, Roger Frech, has been a wonderful father and the other half of the home team.

About the Author

Education and Professional Career

1981	BS Biochemistry, Oklahoma State University, OK
1984	MS Analytical Chemistry, University of Oklahoma, OK

1985	Visiting Research Assistant, Technical University Berlin, Germany
1987	PhD Analytical Chemistry, University of Oklahoma, OK
1987–1988	Postdoctoral Fellow, University of Oklahoma, OK
1988–1989	Postdoctoral Fellow, Max-Planck-Institute for Polymer Research, Mainz, Germany
1989–1991	Visiting Assistant Professor, University of Oklahoma, OK
1991–1997	Assistant Professor, University of Central Oklahoma, OK
1997–2001	Associate Professor, University of Central Oklahoma, OK
2000–2001	Visiting Academic, University of St Andrews, UK
2001–present	Professor, University of Central Oklahoma, OK
2004–2012	Chair, Department of Chemistry, University of Central Oklahoma, OK

Honors and Awards (Selected)

2015	Vanderford Award—Outstanding Teaching, College of Mathematics and Science, University of Central Oklahoma
2013	Fellow, American Chemical Society
1999	Neely Award—Outstanding Teaching at University of Central Oklahoma

Cheryl is serving a 3-year term in the chair succession of the American Chemical Society Division of Chemical Education from 2017 to 2019. She is a consultant on the ACS Society Committee on Education and is the public relations chair for the Oklahoma Section of the American Chemical Society. She most enjoys helping students learn how to learn in her general chemistry courses at the University of Central Oklahoma.

"Are You Always This Enthusiastic?"

Jennifer M. Heemstra

Photo Credit: Jessica Lily Photography

I get asked this question frequently. I can't help it—I love my job. I love the creativity and autonomy of independent research. I love the gratifying feeling of seeing a project brought to completion and published for others to read. Most of all, I love the opportunity to teach and mentor so many talented and unique students. And, I realize how close I came to not pursuing this career path.

J. M. Heemstra (✉)
Department of Chemistry, Emory University, Atlanta, GA, USA
e-mail: jen.heemstra@emory.edu

© Springer International Publishing AG, part of Springer Nature 2018
K. Woznack et al. (eds.), *Mom the Chemistry Professor*,
https://doi.org/10.1007/978-3-319-78972-9_14

Somewhere near the end of my time as an undergraduate, I realized that my dream job was to be a chemistry professor at a major research university. However, I also dreamed of someday having a family, and in my mind, these two dreams were mutually exclusive. As a result, I spent the following 10 years in a sort of career goal limbo, knowing what I really wanted to do, but convinced that it was unattainable.

It's not that I'm afraid of working hard. I just know that there are only so many hours in each day, and there didn't seem to be enough to simultaneously meet the demands of an academic career and raise children. I'm also generally not afraid to take on a new challenge, even if failure is a possible outcome. But, as anyone who has worked in a research lab knows, an academic job is not something that you can simply "try out." Upon accepting an academic position, my department would be investing hundreds of thousands of dollars to get my lab started up, and my students would be counting on me to see them through their PhD studies and to continue to mentor them throughout their careers.

During the first year of our postdocs, my husband John and I welcomed our first son, Evan. Becoming parents was nothing short of amazing, and everything we had been told about the unimaginable and unconditional love you feel for your children was true. And, because I so enjoyed spending time with my family, I became even more strongly convinced that an academic career was inherently incompatible with the work-life balance that I wanted.

As the end of my postdoc approached, I needed to start making real decisions about my career choice. My undergraduate, graduate, and postdoctoral mentors each expressed that they supported me no matter what career I chose. But, they also each knew that I dreamed of a career in academia. I often tell students that they can't underestimate the value of working for supportive and encouraging research advisors, and, in my case, this made a critical difference in my career trajectory. As I agonized over whether to apply for academic jobs, my advisors each listened patiently to my concerns and struggles and then dispensed with wise advice that essentially boiled down to "Don't overthink it. If this is the career that you really want, then just go for it."

The other critical factor in my decision-making process was having a very supportive spouse. By this point, we had been married for over 6 years, which meant that John had tolerated countless discussions involving this difficult decision. John told me that he knew I would regret it if I didn't at least try for my dream job but that he loved me and supported me no matter what the outcome.

I decided to follow everyone's advice and "just go for it." The only problem at this point was that I found myself just 3 short months away from application deadlines, but with no proposals prepared. For the record, this is clearly not the model to emulate when preparing to apply for academic jobs. If you have even an inkling of a notion that you will go into academia, you should start working on your proposals the day you start your postdoc, or even earlier. Fortunately, I was able to cobble together three proposals that, even if they didn't necessarily reflect the exact research I would pursue, were representative of my general interests and showed that I could formulate creative, high-impact ideas.

"Are You Always This Enthusiastic?"

My interview experience was probably pretty typical—nerve-wracking but surprisingly fun. From the first day of my first interview, I realized that this was the job I was meant to do and going on interviews gave me the chance to meet many of my science "heroes" face to face. At the end of my interviews, the University of Utah was my top choice, as the job involved a joint appointment in an interdisciplinary research center on campus, which would provide a collaborative framework that was ideal for much of my research. I was also very drawn to the collegial atmosphere in the Department of Chemistry and the presence of a large number of female mentors who seemed to effectively balance family life with being at the top of their research fields. So, receiving the offer from Utah was a dream come true, and we quickly accepted and packed our bags.

On my first day at Utah, I found myself sitting in my new office overjoyed that I had landed my dream job, but simultaneously terrified. Having made such a late decision to apply for jobs, my proposals were ideas that I was extremely excited about, but they weren't necessarily the very first things that I wanted to start working on. So, as I pondered the generous amount of empty lab space just down the hallway, I was faced with the monumental question of "what research do I *actually* want to do?" While this was indeed terrifying, I highly suggest that every scientist embark upon this soul-searching process at least once in their career. Throughout the interview process, I had been continuing to develop new research ideas, and I was able to spend a few days brainstorming and outlining all of the projects I had been thinking about. I then looked over this list and asked myself which ideas I was the most excited about. The beauty of this process is that it not only helped me to strategically plan the trajectory for our initial research goals, but, after putting all of my ideas together on paper, then narrowing down to the ones I was the most excited about, I was able to step back and say "okay, this is who I am as a scientist." Now, my group annually meets and goes through a similar process to plan for our future research goals and proposals. I've found that this brainstorming and planning is even more fun and enriching when it is driven by the members of my research group.

The best advice I received for starting a lab came from my graduate advisor, Jeff Moore. He told me to simply recruit great people, provide them with the resources they need to be successful, and then set them loose to do exciting research. I've been unbelievably fortunate to recruit a group of creative, talented, and independent-minded students and postdocs to join my lab and am very proud to say that they now run the place. Having benefitted from this mentoring style throughout my education, I feel that the autonomy it affords makes research fun and gratifying, and it prepares students to eventually embark upon their own independent careers. Also, as a side benefit of choosing not to micromanage my group members' projects, I am able to carve out a reasonable amount of time to spend with my family and enjoying my hobbies.

Sometime in my first couple of years at Utah, John and I started to ponder the inevitable question of whether we wanted to have another child. We decided that we did want to expand our family, but this brought up the tricky question of when. During my training, it seemed that the canonical advice for female students and postdocs was something along the lines of "you can have both an academic career and a family ... just wait until after tenure to have kids." Needless to say, I had already gone against this advice once and couldn't help but question the wisdom of having a *second* child before tenure. In a conversation with one of my female colleagues, I cautiously mentioned that I was pondering having a second child, expecting to hear that this was a fantastically terrible idea at that point in my career. Her response surprised me, as she said "there's really never a perfect time, so if you want another kid, just go for it."

John and I were blessed to welcome Owen into our family during my third year at Utah. While my colleague was right that there is no perfect time, and having another child before tenure certainly did add stress in some ways, I can't imagine life without both of my kids. I am also fortunate that my department was supportive through this time. Our University policy provided me with one semester of teaching release so that I could stay home part-time with my son, and I was given the option to extend my tenure clock by 1 year. Importantly, I was universally encouraged by my colleagues to exercise this option, even if I didn't think I would ultimately need the extra time.

As I write this chapter, I am now in the seventh year of my independent career at Utah, where I was recently promoted to Associate Professor with tenure. I cannot imagine a better life—I have a wonderful husband and two kids who are now old enough that their personalities shine through, and we can travel and embark on other family adventures together. My lab is well-funded and growing quickly, and more importantly, my students and postdocs are the ones who drive the success of our research and come up with the best ideas. At least for the moment, I'm even able to devote a decent amount of time to my hobbies of rock climbing, biking, and running. The path to this point has certainly not been without challenges, and I'm sure that this job will continue to require me to seek out creative solutions to finding work-life balance. What follows is my attempt to give an honest picture of the ups and downs I've encountered along the way.

Time

This was certainly my biggest barrier to attempting an academic career. Throughout my education, I had seen the large amount of work that is required of faculty members and just couldn't fathom how I could balance this with raising children. In fairness to my former advisors, they did each seem to achieve a nice work-life balance. However, I always imagined that this was only possible because they possessed a superhuman ability to function without sleep. Based upon my personal experience so far, it is in fact reasonable to run a vibrant research lab of ~15 people, write grants, teach, and take care of committee work, all while still spending time with family and managing to get a reasonable amount of sleep. I'm sure that each person finds different solutions to making this work, but for me, two of the bigger pieces of the puzzle are the ability to write fairly quickly and a refusal to micromanage the members of my lab and their research. But, this is by no means the only solution to this challenge.

To offer an alternative perspective on the time issue, I'm pretty sure that if I polled all of the faculty in my department and asked them whether they felt that they had enough time to do all of the job-related tasks that they wanted to, almost all of them would say "no." No matter how much time you have, there is always another grant application that could be written, another experiment that could be done in the lab, or another research article that could be read. So, whether or not we have children, we all have to draw the line somewhere and leave something undone that we would have liked to do. And, arguably, having less time actually helps to focus and clarify which things are the most important and thus worthy of spending time on.

Discrimination

In an ideal world, this would be a nonissue. Unfortunately, however, this does still happen, so it's good to be prepared for it. One incident that stands out clearly occurred as I was nearing the end of my time in graduate school. I had lined up a postdoc position, and on the date of my thesis defense, I received a phone call from my putative future advisor. During this phone call, the professor explained that because I was female, there was some chance that I might become pregnant and thus not be able to work in lab for a time. So, they had decided to rescind my postdoc offer. I struggled with whether or not to include this story here, partly because it's something I've moved on from and care not to think about but also because it can be distressing to younger female scientists. However, this experience taught me a very valuable lesson that I think is worth sharing. While I was of course traumatized by losing this postdoc opportunity over the fact that I was female (and having this happen mere hours before my thesis defense), in the long run, this was the best thing that could have happened. All of my former research advisors have clearly supported me during my time in their labs and throughout my job search. However, what I took for granted until recently is that I will benefit from their support in many ways throughout my career. Thus, it was absolutely worth going through this emotionally painful experience to learn the valuable lesson that if someone is not supportive of you and your goals, then they aren't worth working for.

Lab Work While Pregnant

My last story brings up the very real and practical issue of how one does go about having children during the lab work-intensive years of graduate school or postdoc. This issue is more specific to "wet lab" disciplines such as organic chemistry, where daily job duties can involve working with hazardous materials including carcinogens and teratogens. Fortunately, overall lab safety has improved significantly in the past few years, but the problem still exists that an accidental chemical spill could be devastating if it causes exposure during pregnancy. I have seen a number of strategies employed in this situation, sometimes involving time out of the lab to write manuscripts, or a dramatic shift in research project to something that minimizes use of hazardous chemicals. I was fortunate that my postdoc lab was largely segregated by discipline, with organic chemists in one wing and molecular biologists in the other. So, even though I was doing mostly organic synthesis, I was able to work amidst the molecular biologists, which meant really only having to worry about hazards from my own experiments. To further minimize risk, I wore extra personal protective equipment, became very skilled at reading MSDSs, and did all of my experiments in the fume hood (even those using only a few microliters of solvent). Also, if a procedure used chemicals that were known carcinogens or teratogens, I simply found an alternative procedure. Going to these great lengths definitely slowed my research progress a bit, but I was able to successfully do my job during this time, which is something that many women in my field are told is impossible.

Maternity Leave

No matter how superhuman you are, everyone needs some sort of maternity leave. So, this may seem like a bit of a no-brainer. However, I have found that maternity leave during graduate school or postdoc is very different from that while a professor. As a postdoc, I was primarily just responsible for my own work, and that work largely revolved around doing experiments in lab. After my first son was born, I was given 9 weeks of paid leave—as a sidenote, I will take this opportunity to thank my postdoc advisor, as I was only technically entitled to 6 weeks of leave and he generously told me to take an extra 3 weeks. Because my project more or less came to a halt while I was on leave, I was able to spend those 9 weeks focused almost exclusively on caring for my new baby. However, at the end of that time, I was immediately back to work full-time, and my son was in day care full-time.

By the time my second son was born, I was in the middle of writing a grant, had committed to teach a class shortly after my due date, and was advising a lab of 10 people. So, while I had some vision that I would take at least 2 weeks completely off from work, I found myself answering emails while in labor, while in the hospital recovering, and basically everyday thereafter. Perhaps other women have stronger willpower than I to simply ignore their email for a time. But, I found that I was actually less stressed knowing that my students were getting timely advice on their research and that I wouldn't be coming back to over 1000 emails in my inbox. While this lack of time completely away from work had its drawbacks, there has been a tremendous benefit to the flexibility that my job affords. Between the teaching release provided by my department and my ability to do much of my work from home while my son took naps, I was able to stay home part-time with him for almost

Fig. 1 Jen, John, Owen, and Evan hiking in southern Utah (2014)

a full year, getting to participate more in all of the fun milestones that come during that time (Fig. 1).

Travel

Travel is a generally fun and pretty much inevitable part of this job. My first bout of intensive travel came when I was interviewing for academic jobs. This was by far the most difficult time of travel, as I was gone anywhere from 2 to 4 days each week for almost 3 months straight. Moreover, Evan was about 18 months old, which was old enough to realize that I was gone, but not old enough to realize that I was coming back. As a result, our relationship did suffer during this time. As luck would have it, my second intensive season of travel, my "tenure tour" came during Owen's toddler years, which was also difficult. I do want to highlight, though, that while travel right now has the downside of taking me away from my family, there will come a time (hopefully soon) when my kids will be old enough to join me for some of these trips. I see this as a tremendous benefit to the job, as it will not only provide extra time

Fig. 2 Evan, Owen, and John in South Lake Tahoe (2015)

together as a family, but my kids will grow up getting to see a large number of fun and exciting places around the world (Fig. 2).

Rejection

Along with travel, this is an inevitable part of working in academia. As an academic principle investigator, I regularly send out manuscripts and grant applications, which are then reviewed by other scientists and judged based on factors including novelty, innovation, significance, and impact. Especially in these financially lean times, only a small fraction of grant applications are ultimately successful, and the inherent subjectivity of the review process can make the sting of rejection all the more great. One of my greatest disappointments with grant writing came during the fall of my third year at Utah. I had spent a month putting together my first grant application to the National Institutes of Health. My lab had some exciting preliminary data, and what I felt were creative and important experiments planned. I had very high hopes for the application and thus was taken by surprise when it received only mildly enthusiastic reviews and was not funded. Upon receiving this news, I felt a crushing sense of disappointment and frustration. I came home that evening and told my family the bad news. My husband, who is also a chemist and knows well the sting of the review process, said that he loved me no matter how successful I ultimately was in my academic career. And, I realized that my then 4-year-old, like most of the general population, didn't even know what an NIH grant is and couldn't care less whether or not mine was funded. He just wanted to have fun and play games together like every other evening. Writing this story, I realize how overly dramatic I might seem—I sound as if the world was ending because one grant was not funded. But, it's hard not to take things this personally—each grant application represents about a month of hard work, and even more, in research, your value is largely determined by your thoughts and ideas. So, a rejection of your research can feel very much like a personal attack. This story highlights what is by far my least favorite thing about my job. However, I saved it for last because it also highlights one of the best aspects of having both a family and an academic career. It is inevitable that I'm going to face both success and rejection along this career path, and while it is especially sweet to celebrate successes with my entire family, a hug and a smile from my kids can almost completely wipe away the devastation of having a grant or paper rejected.

Reflections and Outlook

It is with great excitement that I see the landscape for women in academia rapidly shifting. My hope is that by the end of my career, women will comprise approximately half of all tenured and tenure-track faculty in chemistry departments. There is certainly no lack of talented, motivated, and intelligent women in the PhD pipeline. Rather, a key barrier seems to still be the notion that having a family and pursuing an academic career are inherently incompatible. Fortunately, university and department policies are becoming increasingly supportive and family-friendly, providing increased flexibility in the choice of when to have children. I also hope that my story and the many others in this book demonstrate that it is possible to successfully run a lab while raising children and encourage women who are pondering their career options to "just go for it" if an academic career is what they aspire to.

Acknowledgments I owe a tremendous debt of gratitude to the women who were pioneers in the generation before me, frequently starting their careers as the only female faculty member in their department. I am also grateful to the many faculty and administrators (of all genders) who have fought for institutional change in the area of gender equity. Most of all, I am thankful for John—my husband and best friend—and my children, Evan and Owen, who continue to walk through this journey with me and make life so much fun.

Editor's Note

Between this chapter's submission and the publication of this book, Dr. Jennifer Heemstra has changed institutions. Dr. Heemstra is now an Associate Professor of Chemistry at Emory University.

About the Author

Education and Professional Career

2000	BS Chemistry, University of California, Irvine, CA
2005	PhD Chemistry, University of Illinois, Urbana-Champaign, IL
2005–2007	Research Scientist, Obiter Research
2007–2010	Postdoctoral Research Associate, Department of Chemistry and Chemical Biology, Harvard University, MA
2010–2016	Assistant Professor of Chemistry, University of Utah, UT
2016–2017	Associate Professor of Chemistry, University of Utah, UT
2016–2017	Deputy Director, Center for Cell and Genome Science, University of Utah, UT
2017–present	Associate Professor of Chemistry, Emory University, Atlanta, GA

Honors and Awards (Selected)

2016	W.W. Epstein Outstanding Educator Award
2016	NSF CAREER Award
2015	Cottrell Scholar Award
2015	Myriad Award of Research Excellence
2014	University of Utah College of Science Professorship
2011	Army Research Office Young Investigator Award

Jen is an editorial board member for the journal Aptamers and is active in educational and outreach activities through the American Chemical Society, Research Corporation for Science Advancement, and the International Science and Engineering Fair.

How Motherhood Shaped My Professorship

Jani C. Ingram

Photo Credit: Steven Toya

Growing Up in a Small Arizona Town

I was born and raised in a small town in Arizona, Kingman. When I was a kid, I thought about becoming a bank teller, a middle linebacker for the Pittsburgh Steelers, or a sports statistician. I was good at math and played lots of sports so I think that shaped my world view. I had one chemistry class in high school that,

---

J. C. Ingram (✉)
Department of Chemistry and Biochemistry, Northern Arizona University, Flagstaff, AZ, USA
e-mail: jani.ingram@nau.edu

Fig. 1 Jani's family—picture taken her junior year of high school

honestly, was far from great. In fact, if someone were to tell me that I would become a chemist, I would have told them that they were crazy.

My parents were teachers; my dad taught high school mathematics and my mom taught second grade. I have an older brother and a younger brother (Fig. 1). My mom is a member of the Dinè (Navajo) Nation, and my dad is Caucasian. This makes me a person of mixed race and culture which could explain why I have always been drawn to diversity. I grew up off the Navajo Reservation but spent many summers with my Dinè grandmother and family. I was a bit of tomboy and played all kinds of sports with my brothers. In high school, I played volleyball and basketball as well as running track. I had a few dolls and Barbies, but I would say I preferred playing football or building forts to playing with dolls. In school, I had a bit of a type "A" personality which meant I worked hard to get an "A" in every class, be the best player on the team, and get elected to a position of leadership in all the clubs I joined. I definitely excelled at many things I did in high school, but it was not that difficult since my high school was small.

How College Shaped My Career Path

In my first 2 years of college, I went to a junior college that was about 150 miles from my hometown. The reason I chose this route was because I earned a basketball scholarship. I was also convinced to play volleyball as well. Long ago, even though

Fig. 2 Jani as a college basketballer

there was Title IX, there was less gender equity in college athletics than there is today. At the junior college I attended, the coach recruited half a volleyball team and half a basketball team and then convinced all the athletes to play both sports. This would not have been so bad except I had no idea that I was expected to play both volleyball and basketball in college. The first week I was at school, the coach (who was both the volleyball and basketball coach) began talking to me about volleyball practice. This was very confusing to me since I was a basketball player. However, I did figure out the new expectation, and since I was a naïve freshman, I began practicing volleyball as well as basketball (Fig. 2). As a result, my college days were very long with practice beginning at 6 am and ending late in the evening with studying. Perhaps this was good training for life as a graduate student and university professor.

When I started college, I thought about studying to become an engineer even though I really did not know what an engineer did. I was good at math which, I was told, is important in engineering. Since I was at a junior college, I had the same professor for multiple classes. For example, I had Dr. Adams for all three of my calculus classes and Dr. Barkhurst for all four of my chemistry lecture courses plus all four of the associated laboratory courses. I can definitely say that Dr. Barkhurst is the reason I am a chemist. He was an amazing teacher, and even more amazing was that he was married to a Navajo woman so he understood Navajo students. One clear memory of one of my early interactions with Dr. Barkhurst was a time in his office when I was getting assistance with homework. It was early in the semester of my first year of college, and I was feeling overwhelmed. I was practicing two sports, and I came from a fairly average high school in terms of academic preparation. I asked Dr. Barkhurst about switching from the regular first semester general chemistry course into the introduction to chemistry course. I thought this would be better since my high school chemistry class was weak. He said no. He told me I was doing just fine, and he would help me with any questions I had—and he did. I ended up with all A's in all my chemistry courses. He was definitely a remarkable teacher and inspirer.

In addition to classes and sports, I met my future husband, James, in junior college. He was also a basketball player. He, too, thought he would go into engineering as his father was an engineer. However, he too decided that engineering was not the path for him. Through a discovery phase in college, he settled on an education degree. He was from Las Cruces, NM, and we decided to continue our education at New Mexico State University (NMSU) after junior college. Some of our decision was based on the strong engineering program at NMSU, which we both thought was our career path at the time. My first year at NMSU was in the chemical engineering program; it definitely was a challenge and only somewhat interesting. I applied for and was accepted into a summer internship at Los Alamos National Laboratory. During that summer, I worked on various projects that all focused on chemical research. I really liked thinking about the chemistry aspects of various issues. When I returned to NMSU the following fall, I changed my major to chemistry. It was the right choice for me. I applied for and was accepted into the MARC (Minority Access to Research Careers) program which is a training program funded by the National Institutes of Health. It allowed me to do undergraduate research in chemistry. I loved it! I would go to the lab all hours of the day and on weekends. My project was in the area of environmental analytical chemistry; my mentor was Dr. Gary Eiceman. It was an amazing experience and really helped me to realize that I liked analytical chemistry. I actually really like environmental analytical chemistry, but I was advised that there were not many jobs that focused on environmental analytical chemistry as this was the late 1980s. It was at this point that I realized I would need to go to graduate school in order to get additional training. I decided to look at graduate programs in analytical chemistry as I really enjoyed the application side of chemistry. I remember as a high school student thinking that anyone who went to college past their bachelor's degree was crazy. So now I was going to graduate school in chemistry. A lesson learned for me is that what seems crazy to a high school girl might actually be one of the best decisions you will ever make.

The summer before my senior year at NMSU, James and I were married. For us, it was a great decision although we honestly did not really do a lot of rationalizing about the timing of getting married with respect to our education or careers. We were young and in love and that was our motivation. I would have to say that marrying James was one of the best things I have done in my life as he has been a great partner in so many things. We have grown as individuals as well as a couple, and we have raised three amazing children which is the best thing you can ever do together.

Graduate School: Some Days You Love It and Other Days You Hate It

I went to graduate school at the University of Arizona. It was great experience with days of wanting to quit and days of major triumphs. Additionally, there were many times that I did not think I belonged with my classmates nor the other scientists. I believe this was partly from being different than many of the other students (female and Native American) and partly from attending a public high school, a community college, and a public university—none of which would be considered top-tier schools.

Graduate school is definitely a time of growth for any individual. The growth is a mix of taking classes in your discipline, gaining confidence, and learning how to navigate issues as they arise. Before graduate school, I was a 4.0 student who was able to get good grades by working hard. In graduate school, I continued to do well in my course work, but research was definitely a challenge. Even though I had multiple research experiences as an undergraduate including work at the Los Alamos National Laboratory, at NMSU, and even a summer at Bell Laboratories, I was not an independent researcher. This means that I had a research mentor that pretty much trained me in a particular technique or method and told me what to do. Graduate school was a whole new world; now I was more in the driver's seat guiding the project. This meant a lot of trial and error learning which, as a type "A" personality, gave me many frustrating days and evenings in the lab. Fortunately, my research advisor, Dr. Jeanne Pemberton, was an excellent mentor. She provided guidance and advice but at the same time did not tell me what to do, as was my past experience with research. She told me (after I had graduated) that her litmus test for her students' readiness to defend their dissertations was if she could have a peer-to-peer conversation about the research, not a mentor to mentee discussion. I think of her as my chemistry mom as she definitely was nurturing in terms of developing my research abilities. Dr. Pemberton definitely helped me feel like I belonged in science and not just an outsider.

One experience in graduate school that compared amazingly well to motherhood was studying for and taking my oral exam. At the University of Arizona, the doctoral students are required to write an original research proposal on something in your field of study but not on your personal research topic. The proposal is reviewed by your dissertation committee, and then you defend the proposal in a committee meeting, as well as answer any chemistry-related questions posed to you by your committee. The actual oral exam takes from 2 to 3 hours during which time you are on your feet at a chalkboard answering questions fired at you by five professors. This may seem to have nothing to do with motherhood; however, after experiencing pregnancy and childbirth, I believe there are major similarities. Preparing the proposal and studying for the oral exam takes many months of preparation that has many aches and pains (mainly mentally) associated with it; this experience compares to 9 months of pregnancy. The oral exam itself is quite exhausting and also has some pain (mainly mental anguish) associated with it; again, this is not so

Fig. 3 James and Jani as students at the University of Arizona visiting the Grand Canyon

different than labor and delivery. The outcome, which in the case of my oral exam, was a pass; I experienced tremendous joy and accomplishment. Even more joyful and an incredible accomplishment was the outcome of childbirth, which, for my husband and I, was a healthy, 10-pound baby boy. I experienced the oral exam first so I actually reflected on this experience throughout my pregnancy and delivery of each of my children. It is interesting how one experience can prepare you for a totally different one later in life (Fig. 3).

Idaho: National Lab Life and the Birth of Three Children

As I finished my graduate work, I decided that *I did not* want to become a college professor. Even though I really admired my graduate advisor, I thought that I did not want her life. She was an assistant professor when I started working for her, and I was not sure I wanted to work as hard as she did to gain tenure. She was not married nor did she have children; I was married and knew I wanted to have children. Also, my parents were both teachers, and I really did not want to teach (at least this is what I thought at the time). I interviewed with a number of industrial companies and decided to accept a position with the Idaho National Engineering Laboratory, which is now the Idaho National Laboratory. Besides living in New Jersey for 10 weeks for

How Motherhood Shaped My Professorship · 215

an internship, this would be my first time not living in the southwest. My husband and I moved to Idaho in August and it seemed like a beautiful place to live. To be honest, it is a beautiful place; however, I had never truly experienced winter. My first winter in Idaho included days of subzero temperatures, snow, and winds that drifted the snow. I thought I had moved to Siberia or some other far-off land where you did not dare go outside without a heavy coat, gloves, scarf, boots, etc. It was a bit difficult adjustment for me, but over time, I learned to adapt to the cold. Another wonderful experience that has been useful in my life was living at 7000 feet in Flagstaff, Arizona, where it snows in the winter.

I was a staff scientist at the Idaho lab for 12 years. My training at University of Arizona was in surface analytical chemistry. At the Idaho lab, I became part of a team of scientists developing and applying surface analytical chemical techniques, mainly static secondary ion mass spectrometry, to contaminants on environmental surfaces. It was *awesome*! I got to combine analytical chemistry to environmental problems; I did not think I would get such an opportunity. The research team was a group of mainly older men who were incredible mentors to me. Since I had not done any postdoctoral training, I definitely had a lot to learn in terms of writing grants, managing technicians, and executing research. It was very helpful being a part of a team that included chemists, physicists, engineers, and technicians. We each had our role on various projects as well as opportunities to collaborate with other scientists at the lab. While in Idaho, I worked with microbiologists, geologists, and hydrologists which I found to be very interesting providing analytical chemistry support to various projects. As part of my job, I was given the opportunity to mentor junior staff as well as summer students who were interning at the lab. I found that this part of the job was very satisfying, and it made me begin to think about teaching.

Another adventure that my husband and I began in Idaho is parenthood. We had tried to start a family on my last year in graduate school, which thankfully, did not happen. Getting my dissertation written and searching for jobs was very time-consuming on my last year in graduate school, so I cannot imagine adding a baby. We had our first son, Jordan, after living in Idaho for almost 2 years. We were overjoyed by his arrival, but at the same time, we were not sure what it meant to be parents. As like graduate school, parenthood has its ups and downs, but for us, it has been amazing. We had two more children while in Idaho; our second son was born about 3 years after Jordan. His name is Joshua. We thought we're content with two sons, but I had a strong desire to try for a daughter. Part of my motivation was that I had strong women in my life. My grandmother on my dad's side had ventured to the west from Wisconsin at the age of 17 mainly because she had tuberculosis and needed a dryer climate. She ended up at the Grand Canyon and worked as a Harvey Girl, which meant she worked for Fred Harvey who created Harvey Houses along the railroad line which had restaurants and hotels. She moved west on her own, so as one might imagine, she was very independent. My grandmother on my mom's side grew up on the Navajo Reservation and was forced to go to boarding schools that were away from her family. She was able to return home for the summers, but she too was very independent. My mom was the first person in her family to attend and graduate from college. She is an inspiration to me. I thought, with all these

Fig. 4 Ingram Family in Idaho

independent, strong women in my family, a daughter would have amazing genes. We were blessed with a daughter, Jalisa, who was born 4 years after Joshua. Our family reflects biodiversity in that our oldest son, Jordan, looks like his father, while our second son, Joshua, looks more Navajo like his mother. Our daughter is blond and blue-eyed which I believe comes from my dad's side of the family who are Germans from Wisconsin. I believe the diversity of my children's appearance is an interesting reflection of my interest in diversity (Fig. 4).

My children grew up with a mom for a chemist. They honestly had no choice in the matter. I worked in the lab throughout my pregnancies, right up until the day each of my children were born. Fortunately, I was not working with any toxic chemicals at the time so I was able to do most of the work I had done before being pregnant. For me, working allowed me to focus on my science and not think about the aches and pains that come with being pregnant. I believe that having a mom who enjoyed doing science gave my kids a view of the world that is a bit different that of other children. One example of this assessment came from my son Jordan when he was about 6 years of age. He knew his mother was "Jani Ingram, PhD", although I am sure he really did not know what the PhD really meant. One day I introduced him to one of my co-workers, who was a male, and I happened to say he was a PhD. My son looked at me surprised and said that my co-worker could not be a PhD because he was a man. I found this very amusing since most of the world would think the opposite. The following are thoughts that my children have on their mom (Fig. 5).

Reflections from Jani's Children
Jordan Ingram

When I was younger, my mom came into my class and did experiments. It was inspirational to have her as my mom. She was on us all the time about our schoolwork because she wanted us to succeed. I would credit her as one of the people who established my work ethic and discipline. As a young professional, I am a better person because of her.

Joshua Ingram

When I was young, I remember my mom coming into my classes to do science experiments with our class. My favorite one was making ice cream out of liquid nitrogen. Regardless of what experiment it was, I always enjoyed when my mom came and did science experiments with us because she always blew our minds with the different things she could do with science. My favorite thing about my mom being a chemist is that I can tell she was doing something she loves. It inspired me to find out what I love and pursue that in my career.

Jalisa Ingram

Painting our nails and fixing each other's hair are not how my mom and I have fun. Instead, we test the pH of every liquid in the house or go out sampling in the field. She has shown me that just because I am a girl does not mean I have to steer clear of science labs and big equations. It is okay for a girl to ditch the pretty dress for a lab coat. My mom has given me the opportunity to see what a college lab looks like, what butchering sheep looks like, and what a strong, intelligent, and professional woman looks like.

Becoming a Professor/Chemistry Mom

Although I thought I did not want to go into academia after graduation school, my thoughts changed at the Idaho lab mainly as a result of the limited mentoring I did with junior staff and summer students. I learned that there were a number of different kinds of academic institutions which was definitely a positive revelation for me in my search for an academic position. I applied for a number of tenure-track chemistry positions, and I was pleased to accept an offer to Northern Arizona University (NAU). As many things in my life, I had not wanted to attend NAU as an undergraduate because it was too cold. However, after living in Idaho for 12 years, Flagstaff (where NAU is located) had less severe winters, so it seemed like a great place to live. Again, it is interesting how one's perspective changes with time. My husband and I were excited to return to the southwest; he got a position as a special education teacher in one of the Flagstaff high schools. We also were happy to live a place with a diverse population. For example, the first day of second grade for our son Joshua was an incredible experience for him as there were several boys in his class that were Native American. In Idaho, Joshua was a bit of an outlier in that he

Fig. 5 The Ingram Kids—Jordan, Jalisa, Joshua

did not look like most of his blond classmates. He was happy to see other boys his age that actually looked like him. Overall, it was a great move for everyone in my family, and we still are happy to live in Flagstaff.

My position at NAU started at the associate professor level, mainly because I had 12 years of research experience at the Idaho lab. However, I did not start with tenure since I had never taught a chemistry class (they did not accept my Sunday school teaching experience). My first teaching assignment was teaching the junior level quantitative analysis course and the corresponding lab. I was, again, blessed with great colleagues who shared their lecture notes with me and provided insights on teaching at the college level. I definitely was not a great instructor at the beginning, but I learned the best approaches for both me and my students through trial and error. I found teaching to be a lot of fun, mainly due to the enthusiasm of the students. Even the students who are less than enthusiastic offer some entertainment value with their interesting excuses for late homework or widely varying outside interests. I believe that teaching has definitely provided the opportunity to see the world through the eyes of the younger generation. I constantly try to put myself in their positions as college students trying to juggle classes, work, family, extracurricular activities, etc. I believe that being a mom to three children has made a significant impact on my patience and creativity in working with college students. Since my children are very different in many different ways, it helps me interact with my students in many different ways since they all are unique individuals.

How Motherhood Shaped My Professorship

In addition to teaching, I also developed a research program that is centered on studying contamination from abandoned mines on the Navajo Reservation. The research has provided a way for me to assist the Navajo Nation in dealing with contamination issues as well as work directly with students from a wide range of backgrounds and ethnicities to study the issues. Because I am Navajo and the research focuses on issues on the Navajo Nation, I have worked with a number of Native American students over the years with a majority of these being Navajo tribal members as well. I have learned a tremendous amount from my Navajo students about culture, and this has been a delightful benefit of my job. Additionally, I have worked with many non-Native American students who have made being a professor enjoyable. I feel that my lab is a place where students from all walks of life come together to work as a team for a common goal. I think that the research brings these students together who would have never interacted with one another in most other settings. I feel privileged to be a part of experience, and I am proud to be their chemistry mom.

Final Thoughts

Life to become a chemistry professor has been a great journey for me. I have enjoyed the freedom to pursue research that I find both interesting and beneficial to others as well as spending time with students who are passionate about their work. I have appreciated the opportunity to teach students concepts in analytical chemistry even though many do not want to become analytical chemists. The energy that students bring to class is definitely a perk to the job. However, the most rewarding job I have had for the past 25 plus years has been being a mom to Jordan, Joshua, and Jalisa. Reflecting back to graduate school, I thought that being both a professor and a mom was not possible. Clearly I was incorrect in this thinking, which has been the case for a variety of opinions early in my life. However, I know that being a mom and a chemistry professor would not have been possible except that I was blessed with an amazing husband who is the best parenting partner a person could ask have. Overall, I truly believe that being a mom has made me a better professor (Fig. 6).

My advice to young women interested in pursuing a career as a professor would be to think broadly. I was sure I did not want to become a professor based on my perceptions from graduate school. Since I had not been exposed to the various types of academic institutions and opportunities that exist, I had a narrow perspective on my options. I suggest that students should take advantage of visits to different types of schools, even if it is just a day trip, which can really help to gain perspective on different options that do exist. Additionally, I suggest becoming comfortable being the person in the room who is different—it might be gender, age, ethnicity, or even being from a different part of the country than everyone else. I try to remember that if I can become comfortable in uncomfortable situations, I can use this experience to help my students overcome situations that they find uncomfortable. Finally, I would say that one of the most difficult things for me is advocating for myself. Often, being the person in the

Fig. 6 Ingram Family at the Grand Canyon

room who is different results in many requests to serve in a variety of capacities. These requests can make you feel more accepted, but often the requests take time, energy, and even resources that take away from other activities that are more important. My advice, which I try to enforce for myself, is to learn to say no or, at least, thank you but not at this time. The simple word, no, can make a big difference in terms of success.

Acknowledgments I would like to thank my many mentors, including Dr. Rod Barkhurst from Yavapai College, Dr. Gary Eiceman from New Mexico State University, Dr. Jeanne Pemberton (my chemistry mom) from the University of Arizona, my colleagues from the Idaho National Laboratory, and my colleagues from Northern Arizona University. I am indebted to my students, who have taught me about culture and curiosity. I am appreciative of the National Institutes of Health, National Science Foundation, the United States Environmental Protection Agency, and the United States Geological Survey for funding to do what I do in the lab, as well as the John & Sophie Ottens Foundation at Northern Arizona University and the Haury Foundation at the University of Arizona for funding to work with students and communities. Lastly, I want to thank my family—my parents, brothers, children, and, most of all, my husband, who has been supportive and encouraging throughout my journey.

About the Author

Education and Professional Career

1982	AA, Yavapai College, Prescott, AZ
1985	BS Chemistry, New Mexico State University, La Cruces, NM

1990	PhD Chemistry, University of Arizona, Tuscon, AZ
1990–1992	Scientist, Idaho National Engineering & Environmental Laboratory, Idaho Falls, ID
1992–1998	Staff Scientist, Idaho National Engineering & Environmental Laboratory, Idaho Falls, ID
1998–2002	Advisory Scientist, Idaho National Engineering & Environmental Laboratory, Idaho Falls, ID
2002–2015	Associate Professor, Department of Chemistry & Biochemistry, Northern Arizona University, Flagstaff, AZ
2015–present	Professor, Department of Chemistry & Biochemistry Northern Arizona University, Flagstaff, AZ

Honors and Awards (Selected)

2013	Cal Seciwa Outstanding Faculty Recognition Award (Northern Arizona University)
2012	College of Engineering, Forestry & Natural Sciences Teacher of the Year (Northern Arizona University)
2005	Outstanding Faculty Award for outstanding support of Native American students at Northern Arizona University, received at the Native American Convocation
2001	U. S. Young Observer Travel Award to attend IUPAC General Assembly and Congress in Brisbane, Australia
1996	LMITCO Excellence Award Program Award for Excellence
1990	American Vacuum Society—Nellie Yeoh Whetten Student Award
1985–1990	AT&T Bell Laboratories Cooperative Research Fellowship Program, University of Arizona Faculty of Science Fellowship
1985	American Institute of Chemists Outstanding Senior Chemist
1984–1985	Minority Access Research Careers

Upward Bound to a PhD in Chemistry

Judith Iriarte-Gross

Photo Credit: Middle Tennessee State University photo by Andy Heidt

"Mom! I am the only kid in high school who knows what an NMR is!" My career path to a PhD in chemistry included raising a son—as a single parent. Once I made the decision to go back to college and to earn a degree, I realized that a career and a family life were hard to balance, but it could be done. I wanted a better life for my family and realized that this included getting a college education. This path to a PhD in chemistry started in a small way in high school though I must admit that I did not like high school chemistry. I can say that I embraced a career in chemistry after working with some amazing mentors who helped me navigate undergraduate and graduate school as a student and as a single mom (Fig. 1).

J. Iriarte-Gross (✉)
Department of Chemistry, Middle Tennessee State University, Murfreesboro, TN, USA
e-mail: Judith.iriarte-gross@mtsu.edu

Fig. 1 Daniel practicing titration for his 6th grade science fair project

Before College

I grew up in Prince Georges County, Maryland a suburb of S.E. Washington D.C. I am the oldest of seven children and one of two girls. My mother was always a stay-at-home mom and did not attend college. She did not know how to drive a car. My mother passed away 3 years ago so of course, she still not not drive. My father divorced my mother when I was 13. Imagine seven kids, no transportation, very little child support, and no job. After looking back at this time in my life, I can honestly say that the Daughters of Charity at the Catholic schools that I attended, from kindergarten to grade 12, were my first mentors. They made sure that we received food baskets, toys, and clothes at holidays. I was given a scholarship to attend the girls' Catholic high school and bus money to make sure that I got to school. One year I was asked to work in the high school chemistry lab washing glassware and cleaning up the benches. Little did I know that my first introduction to chemistry would start me on a path to a PhD in chemistry. In fact, at that time, I did not appreciate the beauty of chemistry but did appreciate the beauty of clean glassware! I did not realize that my high school chemistry teacher, Sister Mary Ellen, saw that I was an apprentice chemist. Today I tell my students that mentors have X-ray vision and can see things about you that you cannot see yourself.

During high school, I worked as a paper "boy" for a Washington daily newspaper. Keep in mind that during the late 1960s, there were strictly defined job titles for males and females. I also babysat children in the neighborhood, a traditional job for a teenage girl. As a junior and senior in high school, I worked at Ford's Theater and National Theater in the evenings and on weekends as an usher. It was nice to have some spending money! I was a good student in high school but was not thinking about college. I did not know what I wanted to be when I "grew-up." I knew that I did enjoy science. If I had any thoughts about college, I did not have a clue about how to start the college application process or where I would find the money to pay for college tuition. No one in my family had ever attended college or knew how to apply and I was told that there was no money for college. One day, however, the sisters spoke to my mother about a new program for low-income students who had the potential to go to college. This program even paid the students a stipend for transportation, $10.00 per week! This program was called Upward Bound and was held at Trinity College in Washington D.C. Once again, my mentors provided me with a door to a better future, an opportunity for a college education. I tell my students today that one should always listen to a mentor because his or her advice can take you farther than a car!

Learning About College Opportunities

I attended Upward Bound at Trinity College for 3 years. I was delighted to be able to live on the campus during the summer and to not have to watch over my younger siblings. I looked forward to the Saturdays during the school year, where I could take the bus to Trinity College for classes, tutoring, lunch, and $10.00. Upward Bound showed me the possibilities that a college education could provide to a first-generation student who was eager to learn. Upward Bound showed me that I could succeed in college by providing me with supplemental instruction, tutors, and opportunities to explore majors. During the summers, I lived on campus, ate in the cafeteria, and attended classes. One year, I asked for a science and math tutor and was able to work one on one with a Trinity student who was hired just for me! In addition to academic classes, we learned how to read college catalogs and to complete applications for admission and for financial aid. We took practice SAT exams and learned how to improve our study skills. Each summer, Upward Bound took us on trips to visit colleges and universities. We also took cultural trips to concerts and shows. I remember seeing the Alvin Alley Dance Theater and hearing Ray Charles at a concert. We also attended a Washington Senators baseball game. Life was more than watching my brothers and sister, and I began to imagine the possibility of attending college as a science major.

Hard Decisions to Make

At home, I was still expected to watch my brothers and sister, and this turned out to be a major roadblock to college. So in a way, my family, brothers, sister, and mother did influence my career. I had decided that I wanted to major in a science and had applied to several colleges on the east coast, hoping for scholarships and grants. I found out later that I had received acceptance letters, but my mother did not understand that financial aid offers were sent in separate letters. She turned down offers of admission without telling me because there was no mention of funding. I learned about this several years later. I was accepted to the University of Maryland, a bus ride with two transfers away. I could still live at home and attend college. I could still live at home and babysit. This was not my plan. I desperately wanted to get out of the house, so I married the man I was dating, 2 months after I graduated from high school. I could attend college and live in an apartment and not have to babysit! At the time, this was heaven. But soon afterward, I discovered that this was a mistake. I was pregnant by the end of the fall semester. I dropped out of Maryland; had my son, Daniel; and started a job as a clerical worker at an insurance company in Washington D.C. My college career was over, so it seemed.

Best Decision of My Life

Reviewing insurance applications and ordering new ID cards for clients was not the life that I imagined, but we did get health insurance, the office was on a bus route, and my salary helped to pay the bills. After 2 years of marriage, I realized that my marriage was a huge mistake and moved out with my infant son. One year later, my ex-husband was killed drunk driving. My job at the insurance company paid the bills and provided health insurance; however, I wanted more for myself and my son. I learned to drive for the first time and bought a car. I returned to college as a part-time student with hopes of becoming a premed major.

My first class was held in the evening at Andrews Air Force Base. It was a college algebra class with no more than two or three women in the section. Keep in mind that I was a clerical worker for 5 years. Clerical workers do not use algebra. I knew that I could ask questions, so when the teacher started discussing "slope," I raised my hand and asked him to define a slope. He drew a "slope" on the board and laughed along with the class. I was not happy but I was very determined. I told the teacher after class that I asked a legitimate question and that I would earn an "A" in his class and I did. I tell my students this story and remind them that no question is a silly question and that my job as a professor and mentor is to facilitate their learning.

Finding the Courage as a Full-Time Undergraduate

This class and the few other classes that I took as a part-time student gave me the courage and the time to do the unthinkable: resign my job and go back to college as a full-time student. I needed to know that I could succeed in college. I also needed to know that I could afford college as a single mom. I was terrified of quitting a secure job, but I had to quit in order to move forward in my life and to find a satisfying career. It helped that my son was now in kindergarten and in a Head Start program which reduced my babysitting costs. My Upward Bound training kicked in, and I was accepted as a student at Prince Georges Community College, Maryland. I knew how to apply for financial aid and to search for private scholarships and grants. From the federal government, I received a Basic Education Opportunity Grant, BEOG, better known as the Pell Grant of today. I also wrote to my state senator, Steny

Hoyer, for assistance with finding financial aid for college. Remember that there is no such thing as a silly question! As a result of my question to then state Senator Hoyer, I received a 2-year full scholarship from the state of Maryland. Today, Congressman Hoyer represents the 5th District of Maryland and is the second-ranking member of the House Democratic Leadership. Though I have never met him in person, I am truly grateful that he took the time to provide a constituent with the help necessary to raise a child and earn a college degree. I consider him a mentor who supported my decision to return to college full-time at a time in my life when I had many doubts.

Now that my son and I were both full-time students, we settled into a routine, wake up, pack lunch, eat breakfast, and head to school. After school, relax a little, prepare dinner, do homework, and get ready for bed. During the weekend, we cleaned house, washed clothes, studied, and visited the Smithsonian. I was a college student, premed major, and a mom. Life was hard, but, finally, life was meaningful.

Finding Mentors and Role Models

I enrolled in my first general chemistry course in the 1978 fall semester. I relearned the language of chemistry and fell in love with dimensional analysis. Where was dimensional analysis when I needed it in my math classes? In the spring 1979 semester, I truly enjoyed thermodynamics, balancing oxidation-reduction reactions, and solving chemical equilibrium problems in gen chem II. I still have my gen chem textbook, Nebergall 5th edition, though minus its front cover. My professor, Dr. Pat Cunniff, encouraged me to major in chemistry and suggested that I attended a summer program for women and minority students at the University of Maryland medical school in Baltimore. I attended that program and she was right. I was not a premed student but a chemistry major. I listened to my mentor!

I flew through classes at Prince Georges Community College and transferred to the University of Maryland as a chemistry major. My son was growing up fast and was in elementary school full-time. A major obstacle for any parent is to find an amazing and trustworthy babysitter. I was fortunate that there was such a woman in the next apartment building who took care of neighborhood children like they were her own. How do I know this? Late one night, my son and I were outside watching something in the sky. It was too long ago, and I don't recall the astronomical event that had us up late. It was dark and cold outside and our babysitter saw us standing in the parking lot. She hurried down thinking that her "baby" was hurt, and we were waiting for an ambulance. Great babysitters are precious like gold and make a significant obstacle to work and family balance, child care, turn into a small issue that is easily solved.

Fig. 2 I am very happy to be the first in my family to graduate from college. I earned my BS in chemistry and shared my first graduation photo with Daniel

Where Are the Women in STEM?

I mentioned earlier that there is no such thing as a silly question. I asked my way through my undergraduate program. Who is the best calculus teacher? I asked the chair of the math department and found an amazing calculus professor. How can I survive calculus-based physics? Visit the physics tutoring lab and practice problem-solving. Should I take physical chemistry during the summer? Yes! I took both physical chem I and II during the summer session. I regret that Dr. Sandra Greer was not my physical chemistry professor. She is an outstanding advocate for women in chemistry and in STEM. Today, I consider her a mentor.

What would undergraduate research do to help my career? Undergraduate research will introduce you to opportunities beyond that in a traditional laboratory course. I did find an undergraduate research mentor, Dr. Mike Bellama, who introduced me to basic research methods. My son was in sixth grade and I graduated with my BS in chemistry in 1981 (Fig. 2).

Upward Bound to Graduate School

Life was good and I knew that I wanted more, so I applied for and was accepted in the MS program at Maryland. Life was good but not good enough to go to another university out of state which would mean finding a new babysitter, new school for my son, and new housing. At Maryland, I was eligible for graduate student housing. My son was now 10 years old and convinced me to give him a house key. The chemistry department was a 5-minute drive or 15-minute walk away from our apartment on the College Park campus. I was first nervous about leaving him early in the morning, 30 minutes before the school bus arrived, but several kids in graduate housing caught the bus, and he did not want to be late in front of the other kids. He also checked in when he got back home with a quick phone call to my lab. Life continued to be good.

I conducted my MS research at the National Bureau of Standards, now called the National Institute of Standards and Technology. I synthesized barium and titanium sol-gel coatings and learned how to use an inert atmosphere glove box and how to tune an NMR. However, driving to Gaithersburg every day caused some stress, due to my son being at home in College Park. I mentioned earlier that babysitters are gold though we do not call a caregiver for a 10-year-old, a babysitter! I found other moms in graduate housing who were happy to keep an eye on my son while I was off campus. Though I must say that 95% of these moms were not graduate students but were married to graduate students. As an undergraduate at Maryland, I had one class with one woman professor, and I did not have any classes with women faculty in the M S program. I was fortunate to find mentors at Maryland who understood that I was a grad student and a mom. However, I did start wondering about the absence of women faculty in chemistry and the sciences. I earned my MS in inorganic chemistry in 1984 and with a 12-year-old son; I was ready to start a PhD program out of state.

Questions that Everyone Should Ask

Today I tell my undergraduates to ask many questions as they explore graduate program opportunities. What is your policy on comprehensive exams? Do you have to choose a research group by the end of your first semester? Can one interview both grad students and research directors before choosing a group? Is health insurance provided? How many seminars do you expect a grad student to give? Are there travel funds to present at professional meetings? What other training (proposal writing, IRB, safety) is provided to grad students? Will you pay my travel to visit your program and campus? Ask questions before you make that critical decision about a graduate program. I must admit that I did not ask all of the questions that I should have asked when considering PhD programs. Ask questions that are important to you. I did mention that I was a single parent with a son. One professor told me about the preschool program on his campus. I quickly learned to mention that my son was in 8th grade and that I was not the typical grad student with younger children. After many questions and three campus visits, I decided to attend the University of South Carolina where I could afford to live on my graduate student stipend and raise a growing boy.

Being a Mom and a PhD Student

Our move to Columbia was not without problems. I discovered that I had to have knee surgery and postponed my PhD program to January 1985. Once we arrived in town, my son was very nervous about starting a new school in a new state in the middle of the school year. He broke out in a rash, and I had to find a local doctor during the first week that we were in town. He spent the first 12 years of his life in the Washington D.C. area and was scared to leave all of the family behind. He also thought that he would be called a "Yankee." He was worried about a class that he (and all students) was required to take in 7th grade: a quarter semester of sewing, shop, cooking, and Spanish. He did not want to be teased for taking sewing and cooking! He could deal with shop and Spanish. He was worried about his math skills and if he had to read extra literature books to catch up. His worries were legitimate but turned out to be groundless. He quickly made friends and, to my knowledge, was never called a "Yankee" or even a "Damn Yankee." He was ahead in math and helped his new friends understand "The Hobbit" which, according to my son, was disliked by all. Remember that this was over 25 years ago before the Hobbit films that we love today. We both agreed that the spring 1985 semester was a tough one without family and longtime friends in the neighborhood, but we survived. Thinking back on this major change in our life, I would not have done anything differently. I took the time that I needed to be "Mom" apart from my being a graduate student. Make time for your child because he or she will grow up so quickly. Your research will still be there for you.

Learning About Life as a PhD Student and Mom

We settled into a routine once that spring semester was over. During the summer, my son hung out with his new friends, and I hung out in the Odom research lab. I took my time to find an understanding research advisor who accepted that I was both a single mom and a graduate student. It was important that whomever I worked for would recognize that sometimes I had to be a mom, especially a mom to a teenage son. I decided to work for Jerry Odom for several reasons. I was intrigued by the vacuum line chemistry at NIST and thought that I would enjoy learning new techniques. I also wanted to move to the p block of the periodic table. I wanted to learn new chemistry and found that selenium and tellurium chemistry was the focus in the Odom research group. I could learn new synthetic methods using my personal vacuum line! The Odom group also used NMR extensively to characterize products and to follow reactions, but this was not the usual ^{1}H or ^{13}C NMR. We played with ^{29}Si and ^{73}GE as well as ^{77}Se and ^{125}Te. What fun! My son and I were finally at home in South Carolina. Daniel was happy at Irmo High School, home of the world famous, so I was told, Okra Festival. I don't recall eating okra except if it was hidden in soup. I never ate the favorite southern breakfast food, grits, while living in South Carolina. This was our life in South Carolina!

My classes and comprehensive exams were over and my focus was on my research. I spent many long weekends running NMR experiments. My son would come with me and do his homework while I set up my experiments. He often commented that he was the only high school student in South Carolina who knew what an NMR experiment was! Afterward we would go shopping, see a movie, or

Upward Bound to a PhD in Chemistry

visit friends. This worked great until Daniel was 15, that age when children become very vocal about driving and wanting a car. However, I was a graduate student raising a growing boy on a graduate student's salary. "Mom, I need a car," was his favorite sentence though sometimes it changed to "Mom, I need a Mustang." He got his learner's permit and during the weekends, I taught him how to drive on empty parking lots on campus. He learned to drive on interstates by driving me to campus, again on the weekends, when I needed to set up experiments. One morning he was trying to merge onto the highway and saw a line of tractor trailers quickly approaching his lane. I laugh at this now because he was not asking for a car at this moment, instead he said over and over again "Trucks are coming!" After we arrived at campus and he realized that he survived highway driving, he still wanted a car!

My new saying became "Get a job if you want a car." He did and started working at a veterinary clinic on weekends and sometimes after school. We moved to South Carolina with one cat, Garfield, and one car. During our time in Columbia, our cat family grew to include Doris (named after Daniel's grandmother who did not like cats), Oreo, and Patches and a second car. There was peace at last for a while.

Time passed quickly and I was soon writing. Daniel was happy that I had reached this stage of my program and that my benchwork was done. Raising a teenager while in graduate school was challenging and at times very funny. He did have a car but he also wanted a mom who did not smell like rotten garlic, the smell of selenium compounds. Each evening, I had to change clothes after a day in the lab because of the "garlic" (selenium) smell. We were in the grocery store one day and the clerk remarked that she smelled something funny. Of course, my son assumed that it was me and was totally embarrassed. He did not want to be seen in public with me in my fragrant lab clothes. He took over the grocery shopping and often made dinner during the week. I taught him how to use the best kitchen appliance in the world: the Crock Pot! He learned to make chili, soups, and became quite good with Crock Pot chicken and potatoes. We were a team with a common goal: graduation for both of us!

I completed my dissertation and was preparing to defend when a hiring freeze was put into place on campus. Jerry asked me to put off defending my dissertation till the very last day since he would not be able to pay me with the hiring freeze in place. See what I mean about finding the best research mentor! The day finally came and Daniel was in the audience. The presentation went smoothly and the questions began. I could see Daniel squirming in the back of the room. He did not understand the "defense" process and was getting upset about the nonstop questions. I reached a point where I could not answer a question and the defense was over. A lesson learned is to explain to your older children about the process that you are going through to earn that degree. It requires long hours in the lab, in library, and writing. Grad students must be able to communicate their chemistry as a teaching assistant in the undergraduate lab, giving a departmental seminar or presenting at a professional conference. Tell your children that being a grad student or research scientist is not the same as a 9 to 5 job but depends on the chemistry.

Fig. 3 Dreams do come true. Daniel graduated from high school, and I earned my PhD in Inorganic Chemistry from the University of South Carolina

Graduation at Last

My son graduated from high school 3 days after I passed my defense and received my PhD in inorganic chemistry (Fig. 3). I would have never thought that one day I would earn a PhD in chemistry when I was growing up in Washington DC. Even today, I still am amazed that I have a PhD, but this should not be a feeling of amazement but one of confidence. This is another lesson learned that I share with my students. You can do anything if you have the determination and drive to succeed. Find mentors who understand that work (both undergraduate and graduate school) and family need to be balanced in a way that is best for you. Keep your mentors informed even when chemistry has to be pushed to the back Bunsen burner. Don't be afraid to ask questions. If you can't feel comfortable talking with your mentor, then that partnership will not succeed. Remember the goal and do what is best for your family and for your science (Fig. 4).

Upward to a Career in Higher Education

I moved to Dallas Texas as a postdoctoral research associate in the lab of Patty Wisian-Neilson at Southern Methodist University. Continuing my tradition, I joined her group to learn new skills and polyphosphazene chemistry. I moved to Texas with four cats! Cats were not welcome in residence halls! Daniel stayed in South Carolina

Fig. 4 Daniel did not receive a car for his high school graduation. He did receive this t-shirt!

for college. He studied finance and now lives in Maryland with Bonnie, his wife, and their three children, Elaina, Waverly, and Logan.

I was still trying to find my chemical career path after my postdoc. I worked at the FDA lab in Dallas where I was an analytical chemist. It was an interesting job with great benefits and never boring. After a day identifying pesticides in cantaloupe or measuring how much mercury is in a shark, I was teaching intro chemistry at a community college twice a week. I really enjoyed interacting with students more than interacting with a GC-MS. I was not an analytical chemist, so I found a job in Fort Worth as a synthetic sol-gel chemist and lab manager. I soon discovered that a job in industry was not for me. After a search for a tenure track position, I joined the faculty of the Department of Chemistry at Middle Tennessee State University in 1996. I am delighted to say that my husband, Charles Gross, a native Texan (whom I met while I was in Texas), and my four senior cats, followed me to Tennessee.

Today I am a full professor and director of the Women In STEM Center. I am an advocate for our women students who sometimes still question if they can have a career in chemistry and a family. We share stories, discuss options, and offer possible solutions. I encourage my students to consider all opportunities such as undergraduate research, internships, and professional presentations which will enhance their resume. I ask women students to step outside of their comfort zone and take on leadership roles on campus. Women can juggle and balance both a career in chemistry and a family. I know because I have a wonderful career in academia and a son who does know an NMR!

Advice and Lessons Learned

Find a mentor or mentors for every step of your education and career!
Ask questions!
Do not give up!

> **Reflections from Judith's Cousin**
> **Crystal Wood**
>
> Dr. Iriarte-Gross is not only my cousin Judy but also my role model. As a young girl I can always remember being fascinated by the thought of having a college professor as a cousin. That was back when I was really young and still under the impression that teachers didn't have lives. As I got older, I would turn to Judy for science fair project ideas and later in high school chemistry homework help. It was either when I was in middle school or high school she told me her story, the thing that gave her the drive to succeed in her studies. When she was in college, a professor had told her that women had no business being in the sciences, or something to that effect. Judy proved that professor and a lot of other people wrong. She has been a nonstop force in paving the way for girls in science. She is my own personal cheerleader as I have gone through my own trials and tribulations in my own educational journey and a constant supporter of all things happening in everyone's lives around her. Anyone who knows Judy is sincerely lucky and those who have been taught by her are truly privileged. Today I am 33 and I can still say she is still one of my main role models.

Acknowledgments First of all, I want to thank my family of which there are too many to count! Special thanks goes to my son Daniel, for having a Mom the Chemistry Professor and my cousin Crystal for having a Cousin the Chemistry Professor. I would not be successful today without the ongoing support and love of my beloved husband, Charles, aka Santa to my students. He is my Sherpa and Cat Dad. I also want to give a shout-out to some special "chemistry" mentors: Dr. Pat Cunniff, Dr. Sandra Greer, Dr. Jerry Odom, Dr. Ann Nalley, Dr. Diane G. Schmidt, Dr. Donna Dean, and Ruth Woodall. Finally, I acknowledge the love and support of my students and friends who make life as Mom the Chemistry Professor a true joy!

About the Author

Education and Professional Career

1981	BS Chemistry, University of Maryland College Park MD
1984	MS Chemistry, University of Maryland, College Park MD

1990	PhD Inorganic Chemistry, University of South Carolina, Columbia SC
1990	Postdoctoral Research Associate, Southern Methodist University, Dallas TX
1991	Analytical Chemist, Food and Drug Administration Laboratory, Dallas TX
1992	Polymer Chemist and Lab Manager, Fresnel Technologies, Inc., Fort Worth TX
1994	Chemistry Instructor, Texas Wesleyan University, Fort Worth TX
1996	Assistant Professor of Chemistry, Middle Tennessee State University, Murfreesboro TN
2000	Associate Professor of Chemistry, Middle Tennessee State University, Murfreesboro TN
2001	Tenured, Middle Tennessee State University, Murfreesboro TN
2003	HERS Leadership Development for Women in Higher Education, Bryn Mawr, PA
2007–present	Professor of Chemistry, Middle Tennessee State University, Murfreesboro TN
2009–present	Director, Women In STEM (WISTEM) Center, Middle Tennessee State University, Murfreesboro TN

Honors and Awards (Selected)

2017	ACS National Award for Encouraging Women into Careers in the Chemical Sciences
2016	William E. Bennett Award for Extraordinary Contributions to Citizen Science, NCSCE
2016	Fellow, American Chemical Society
2016	Fellow, American Association for the Advancement of Science (AAAS)
2015	Woman of Influence, Nashville Business Journal
2015	E. Ann Nalley Southeast Region Award for Volunteer Service to the American Chemical Society
2015	Science Educator of the Year for Higher Education, Tennessee Science Teachers Association
2014	First Athena International Leadership Award, Rutherford CABLE
2010	TRIO (Upward Bound) Achiever for Washington DC
2009	Association for Women in Science (AWIS) Fellow
2008	Science Education for New Civic Engagement and Responsibilities (SENCER) Leadership Fellow
2008	Southeast Regional Stan Israel Award for Advancing Diversity in the Chemical Sciences

Dr. Iriarte-Gross is nationally known for her advocacy for encouraging girls and women in the sciences. She introduced Tennessee to the Expanding Your Horizons (EYH) Network and is the EYH Regional Consortium Chair for the southeast. She established the Women In STEM (WISTEM) Center on her campus. She is active in many organizations including American Chemical Society (ACS), Iota Sigma Pi, Association for Women in Science (AWIS), American Association of University Women (AAUW), and Sigma Xi. She is chair of the Women Chemists Committee for the Nashville Local Section.

I'm.a.Gene: Destined for a Career in the Sciences

Margaret I. Kanipes

Photo Credit: Howard Gaither Photography

Please don't laugh, my parents named me Margaret "Imogene" Kanipes. Imogene is pronounced as "**im**-*uh*-jeen." Webster's Dictionary defines a gene as "a unit of DNA that is usually located on a chromosome and that controls the development of one or more traits and is the basic unit by which genetic information is passed from parent to offspring." So you see, I was destined for a career in the sciences. Bob Goshen said, "Leaders should...influence others...in such a way that it builds people up, encourages and edifies them so they can duplicate this attitude in others." I am who I am today, because of great mentors throughout my life and career. My goals have

M. I. Kanipes (✉)
Department of Chemistry, College of Science and Technology, North Carolina Agricultural and Technical State University, Greensboro, NC, USA
e-mail: mikanipe@ncat.edu

© Springer International Publishing AG, part of Springer Nature 2018
K. Woznack et al. (eds.), *Mom the Chemistry Professor*,
https://doi.org/10.1007/978-3-319-78972-9_17

always been to mentor and shape young people into great scientists. I feel that it is so important for me to pay it forward and hopefully my mentees will do likewise.

Role Models and Mentors

My three siblings and I grew up with a "science dad," who always asked us science trivia, and a mom who taught us discipline and focus. Our father, an outstanding educator, majored in chemistry and taught junior high school science. Dad would take us to school with him in the summers to prepare his room for the next academic year. His classroom was filled with jars of animal specimen (I thought they were cool and pretty gross!!), rocks, minerals, thermometers, science charts, and books. He watched tons of shows such as those on the Animal Kingdom that featured animal wildlife and interesting life and earth science facts. Even upon his deathbed, he quizzed the doctor about interesting facts such as did he know the difference between a predator and a prey? I think it is my name and science dad's influences that are some of the many reasons why I have always been so inquisitive and loved science. My dad wanted to go into dentistry; he always talked about it and the impact that fixing people's teeth had on people's faces and their ability to live life through their smiles. He never applied to go into dentistry because he knew it would cost more time and money that his family did not have. In my junior and senior years in high school, I decided that I wanted to be a dentist. I had an opportunity through my high school to shadow two dentists and was absolutely certain that this would be the route I would take in college. I also had a fun but challenging high school chemistry teacher, Mrs. Blakeney. Of course, I also would major in chemistry like my dad. Besides, chemistry was the one science where I really excelled and that I really loved. I chose to major in chemistry and attended my parents' alma mater, North Carolina Agricultural and Technical State University (NC A&T).

When I arrived in college at NC A&T in 1986, I came with a Dental Admission Test (DAT) book (purchased by my parents). I had my life mapped out. I was going to attend NC A&T for 2 years and then matriculate to Howard University's dental school. I would start out in the chemistry program at NC A&T and complete the necessary requirements before heading to the dental school. They listed a 2+2 program in their catalog. Little did I know that my life was going to be forever changed when I met the Chemistry chair, Dr. Walter Wright. Dr. Wright set up my first internship at Rohm and Haas Chemical Company during the summer after my freshman year. I was the first chemistry major to have received a summer internship after the freshman year from that department. I remember that my mom was very hesitant about letting me go to an internship that was outside of NC. Rohm and Haas graciously agreed to fly my parents and me to the facility in Pennsylvania so that they could see where I would be for the summer. My father and I went; my mother does not fly!!! I also completed internships at MIT for two summers in organic chemistry and toxicology. My last internship was at DuPont Chemical Company in Delaware. It was these experiences that made me feel like I could be a professor as a college student. I loved being in the research lab and working collaboratively with other students, postdocs, and research associates. I saw myself as a professor, someone who could lead a research lab to address important scientific questions. I also told myself that I would return to my alma mater to work with students and mentor them into pursuing PhD careers in Science, Technology, Engineering and Math (STEM). The thought of dentistry faded away.

I enjoyed my chemistry classes. I was asked to assist students by tutoring in the study room at Hines Hall, and helping others was exciting to me. I also later became a Minority Access to Research Careers (MARC) scholar that was under the leadership of Dr. James Williams. The MARC program was for students who had the desire to pursue a PhD in biomedical sciences and related fields upon completion of their BS degree in the sciences. Dr. Williams would become a lifelong mentor and would guide me throughout the early years of my career.

Dr. Wright also introduced me to a new NC A&T chemistry professor, Dr. Lynda Jordan, who had received her PhD from the Massachusetts Institute of Technology and completed a postdoc at the Pasteur Institute in France. I participated in my undergraduate research with Dr. Jordan on a project titled "The Isolation and Characterization of Phospholipase A2 from Human Placenta." I spent time at both NC A&T and Wake Forest University conducting my research.

On one of the scientific trips that I attended with Dr. Jordan, she gave me a picture of a woman carrying a basket and bags of various sizes in her hands and on her shoulder. When I look at the picture, I often ponder, where is this woman going? Is she going to work or is she going home after a long day at work to feed her family? The picture is in my office and is a constant reminder that the road we travel as women in academia is not always an easy one. Like this woman, we sometimes carry lots of baggage. We can become weary under the weight of our responsibilities. We carry self-doubt and worry about our capacity to succeed in our careers. We lug our insecurities that we are not paying enough attention to our personal relationships with family and friends (roles as mom, wife, daughter, and sister). We wonder how we can manage to tote all of the things required to be a good mom and wife and be successful professionally. However, we must press on, and we should share our experiences rather than haul them around.

Upon completion of my BS degree, I entered Carnegie Mellon University (CMU) in Pittsburgh, PA. I had a supportive graduate advisor, Dr. Susan Henry (Fig. 1). Through Susan, I learned what it meant to be a great mentor and advisor. Susan always knew our strengths and weaknesses and helped us to reach our full potential scientifically and as leaders. During my graduate years, I had lots of support from not only my advisor but my family, supportive church members at St. James African Methodist Episcopal Church in Pittsburgh, and new friends at CMU. With the help of the Dr. Barbara Lazarus in Academic Affairs, I started a networking group for students of color while at CMU. The purpose was to support and foster a nurturing environment for each other in our graduate programs so that we could succeed in our PhD programs at CMU. At the time, there were only about 11 African American students who were pursuing their PhDs on the entire campus. In 1996, I was awarded the CMU Graduate Student Service Award and inducted into Phi Kappa Phi National Honor Society. Upon the completion of my graduate program with two publications, I pursued a postdoc in Biochemistry at Duke University Medical Center in Durham, NC.

Fig. 1 Margaret and her graduate advisor and mentor, Susan Henry (2016). Picture courtesy of Dr. Sepp Kohlwein

I had a challenging postdoc experience; I would say it was one of the most challenging experiences of my career. I was the first African American woman in the lab, and I sometimes think that my advisor doubted my results even after showing every possible control. I also felt very isolated at times although I eventually found a great friend in the lab. I remember wanting to quit my postdoc. I had been flirting with the idea of becoming a program officer at the National Institutes of Health (NIH) or becoming a Policy Fellow with the American Association for the Advancement of Science (AAAS). I love Washington, DC!! If it had not been for great mentors in my life, I might not have realized my academic career potential today. One of my undergraduate mentors and my graduate advisor both told me that I should not give up on my dream of pursuing an academic career. My graduate advisor invited me back to CMU for a presentation to share my research from my postdoc. It was during this visit that she spent quite a bit of time with me thinking about projects to consider as a junior faculty member to start my career as an assistant professor. She told me that I was not going to give up on my dreams of becoming an academic professor. When I returned from this trip, I had chosen to complete my postdoc. What changed? I decided that I needed to speak up for myself more and let others know that I was feeling uncomfortable. I decided that I needed to change my attitude to one where I was not defeated but able to win. It was not about me, or what people thought I was capable of doing, but it was about the science. To everyone's surprise, I was invited to present my research on "A Calcium-Induced Enzyme that Modifies the Outer Kdo of LPS in *Escherichia coli*." at the Sixth Conference of the International Endotoxin Society meeting in Paris, France, in August 2000. At this meeting, I was presented the Young Investigator's Award. I *completed* my postdoc studies in 2001 with four job offers for assistant professor positions.

Family

My first faculty position was in the Natural Sciences Department at Fayetteville State University (FSU). There, I taught many courses which included biology, chemistry, and biochemistry. I also married my longtime friend, Jay, in 2001. We decided to

I'm.a.Gene: Destined for a Career in the Sciences

wait a few years to start our family that we thought would eventually include two children. We also wanted to get settled in our marriage, our work, and lifestyle. That is exactly what we did! However, I had created a mental timeline of my tenure clock, and my strategy was to be in a good place to secure tenure before becoming pregnant. I also wanted the timing of a pregnancy to be just right so that our first child would come at the end of an academic year. In this way, I could enjoy motherhood and not have the worries of tenure and my teaching responsibilities.

In 2002, I was accepted into a faculty summer research fellowship program in the Enterics Division at the Naval Medical Research Center. I participated in this program for 2 consecutive years while at FSU. As a newly married professional, my husband and I met on weekends and short trips while I took advantage of this opportunity to publish articles, write proposals, and solidify my research program. At FSU, I worked on my research with amazing undergraduate students. One day while I was at FSU, I was called by a friend to let me know that a biochemistry position at NC A&T had been posted. After 3 years at FSU, I joined the faculty in the Department of Chemistry at my alma mater, NC A&T, as an Associate Professor of Chemistry.

After the first year at NC A&T, my husband and I were ready to begin our family. At 38 years old, we were finding it difficult to get pregnant. I was now very worried that while things were going very well professionally that we had waited too long to start our family. However, life as a young professor was great! At NC A&T, I worked hard to develop my research laboratory, write proposals, and teach biochemistry courses. With regard to research, I participated in research as well as directed the research of both undergraduate and graduate students. We were also living in Greensboro and had great family support nearby to help us with rearing our child. I also was in a good place with regard to being able to go up for tenure at the required time.

During the summer of 2005, I returned to the Naval Medical Research Center in Silver Spring, Maryland. My husband and I continued to juggle our marriage and work as a united team. I am very grateful to have his support throughout my career. I returned to the university and was directing the research of three masters and three undergraduate students. I presented my research at national meetings such as the American Society for Microbiology and published three refereed journal articles since arriving at NC A&T in fall 2004. Also, I submitted over six grants where I was the principal investigator (PI) or Co-PI. I had two proposals funded, one from the NIH and one from the Department of Defense. At the end of the 2005 semester, I was excited to find out that I was finally pregnant!

Joshua was born in the summer of 2006. One of the things that I can say is that motherhood is one of the greatest accomplishments in my life. Joshua was one of the most beautiful babies I have ever seen. Due to the age of my first pregnancy and because of medical reasons, I decided not to have any more children. At the time, I had a lab full of students both undergraduate and graduate. My colleagues at the university were very supportive, and while I had no classes, I still monitored my lab. I was thankful for a great technician. I was determined that being a new mom would not bring my lab to a halt or delay my submission of my tenure packet in 2006. I had tons of family support. My husband, Jay, is from Greensboro, NC, and his family helped us keep Joshua until we placed him in daycare. I am also grateful that my mother traveled with me to meetings to present my research findings or other conferences when needed.

At the university, I took on more responsibilities as an associate professor at the college and university levels. In 2011, I stepped into the interim chair position of chemistry. I must say that this is one of the hardest jobs in an academic institution. Juggling home and my new position were challenging. My husband and I always wanted more children, and in 2013, we decided that we would adopt a child younger than our son, Joshua. I wanted a daughter, but God sent me two handsome boys, Journey and John. As my younger son says, God is Amazing! The three boys keep us very busy with sports, boy scouts, school, and church activities, etc. Honestly, I don't know how I make it on most days! My evenings are crazy. Dinner needs to be cooked, and homework must be completed. Two of the boys are chatty in the evenings sharing all of their events of the day, and the other one likes to talk about his previous day in the morning as we are running out of the house. However, these moments are so precious, and they go by fast. How do I juggle all of this? Lots of prayers, support (especially from grandparents), meal planning, and keeping an updated calendar!!!!! This works for me. I wish that I was like my mother who cooked full dinners (meat, two vegetables, some kind of bread) on most days. However, I love the Crock-Pot, a breakfast meal at dinner, and I laugh at this one, "no cooking on Friday dinner policy!!" Friday evenings are when we have family time. Unless it is an urgent deadline, I try not to let anything interrupt our family nights. On certain occasions, the boys have been in my office when I need to complete reports or need to work with a student in the lab. I have also tried to instill in them a love for science, and we have been to the university to complete their science fair projects. I am hoping that I will have one scientist in the family, and it looks like it may be my baby boy, John. He collects everything including rocks, acorns, and pinecones. He told me on one of his visits to the university that he was going to work in my office so I would need to get a new space. I pray that I am a good mother to my boys despite my busy schedule. I used to worry about everything including how they looked, how they dressed, superficial things. Would people think I was a bad mom if they looked tattered? Now, I don't worry about this as much. Someone is always going to spill a drink or drop food on his clothes or not have on any socks. It is about treasuring each precious moment and knowing your limits (Fig. 2).

Can We Control Our Lives?

Laughing now, I think about how I thought I could control everything in my life. Earlier I wrote that I had created a mental timeline of my tenure clock, and my strategy was to be in a good place to secure tenure prior to becoming pregnant. In my mind, I thought I would care for my baby and then submit my tenure packet a year

Fig. 2. Margaret and her family (2015). Picture taken by Blue Box Photography

later. I thought I would be more settled and would enjoy motherhood more without some of the stresses that come with an academic position. How would I juggle marriage, motherhood, teaching, research, and the many committees I had been assigned while preparing to go up for tenure? I didn't plan on Joshua coming near the beginning of the next academic year or in the same year that I was submitting my tenure packet! Even later in my career, how does one juggle three young children while serving in a demanding administrative position?

As women, we drive ourselves crazy because we are under the assumption that we can control everything. I remember one of my graduate students saying to me that she didn't want to go the same route that I did in my career with regard to my timing of children. She said that I was too old, and she did not want to start her family so late in her life. She wanted to start her family in her 20s and if going to get a PhD meant that she had to wait then she would seek other career options. Wow, I thought! Is that how I am seen by young women that I mentor? We don't share our stories enough about the choices we make and why we made them. Or how we can be successful in climbing the academic ladder or completion of our graduate programs at whichever point we are in at that moment in our lives.

It is important to know that we have many options on when we decide to approach motherhood. We must also realize that even on our best day, we are not in control of our lives. Having a strong support network consisting of family members, friends, and colleagues was important for me in making the decisions that we did. There is no way that I would have been successful without the persons who stepped in and up for me.

A New Chapter in My Life

Sometimes, we have to step back to move forward especially when things do not go as planned. I lost my father in August 2015 and was spending time with him in his last days; I realized what is really important in my life and career. After returning to the university after my father passed, I decided that I needed to take care of myself professionally. I was not going to be the chair of the department, and I had served as an interim chairperson long enough. I also decided to submit my dossier and application for full professorship. In December of 2015, I stepped down as interim chair. I am moving forward with a strong realization of who I am and that God is first, my family second, and then my career. I am currently serving as the STEM Center Director of Excellence for Active Learning. I love my new role as director, where I have an opportunity to work with faculty to advance evidence-driven STEM education teaching, learning, and student success innovations/interventions. As a woman in academia, I have learned that I can be a mom, wife, and caregiver and still have a rewarding career. I have more time with my family now at a time when my children are still young.

I found out in April 2016 that I would be promoted to full professor, the highest milestone in one's academic career effective July 1, 2016. I am a mother, wife, and *professor*. This is pretty awesome! I am looking forward to a new chapter in my academic career.

Reflections, Advice, and Lessons Learned

One of the things that I asked myself while writing this book chapter is what advice would I have shared with my younger self while I was navigating my career. My top six pieces of advice would have been:

1. *Know that there are many routes to being successful and staying on track while pursuing an academic career.* What worked for me may not work for you. Be sure you have a support network. I have had lots of family support, but I have known other women who may not live near their families, and in these cases, they relied on friends, church members, etc. I also have known women who decided not to work while their children were young. Know that there are many ways that you can jump back into a career in academia and you can seek advice from mentors and former colleagues when ready;

2. *Never let obstacles get in your way; figure out how to get over, around, or through them.* Use your mentor to help you put together strategies for where you want to go;
3. *Find a role model and mentor and keep them in your pocket each step of your career ladder.* My first role models were my parents. I talked about my father, but my mother has been a great influence on my life as well. She has shown me the value in giving back and inspiring greatness in my life and others. I also have had role models such as my undergraduate and graduate advisors, senior professors in my department when I was an assistant and associate professor, and my former dean throughout each step of my life and career. As an African American woman, I have learned that my role model/mentor does not have to be someone that looks like me or even be my same age. A role model is someone who I use as a guide to support me as I ponder or become prepared for my next career move;
4. *Be aware of leave policies and programs on your campus that support faculty career life situations.* When I adopted my boys, I did not know that there were policies on campus where I could take personal/time away to get my sons adjusted in their new environment. I think it is important to first review work-life policies before you need release time. Next, I would suggest discussing you needs with your supervisor and the designated HR personnel. It is important to be proactive if at all possible. Don't be afraid to ask for release time from teaching, a flexible workload or possibly a tenure clock pause or extension;
5. *Never play the "I should have done this or wish I had done that game."* Learn from your experiences and move forward;
6. *Always take care of Yourself first.* The reason I say this is sometimes we become complacent and lose sight of what is important. While I enjoyed serving as interim chair of chemistry and being involved in many university committees, I moved farther and farther away from staying productive in my research. With so many women **and** men stuck as associate professors in the academy, how can we step into leadership positions in higher administration if we desire to do so? Especially, if your institution requires that positions at the chair level and beyond be held by full professors. Many important decisions are made in positions beyond our classrooms. It is important that we as women are a part of local, regional, and national decision-making that enables the success of our academic institutions and future scientific leaders.

Acknowledgments I would like to thank all my family and friends who supported me throughout the writing of this book chapter. I also would like to acknowledge one of my dearest and closest friends, my mother, who always keeps me on track and loves me unconditionally. Thanks Jay, Journey, Joshua, and John D. You bring out the best in me and for that I am grateful. Lastly, I acknowledge all my mentors who have pushed me out of my comfort zone and encouraged me to soar high and reach my dreams.

About the Author

Education and Professional Career

1990　BS Chemistry, NC Agricultural and Technical State University, Greensboro, NC.

1997　PhD Biological Sciences, Carnegie Mellon University, Pittsburgh, PA.

1997　Postdoctoral Fellow, Biochemistry, Duke University, Durham, NC.

2001　Assistant Professor of Biology, Department of Natural Sciences, Fayetteville State University, Fayetteville, NC.

2004　Associate Professor of Chemistry, Department of Chemistry, NC Agricultural and Technical State University, Greensboro, NC.

2011　Interim Chair, Department of Chemistry, NC Agricultural and Technical State University, Greensboro, NC.

2016　Director, STEM Center for Excellence and Active Learning, NC Agricultural and Technical State University, Greensboro, NC.

2016　Professor of Chemistry, Department of Chemistry, NC Agricultural and Technical State University, Greensboro, NC.

Honors and Awards (Selected)

2016　Faculty Mentor—Minority Access National Role Model Award; American Society of Biochemistry and Molecular Biology Education Fellow

2010　Navy-American Society for Engineering Education Summer Faculty Research Fellow

2007　Young Investigator of the Year Award-North Carolina Agricultural and Technical State University

2000　Young Investigator Award, Sixth Conference of the International Endotoxin Society

Margaret is active member in many associations, including the ACS, the American Association of University Women (campus coordinator), Iota Sigma Pi National Honor Society for Women, and Beta Kappa Chi National Scientific Honor Society (national treasurer).

It Always Seems Impossible Until It Is Done

Sunghee Lee

KrISTeN represents any hypothetical young woman who plans her career in a demanding field such as Chemistry and carries out her motherhood at the same time.

(continued)

S. Lee (✉)
Department of Chemistry, Iona College, New Rochelle, NY, USA
e-mail: SLee@iona.edu

> Dear KrISTeN,
> I couldn't stop thinking of our conversation the other day, about the dilemma you're facing in the career path you want to explore. You want to plan it "right" so that you can perfectly balance a career and family all together. So, here let me rewind my own experience (in no particular order) to recollect my path of the past 20 years or so. When I started writing this, I could not help thinking of Nelson Mandela's quote: "It always seems impossible until it's done."

Starting Graduate School as a New Mom

Chemistry and motherhood are intimately intertwined to me. When my husband and I started our journey to graduate studies in chemistry at Brown University, my 2-month-old son was on the plane with us heading to Providence, RI, from Seoul, Korea. On that plane, I also had the good fortune of having my understanding mother sitting next to me, my relentless supporter for any decision I made in my life. Originally, my mom came along with us intending to stay and help out for just a few months until we settled into the new environment, but it didn't take long for her to realize that she cannot leave the care of her newborn grandchild in the hands of a busy chemistry graduate student couple. Since my mom could not stay for an extended period of time in the USA with only a visitor's visa, she had to frequently leave the USA and go back to Korea and come back again. I still remember the very friendly staff in the Foreign Student Office at Brown, who would write a letter for my mom to make sure that her frequent reentries to the USA were not scrutinized. My mom's frequent trips back and forth across the Pacific, enduring nearly 20-hour flights, lasted until my son was old enough to attend preschool. As you can imagine, every aspect of my life revolved around my mom's presence and absence. Rather more precisely, every aspect of my *mom's* life was planned around my busy schedule of study and research. Therefore, it is not an exaggeration to place my own mother front and center of what I have managed to accomplish.

Still, even with the tremendous support from my mom, it wasn't always easy. For example, my graduate study often involved trips to Brookhaven National Lab in NY to use the Synchrotron Light Source. For each trip, I left town for up to a week, and I found this away period to be very difficult, even knowing that my son was in good hands with his grandma. I kept a notebook filled with notes to my son while I stayed nights collecting data. This was my way of focusing on task and getting things done in a productive way, constantly reminding myself that I can't permit any of these efforts to go to waste. Still this day, I become very emotional when I look at the notes I wrote to James during my times away.

Fig. 1 James (son) in 1992, 3 months old

As you can imagine, entering graduate school with a 2-month baby is not a common scenario, but I was quite naive enough to be not fully aware of what it means to be simultaneously mother and PhD student in chemistry. To be frank, the decisions surrounding my career path and starting a family were not calculated or even carefully considered but rather followed naturally as life itself unfolded. I just let it happen. I met a man who was a fellow chemist, and we got married and had a baby and decided to continue study abroad together. I loved my son, and I loved the idea of continuing to study what I loved, chemistry! Nothing was fully planned ahead of time (Fig. 1).

Luckily for us, the environment at Brown University was nothing but welcoming for a graduate student couple, and I have no single memory of feeling awkward or feeling like a misfit during my entire graduate studies. My son was an integral part of my professional career, and we grew up together, as my knowledge of chemistry deepened. Instead of holding stuffed animals in bed, I often found him hugging and playing with a chemistry book and falling asleep, as those would be the objects hanging around and easily accessible for him. In his toy chest, along with Lego sets, you could sometimes find molecular modeling sets. He became an essential member of every holiday party and group picnic all throughout my graduate school and postdoctoral studies. He had started attending ACS National Meetings at the age of 3 and lasted until he graduated from high school.

Although I had very limited free time, I did organize play dates for my son as best as I could, took him for swimming sessions, arranged for violin lessons, and taught him and other children at the weekend school so that he would be occupied with

Fig. 2 PhD Graduation at Brown in 1996, with James

many activities after his school day ended. My calendar was tightly organized and scheduled for every hour of the day including weekends. Every moment when I was not with him, I would be in the lab working. The worst feeling I got was to realize from time to time that I had allowed my time and efforts to be unproductive—"and I am not even taking care of my son at this moment." As a result, I developed a habit of thinking, "I do not have a second to waste." I carry this habit of treasuring every single minute even now. Largely thanks to my mom, and the understanding community at Brown, we graduated with a PhD in chemistry when James turned 5 years old, with myself being given the best PhD Thesis (Potter Prize) Award in Chemistry for that year (Fig. 2). We moved from Providence, RI, to College Station, TX, and then to Durham, NC, for our subsequent postdoctoral studies. By the time we settled in NY, my son had attended six schools in four different states. His world had been always revolved around the college, from Brown to Texas A&M and Duke University. However, I always felt bad knowing that my son was heartbroken whenever he had to move leaving his friends behind, and he wasn't able to experience the beauty of lifelong relationships with his early childhood friends.

Starting an Independent Career

Fast forward almost 20 years, nowadays, James and I talk about class management and how to engage students in a classroom, for he is now teaching Physics in a high school! James had not expressed a desire to become a teacher like his mom, at any point in growing up. Rather, he would occasionally complain about how I appeared to care more about my students and pay more attention to them. Nowadays, our conversations are often anchored around our students. I am sure that he now understands being an educator is not a 9 to 5 job, but one must be immersed in students' lives, beyond just classroom teaching and curriculum delivery. In other words, a great educator is the one who understands students and has patience, experience, and discipline and the one who would express plenty of care and love yet is not afraid of correcting their mistakes, just like any mother would do for their children. What previous experience could possibly prepare one more perfectly for this tremendous role, other than being a mother oneself?

My career path has taken me to a small predominantly undergraduate institution, where I try to accomplish a nearly impossible balance among full-time teaching, student advising and mentoring, and doing publishable research. Whenever someone describes what I do now as an impossible task, I remind myself how I managed my times at graduate school with an infant; I am reminded of times I felt badly to leave him alone and go away for my own professional development when he was only a small child; and I think of all the times that I chose to stay in the lab instead of joining other moms taking him for fun play activity. "It always seems impossible until it's done." What I am managing to do now is not nearly as impossible a task, compared to that. It does not involve heartbreaking decision. I just have to use my efforts efficiently. Being a mother and a full-time chemistry professor has only made me stronger and taught me how to effectively utilize every single hour of the day. As you can surmise, in a way, I am a mother to all my students with whom I interact. Seeing students' eyes widen upon learning something mind-blowing about chemistry is really a rewarding experience. But, the most rewarding part of my career is interacting with students and watching how they can take their beginner level of skills, knowledge, and confidence and steadily grow as they do research in my lab. It is not much different from raising my own child (Fig. 3).

Balance

Balancing work and life comes down to knowing what the right priorities are. We always have to make a choice about what to do, when to do, and how to do. My choice has always been simply following my heart as a guiding light and doing the most important, meaningful thing first at that very moment. If you accept the fact that all the things you wanted don't have to be done at the same time, then with appropriate prioritization, you can do everything what you want without much

Fig. 3 At the NCUR 2015 (National Conference on Undergraduate Research) with undergraduate research students, from left: Melissa Morales, Jaqueline Denver, Sunghee Lee, Michelle Muzzio, Omoakhe Tisor

sacrifice, just not at the same time. There will be a time when you are more of a PhD chemistry student or researcher or professor than a mother. There will be a time when you are more of a mother than a chemistry professor. The important message is to remember to be present fully, and be immersed in that moment sincerely, and give the best performance you can give at any moment. After all, this could have been the time that you could have spent with your child if you could—would you want to waste time just dwelling on that thought or rather get the most out of your day at work and be as productive as possible?

> KrISTeN, I will tell you once again to follow your heart and be confident about your choice. You can only engineer so much about a successful career or life path. Rather, simply follow where your heart leads you and thoroughly enjoy the journey. Trust yourself and your decision along the way. Before you know it, you will be in a place where you have not imagined to be, but even better, "It always seems impossible until it's done."

Reflections from Sunghee's Son
James H. Park (Age 25)

You'd think that growing up with a chemistry professor as a mother would be tough, right? Right. I can't even begin to explain what the dinners in my house were like. She would frequently quiz me to recite the names of molecules in our food. Our salt shakers were labeled as NaCl instead of, well, salt. Furthermore, quite often the conversation around the dinner table evolves around experiments she did on that day, or recent articles she read, anything that chemists apparently find interesting. Needless to say, I became a physics major. Nothing personal mom, it's just that when you start to see your food as tiny molecules vibrating on a gelatinous mass of hydrophilic bonds, you tend to lose your appetite. And I like eating.

Still, I owe her a lot. Without her, I probably wouldn't have made it as a physics major in college, and I definitely wouldn't have become a high school teacher today. And it wasn't just her love for chemistry that did it. It was her love for me. Ever since I was a child, she pushed me harder than perhaps any other kid at that age to pursue my studies and to push my capabilities to the limit. Like herself, she wanted me to strive for greatness, to be the best in everything I tried. I still recall the motto she taught me to live life by: be ambitious, be organized, and have fun. I remember sitting down with her after school and learning precalculus in the fourth grade. I remember my weekly trips to the library, always eager to check out a new stack of books. I remember all the science fairs and all the geography bees and violin competitions I won. And it was all thanks to her. There was no way I could've mustered up that much motivation as a kid to work toward my goals without her. And none of it could've been possible without all the time and sacrifices she put into me.

Only recently I recounted the last part of my mom's motto. Have fun. I never truly believed that last part as a kid. For a while, I assumed she said that just to try and trick me into thinking doing math problems over having water gun fights with my friends was fun. But then I started to pay closer attention to her. I always knew how invested she was in her job, but the truth was I never understood why. Until now. And all it took was me finally taking the time to notice.

In conversations with my mother, it was immediately apparent how much she treasured her job by how fondly she would speak of her teaching experiences, even the more unpleasant ones. I saw how earnestly she would work on research projects with her students and witness firsthand the enthusiasm she would impart upon them. I could sense how much she cared for her students, how she constantly pushed them to work harder to achieve their goals. And it reminded me of something. It reminded me of my childhood, all those late afternoons spent with nothing but my mom, a pencil, and a notebook. It reminded me of the super-elated giddiness I felt every time I won an award, and the incredible joy I felt for so many years, every moment I was at school surrounded by friends and teachers who loved me for who I was. It reminded me that who I was only existed because of one person: my mom. In the end, it

(continued)

wasn't about her passion for chemistry or teaching. It was about doing what she loved, no matter what it was. Whether it was coming home early from work to help her son learn new things, or drawing molecular structures on the kitchen whiteboard during dinner as a conversation starter. It was all about having fun.

My mom has always been the best at everything and will always be the best in everything. Now, with a teaching job and a physics degree under my belt, I know exactly what to do. I'm going to become the best high school physics teacher in the world. And my inspiration is due to the strongest, smartest, and most loving mother in this universe and all parallel universes. My mom, the chemistry professor (Figs. 4 and 5).

Fig. 4 Sunghee Lee's son James at his university graduation (He is now taller than me)

Fig. 5 Sunghee Lee and her mom in Paris, 2013

It Always Seems Impossible Until It Is Done

Acknowledgments Each and every unfolding event in life can be traced back to its origins: what I enjoy today is the result of yesterday. For that matter, I am grateful for everyone with whom I have crossed paths, from my teachers who always provided encouragement to me to believe in hard-work; my friends from graduate school who made such challenging times be rather fun; and my family, who have had infinite understanding about my distorted priorities at times. I am thankful for all my past and current students at Iona College who allowed me to practice what I preach, allowed me to enter into an important point in their lives, and gave me a chance to pass the wisdom I learned along life's journey.

About the Author

Education and Professional Career

1989	BS Chemistry, Sung Kyun Kwan University, Seoul, S. Korea
1991	MS Physical Chemistry, Pohang University of Science and Technology, Pohang, S. Korea
1992	Pohang Iron & Steel Co. Inc., Materials Science, Pohang, S. Korea, Research Scientist
1996	PhD Inorganic Chemistry, Brown University, Providence, RI
1997–1998	Postdoctoral Research Associate, Texas A&M University, College Station, TX
1998–2002	Postdoctoral Research Associate, Duke University, Durham, NC
2003–2004	Bergen Community College, Division of Science, Paramus, NJ, Science Faculty
2004–2008	Assistant Professor, Iona College, Department of Chemistry, New Rochelle, NY
2008–2013	Associate Professor, Iona College, Department of Chemistry, New Rochelle, NY
2010–present	Department Chair, Iona College, Department of Chemistry, New Rochelle, NY
2013–present	Professor, Iona College, Department of Chemistry, New Rochelle, NY
2014–present	Board of Trustees Endowed Professor in Science, Iona College, New Rochelle, NY

Honors and Awards (Selected)

2017	Iona College Honors Program Faculty Member of the Year
2016	Br. William Cornelia Distinguished Faculty Award, Iona College
2014	Board of Trustees Endowed Professor in Science

2013	The Rising Star Award, American Chemical Society (ACS) Women Chemists Committee
2013	The Distinguished Scientist Award, Westchester Chemical Society, New York ACS
2013	Br. Arthur Loftus Faculty Award for Outstanding Student Research, Iona College
2013	Woman of Achievement Award, Iona College
2011	Honors Program Teacher/Advisor of the Year Award, Iona College

Sunghee is a passionate teacher-scholar, now leading a very successful undergraduate research group at a predominantly undergraduate institution. Her research group, nicknamed "Project Symphony," consists solely of undergraduate students, many of them female. She has spearhead many STEM initiatives at Iona College and the wider community to emphasize the significance of STEM education as well as the foundational role played by chemistry in all neighboring sciences.

On Breastfeeding, Supramolecular Chemistry, and Long Commutes: Life as an Associate Professor, Wife, and Busy Mother of Three

Mindy Levine

Photo Credit: URI Photography

Introduction

My story begins before I became a mother, before I was pregnant, right around the time that my fiancé (now husband of 10 years) and I were talking about the future. I was a third-year graduate student at Columbia University in organic chemistry; he was a third-year PhD student studying electrical engineering at Princeton. I knew I

M. Levine (✉)
Department of Chemistry, University of Rhode Island, Kingston, RI, USA
e-mail: mlevine@chm.uri.edu

© Springer International Publishing AG, part of Springer Nature 2018
K. Woznack et al. (eds.), *Mom the Chemistry Professor*,
https://doi.org/10.1007/978-3-319-78972-9_19

wanted to have children, and probably several of them, but not yet. The "not yet" part was not something that could be taken for granted in our Orthodox Jewish community, where rabbinical permission is often sought for any form of birth control. "We need to wait until I am closer to finishing my PhD at least," I told my fiancé. "I don't know if you think we need to discuss this with a rabbi," mentally bracing for potentially awkward conversation. Much to my relief, he did not think rabbinical discussion was required, and the conversation ended there.

The conversation continued when I was finishing my PhD work in the spring of 2008. It was March, so that meant if I got pregnant right away, I would have a child in the winter of my first year as a postdoctoral fellow. That would be ideal, we thought, as I could take 8-week leave on my NIH postdoctoral fellowship at MIT and still be back with ample time to apply for academic jobs in the Fall 2009 cycle. I wasn't actively doing lab work for my PhD anymore, so we had eliminated the risk of chemical exposure to me and my unborn, hypothetical child. We had been married more than 1 year, practically no time at all, but more than enough time in the Orthodox Jewish community to warrant the decision to stop using birth control. We decided it was time.

It was March, then April, and then May, and I still wasn't pregnant. In the meantime, I was preparing to defend my dissertation, my husband was preparing to defend his, we were looking for places to live in the Boston area, and he was still looking for jobs. We were busy. Maybe it was stress, we thought, still not too worried. Maybe I was exercising too much or too little. Maybe I was working too hard. I bought "Taking Charge of Your Fertility" on Amazon and studied it like the good student I had been trained to be. We took a vacation together. It was August. I still wasn't pregnant.

It was October of 2008 when we decided to seek medical intervention. "I'm not pregnant," I told the nurse practitioner at MIT, a woman whom I had met for the first time just 5 minutes previously. "You're young," she said, "and it hasn't been a year yet. Try being patient." I tried that last spring, I wanted to protest, and now I'm done being patient. "We need medical intervention," I told her. "Please believe me."

She did, fortunately, and just a few weeks later my husband and I were undergoing a barrage of diagnostic tests through the Reproductive Science Center of New England. The tests were invasive, sometimes painful, and always at inconvenient times. "I'll be in a bit late today," I told my colleagues at MIT, where I was trying to start a research project, form important professional relationships, and gain some crucial chemistry expertise. "I have a doctor's appointment." When I showed up at 11 am doubled over in pain from cramping after a hysterosalpingogram (HSG) test, I took four Advils and tried to work. I don't think my MIT colleagues noticed anything amiss. Why would they? They barely knew me.

The results were in a few weeks after that, and the recommendation was clear: straight to IVF for us. This news was delivered matter-of-factly by our reproductive endocrinologist. My husband and I were dumbfounded. "We're young!" we protested. "Aren't there other options?" The doctor shrugged. "You can try intrauterine insemination (IUI)," she said, "but it probably won't work."

The IUI did work that time, right on the very first cycle, much to our and our doctor's pleasant surprise. After receiving the news that my pregnancy test was positive, I left my reaction stirring on the hot plate and went straight to my boss's office. "I'm pregnant," I told him, barely able to contain my excitement. "Congratulations," he said. "Was it planned?" Later that day, I remembered to call my husband.

That pregnancy was spectacularly uneventful as far as these things go, with the only real annoyance being the respirator gas mask that I diligently donned every day before entering the laboratory. It helped that one of my colleagues, a fourth-year graduate student, was also pregnant at the same time. We took pictures in our matching respirators, compared our growing bellies, and waited impatiently for the children to arrive. I traveled only once during that pregnancy to the Fall ACS meeting in Washington DC to present at the Academic Employment Initiative as I prepared to apply for academic jobs. The baby was due in September, right in the middle of job application season, exactly what the original "plan" had been designed to prevent, but I was committed to being ready. While I was at the meeting, I also gave an oral presentation on my postdoctoral research at MIT. I submitted that abstract when I was already pregnant, with the words of my advisor ringing in my ear, "Are you sure you are going to be able to present while pregnant? Because once you submit, it is really important that you follow through." I was 36 weeks pregnant at the time of the conference, but I followed through.

Job Application Process

I wasn't quite ready to submit job applications when my child was born, even though he was 3 days late. I still had to finalize my third and final research proposal. As a result, 2 days after my infant's birth, I made the decision to go straight from being discharged at the hospital back to MIT, where I sat in the basement of Building 18, working to finalize my research proposals that I would submit with my job applications for tenure-track faculty positions. My newborn baby waited in the car with my husband, while I talked about science with my postdoctoral colleagues in the Swager Group. They were helpful, of course, but did seem to be somewhat incredulous that I was there and actually working. They asked about the baby, more than once, and I had the unusual role, as a mother of a 2-day-old infant saying, "Can we please focus on the science? Please? Pretty please with a cherry on top?" (That last part may have been only in my mind.)

The entire time I was in that basement, trying to focus on crafting a competitive research proposal and job application, the voice in my head kept saying, "Whatever you do, don't think about the baby. Don't leak breast milk. Should I have worn a darker shirt? Used extra nursing pads? Done something different? *Don't think about the baby or then you will definitely leak.* Is the baby still sleeping? What if he is screaming hungry and wanting to nurse *right now?* Would my husband come get me? I bet he's not going to come no matter how much the baby screams." Smiling outwardly, I turned to one of my colleagues and said, "So, about that siRNA idea..." and continued the discussion. My work was completed in an hour, and

I returned to the car, to my husband, and to my newborn baby. The baby had slept the entire time.

I submitted my final job applications a few days later from my parents' house, where we had gone for the circumcision ceremony, baby naming, and seasonal Jewish holidays. In addition to having a newborn baby, and being in the middle of job application season, it was also Jewish holiday season. That meant 9 days in the span of 3 weeks that I couldn't write, couldn't use electricity, and couldn't make any progress on my applications. Nine days in addition to the 25 hours every Friday night and Saturday that are always off-limits for work. They were useful in general, this religiously mandated recharging and resting, but in the context of these time constraints and deadlines, proved brutal. There are traditional meals associated with those holidays as well, requiring significant amounts of time in advance of the holidays to adequately prepare. To best manage that year, my husband and I made the decision to move into my parents' house. Four hours in the car, with a newborn baby, because that was the least overwhelming of our options at that point in time.

There were only six job applications, because I was only willing to apply to jobs that would allow my husband to keep his current position at Intel, the one he had obtained when we decided to move to Boston for my postdoctoral fellowship at MIT, the one that would ultimately be the primary source of financial support for our family and young children. Six job applications, in the Fall of 2009, was laughable. I had friends who were applying for 70. With six job applications, I knew there was a nontrivial chance that I wouldn't get any jobs that year and that I would be back in exactly the same position the following year. Fortunately, my NIH postdoctoral fellowship guaranteed funding for 3 years, so I wasn't worried about being unemployed. Mostly I was worried about being too tired and disheartened and burnt-out to do this all over again.

A few weeks later I was back in Boston with an exceptionally squirmy infant. One day, I was in the middle of changing my infant's diaper on the dining room table when the phone rang. It was an unrecognized number, and I had attended enough workshops to know that I wasn't supposed to answer unrecognized numbers during the job search process unless I was fully prepared to have a telephone interview at that very moment. I was not prepared for a telephone interview, merely curious, so I answered the phone anyway. "This is Bill Euler," the voice said. "I am calling to invite you for an interview at the University of Rhode Island. (pause) Is this a bad time?" It was a perfect time, of course, is what I told him, while I held my infant son, naked from the waist down, and tried to keep him quiet for the duration of that conversation.

The next hurdle was the interview, a grueling 2-day affair, where each of the days was nearly 12 hours long. I had heard about such interviews and was sufficiently scared by the very idea of the interview that I weaned my infant completely when he was 7 weeks old, rather than trying to pump breast milk during the interview process. It has been more than 7 years since that time, and I still remember when the weaning was done. I sat at a rest stop along the Massachusetts Turnpike holding my screaming infant son, on the way home from a trip to New York. He was hungry, and because of a work-related decision I had made, my body was no longer able to provide him with what he needed. "This was a mistake," I thought. "Next time I'll do it better."

On Being an Assistant Professor and Going Through IVF

The next time started in the middle of my first year as an assistant professor, at a time when I was pretty sure I wasn't going to be able to sleep ever again, at a time when preparation for an hour-long lecture would routinely take 4 hours or more. I was right in the middle of this, trying to figure out how to buy monkey bars for my fume hoods in my empty laboratory, although I was fairly sure that they weren't called "monkey bars" and couldn't be found in the Fisher catalogue under that name. I had no results, no graduate students, no research program, and no time. "You want the children to be close in age, don't you?" my husband said. Well yes, I thought. Somewhat reluctantly, I agreed.

This time the first IUI didn't work, and neither did the first IVF cycle. Nobody at my work knew we were doing IVF, but it was becoming harder and harder to make excuses for my repeated lateness and erratic schedule. "I'll be in soon," I emailed a colleague. "Just an appointment." I arrived late often, with Band-Aids and bruises up and down my inner arms from repeated blood draws and repeated failed attempts at blood draws. Failed attempts, because it turns out that I was a phlebotomist nightmare: bad veins, needle phobia, and hypoglycemia combined to make routine, sometimes daily, blood draws into highly traumatic experiences. I was traumatized by the needles, exhausted by my job, and terrified that my colleagues and students would find out that I was going through IVF. Terrified of what in particular, I have no idea.

The second IVF cycle worked, and the results of the positive pregnancy test came as we were driving to New York again, for yet another Jewish holiday. This time it was Passover, which required our home to be completely cleaned from all possible crumbs of bread and other grains. Once again, the idea of managing the Jewish responsibilities while taking care of a toddler, working as a first-year assistant professor, and going through multiple IVF cycles was simply too daunting to contemplate. In addition to Passover, though, we had the added benefit of my youngest brother's Bar Mitzvah celebration, all rolled into one big happy family gathering. A bigger gathering than most people might expect, as I am the second born of seven siblings. At the time of the Bar Mitzvah, two of my siblings were also married, and there were four grandchildren, all under the age of 3. I had just been through two IVF cycles, I was nearing the end of my first year as an assistant professor, and I had a high-energy toddler to manage. "Smile for the pictures," my mother instructed us. All I wanted to do was sleep.

Again the pregnancy was uneventful, and again my baby was born 3 days late, this time in the middle of the winter break of my second year as an assistant professor. I know there are no perfect times to have children, but there are certainly times that are less than ideal, and this was one of those times. I had no grants, no publications, and no students with the experience necessary to make progress in their research without me there to supervise and interpret their every thin-layer chromatogram. I was home on maternity leave for 8 weeks, which was the same amount of time I had been home with my first baby, but this time the infant in question was colicky, eczema-prone, unable to sleep, and, as we later found out, allergic to multiple foods and other environmental triggers. There was also the high-energy 2-year-old in the house (although still at full-time daycare), whose presence hampered the relaxation, recovery, and infant bonding efforts only slightly. I went back to work on a Monday morning. By Tuesday afternoon, I was teaching my class, graduate physical organic chemistry.

The Second Child

When my infant son was 3-month-old, I took him with me to an interdisciplinary NSF review panel, which I decided was a great professional opportunity and worth the logistical inconveniences associated with such travel. At the end of each day, I picked up my son from a strange daycare center and found his scalp bleeding from eczema, exacerbated by my increased nut and dairy consumption while traveling. It turns out that he is allergic to both dairy and nuts, but at the time, we had no idea. While the other panelists went out for dinner and stayed up late writing recommendation summaries, I went back to my hotel room with my infant and begged him to please, please, please stop crying. Please let me sleep at least a little. My begging fell on deaf ears. I swore never to travel with an infant again.

This time I was committed to getting the pumping and breastfeeding right, even when I ended up on an elimination diet for my baby's allergies, but it wasn't easy. I wanted to be able to pump in my office, which had locks on the doors, but the young research group of students kept trying to come in to ask me, "Just one very important question." I pumped on the overnight Amtrak Northeast Regional train to Washington DC, where I was scheduled to meet with my NSF program officer to discuss my second attempt at an NSF CAREER proposal. I pumped in the bathroom of Union Station, standing at an empty sink, trying to ignore curious and prying and not-so-friendly eyes. I yelled through my office door at a potential collaborator to "come back later" when he arrived early for our meeting while I was in the middle of pumping. He must have found someone else to collaborate with, because he never came back.

One More Time

Two years later I was ready to be pregnant again, ready to continue on the path to building a large, chaotic Jewish family, not unlike the family I had come from. My husband was not. After some amount of discussion and negotiation, the decision had

On Breastfeeding, Supramolecular Chemistry, and Long Commutes: Life as... 263

been made to have one more child, just one, and then be done. I returned to the Reproductive Science Center, and on the first day of bloodwork and monitoring, the woman next to me shared that she was 38 years old, had no children, and was starting her fifth IVF cycle. "What about you?" she asked. At 30 years old and with two young children already at home, I said nothing. I closed my eyes and tried to be invisible.

I was scheduled to start getting hormonal injections on the same day that I was traveling to Mobile, Alabama, for a Gulf of Mexico research conference in January 2014. Through all the previous fertility treatments, I had never given myself shots, relying on my husband to administer all necessary injections (see needle phobia, above). We talked about delaying the IVF cycle so that it wouldn't conflict with my work travel and talked about finding somebody in Alabama who would be trained and capable of administering the injections. Interestingly, we never talked about skipping the conference itself; priorities, after all, were clear. After much discussion, I decided I would be able to self-administer all injections during the trip. Unfortunately, the first injection was in a public bathroom in the Charlotte, North Carolina airport, after I had run the entire length of the airport to make a connecting flight to Mobile. Sweating profusely from the exercise, hands shaking from the fear of needles, and with the boarding announcements for the Mobile flight playing overhead, I did it. I successfully injected myself twice. Afterward, I remember looking around and feeling like somebody should be giving me a medal for my bravery. No one was there.

The rest of the injections on the trip were uneventful, but the trip itself turned out to be more exciting than anticipated due to an ice storm that shut down Mobile for 3 days. My flight back to Boston was canceled, the entire airport was closed, and I was at risk of missing my first monitoring appointment to determine how my body was responding to the hormonal injections. I was also spectacularly hungry, as due to kosher dietary restrictions I had severe limitations on the food I was able to eat. I had brought some food with me, as I tend to do in general when traveling, but that food supply had run out, and I was still in Mobile. "I want plain salad," I told the waitress at the hotel restaurant. "I'm a bit confused, ma'am," she responded, the epitome of Southern politeness. "No bacon? No shrimp? No cheese? Nothing?" I missed a good fraction of that conference while panicking on the phone with travel agents and airlines and friends and anyone who might be able to help me get home on time. I eventually got out of the South by hiring a private limo to drive me to Gulfport, Mississippi, and flying from there to Charlotte and back to Boston. I made it to my monitoring appointment on time. My body was responding well.

A bit too well, I soon found out, as soon after my egg retrieval procedure I started to develop symptoms of ovarian hyperstimulation syndrome. The syndrome is usually benign enough, albeit somewhat unpleasant, but the medical recommendation is to stop the IVF cycle and try an embryo transfer the following month. In a moment of stubbornness, impatience, and just plain stupidity, I declined to inform my doctor about my symptoms, underwent a procedure to implant two embryos, and developed full-blown ovarian hyperstimulation syndrome. The syndrome resulted in my nearly passing out in the middle of teaching a lecture, lying flat on the floor in my office in the dark, and taking a week off from work to recover. Oh and by the way, I was pregnant with twins.

It was the spring of 2014 when I found out I was pregnant with twins. By this point, my lab had some research funding, I had students with experience, and I vowed that this time I would arrange my maternity leave and teaching schedule in a way that suited my needs. I told my department chair I would need more time off from teaching, and he arranged for me to be off in the entire fall semester in exchange for a somewhat heavier teaching load that spring. It was an imperfect situation, I knew, setting me up for an overwhelming spring semester right before I was going to apply for tenure, with young infants at home, and two older children, but I took it. I figured that the flexibility it would give me at the end of my pregnancy with twins would be worth whatever the trade-offs would be.

That pregnancy turned out to be a nightmare, with one of the twins diagnosed with a rare but fatal genetic condition. He passed away at 32 weeks in utero, and my surviving twin was born by emergency C-section at 35 weeks and 5 days. She is miraculously healthy, happy, and precociously advanced, now at 2.5 years old. My 3rd and final attempt at a CAREER grant was submitted in July of 2014, right in the middle of all the doctors' appointments, scans, and tests that led to my son's fatal diagnosis. I thought about not submitting the grant, not then, but couldn't fathom waiting another full year (see impatience, above). The phone call from my program officer with preliminary notification of the award came 6 months later, in December, when I was already back at work and pumping milk for my daughter. I stopped pumping to take the call.

I traveled during that pregnancy probably more than I should have, including a cross-country trip to San Francisco, while 7 months pregnant with twins to present my research at the ACS Organic Division Young Academic Investigators Symposium. The opportunity was worth it, I decided, but I had some difficulty convincing my high-risk OB of that fact. I had some challenges navigating that conference as well, never really sure what to say to people who asked about my pregnancy. I knew that my son was sick by then, with an uncertain long-term prognosis. What can I say when people tell me how wonderful twins are going to be? Do I tell them that my daughter is unlikely to grow up with her brother? Do I say nothing about twins at all and just let them think I am carrying a very large single pregnancy? Do I comment on their size/weight/body type? And by what right do they think that they can say things about mine? *Can we all please just focus on the science? That's why we're here, right?* Contacts made during that San Francisco trip led to two of my ten stops on my tenure tour the following year and to at least one external letter writer for my tenure application.

On Losing a Child

It is two and a half years now since I gave birth to a healthy daughter and a stillborn son, and I still have difficulty articulating any of the complex emotions surrounding that experience and the resulting aftermath. It is long enough now that most people I meet professionally don't know that my daughter is (was?) a twin, and most of the time, I don't tell them. Occasionally I do, like when I visited a university recently and one of the faculty members there told me that she thinks she would like to have a third child, but she is so nervous that they will have twins, and then what will she do? "I had twins," I told her matter-of-factly. "One died."

Two and a half years later, I am grateful for all the messages of support I received, and re-reading those messages brings me to tears every time. I am grateful for the support of my colleagues and students and the entire department. I am grateful for a mother who is flawlessly useful in a crisis situation, who provided all the support that she could. Being part of an extended Jewish community provided my husband and me with additional support as well: from the rabbi, who came to visit when we told him of our son's passing and spoke so eloquently at the funeral, to the friends and fellow synagogue attendees, who cooked meals for us, offered to watch our other children, and gave us time to heal. Mostly I am grateful for the gift of time and for the human capacity for healing, which leaves me now able to function almost like nothing ever happened.

Mostly I want to say that if you are reading this and you have suffered a miscarriage, stillbirth, or the loss of child at any stage, know that you are not alone. Reach out to me or to someone you trust. It can be so isolating to go through that experience and to deal with colleagues, friends, and relatives who never seem to know what to say. Trust me also that you will learn to be happy again and you will find a place for this loss in your life that doesn't take over your life. Here I am, two and a half years after the stillbirth. Some days I don't think of my baby Gavi at all.

This has been the hardest section of the chapter to write, this part about my baby who died, and I am not sure I am doing any justice to what that tragedy was like for me and my family. I am incredulous that we went through such a horror, the death of a son, and now we are fine. I know that my understanding for how things can go terribly wrong is greatly expanded, and I hope that my capacity for empathy for my students and colleagues is increased as a result. I hope that I live the rest of my life in a way that gives tribute to my son who will always be a baby.

The Tenure Process

Three weeks after my emergency C-section and the burial of my son, I was back in the office, this time for a faculty meeting to decide whom we would interview for an open tenure-track position in our department. I sat there holding my daughter, feeding her pumped breast milk from the bottle rather than risk offending my colleagues by breastfeeding right in the conference room. I didn't have to go to that meeting, I knew, but I wanted to.

I was back at work full-time when my baby was 7 weeks old, this time with the promise of being able to take an additional 3 weeks in June and without the pressure of teaching a class until the end of January. This time, this third pregnancy and maternity leave, I knew what I wanted, I was able to clearly articulate that request, and I successfully secured what I needed to be successful. As a result, the transition back to work was markedly smoother than with my other two pregnancies. The breastfeeding and pumping were fine, and my pumping sessions in my office stopped being interrupted the day I put a sign on my door that said, "Pumping breastmilk. Please do not disturb." I was getting better at being assertive, and my work-life-children setup was starting to look more like something I crafted and less like something imposed on me against my will.

Things got very busy very quickly when the spring semester started, and 192 undergraduate premedical, predental, pre-veterinary, and pre-pharmacy students filled the main lecture hall to learn organic chemistry from me 3 times a week. These students didn't know how to respect a "Do Not Disturb" sign, even one with the word "breast" in it, and so I had to leave my office to pump in a designated pumping room a short walk away. I had organized a tenure tour, but without the family situation that allowed for travel during the fall semester, all of the tenure travel occurred in the spring and overlapped with this highly demanding teaching load. At every school I visited, I had to tell someone, "I need breaks in my schedule for pumping." The first time I wrote that in an email, my host wrote back, "Pumping what?" I got more explicit after that, very quickly. More often than not, my host was male. Usually though, he was also a parent and was sympathetic and accommodating of my pumping needs.

I am, fortuitously, part of a committee at my university that established a Professional Family Travel Fund to enable people with significant caregiver respon-sibilities to travel for work. This fund was established with a fairly general purpose but in its first year was used almost exclusively by female assistant professors with breastfeeding infants.

Similarly, I was able to use the fund to travel with my au pair and my infant daughter to all kinds of interesting places, including a visiting seminar at the University of North Carolina at Greensboro, which, because of the seminar sched-uled on Friday afternoon and my religious Sabbath observance, required me to stay in Greensboro for extra time before heading back to Massachusetts. There was a snowstorm at the time that led to an 8 hour layover in the Philadelphia airport en route to Greensboro. Although such a layover may have been challenging at the best of times, it was particularly interesting with a 7-month-old daughter and Peruvian au pair as travel companions. I remembered that I had sworn to never travel with a baby again and wondered through some sleep deprived and frustrated fog what had happened to that promise. While in Greensboro, I arranged for the Chabad House to deliver kosher Sabbath food to the hotel, but because of my Friday work schedule, arranged for my au pair to accept that delivery. When I returned back to the hotel, she asked me, "Is it usual in this country for a rabbi to meet you in a strange city and give you all kinds of foods?" Yes, I told her, wherever there is a Chabad House, it is usual. The Chabad tradition is known for being remarkably generous in such matters.

My most memorable use of the Professional Family Travel Fund was to enable me to attend a full-week workshop sponsored by NSF, "Advancing Our Under-standing of Supramolecular Chemistry in Aqueous Solutions." This workshop required that all participants agree to stay for 15 hours a day for the entire week;

with a breastfeeding 9-month-old baby, I would not have been able to attend the workshop without bringing my baby and au pair with me. I showed up at the opening reception with my baby and received all kinds of favorable comments from the other attendees, including many well-known supramolecular organic chemists who said things like, "We need to see more babies at these sorts of things," and "Kudos to your university for enabling you to do this." NSF itself indicated that future workshops of this nature may include childcare accommodations. At least two of my external letter writers for my tenure application were people whom I met at this conference, baby girl in tow, attempting to manage both my caregiving and professional responsibilities and to excel at both.

I recently went through the tenure process at my university and came out the other side as a newly tenured associate professor. As part of that process, my tenure evaluation committee was expected to evaluate my teaching. To that end, I walked into the building one day and met a member of the committee who told me, quite matter-of-factly, "I will be visiting your class this morning." Lovely, I thought, as I held my daughter in one arm. The previous evening, my au pair had badly sprained her ankle, and without any other childcare in place, my husband and I had decided that I would take the baby with me to work. That particular morning, in lieu of preparing for my lecture, I was attempting to entertain my daughter, figure out childcare arrangements for my sons, and determine the degree of medical attention that my au pair required. After hearing that my colleague would be evaluating my teaching on that day, I ran to find a graduate student who could watch the baby for a while, gave her strict instructions, *"Do not under any circumstances bring me the baby while I am being evaluated,"* and went back to teach.

After my colleague left, I did in fact bring the baby to class and taught the remainder of the class with my daughter "helping." I put the photo that my student took of this class on Facebook and am sharing it here as well, with the goal of increasing the visibility of chemists who are parents and how we are often (always?) both at the same time (Fig. 1).

Post-tenure

I'm tenured now, and I am sorry to share that it has not changed my workload to any substantial degree. If anything, I would say that the workload is heavier now that it was pre-tenure. What has changed is my ability to manage that workflow more efficiently and my understanding of what tasks require maximum effort, what requires good enough effort, and what does not need to be done at all. Moreover, my willingness to vocalize my opinions at the departmental, college, and university level has increased dramatically with the successful granting of tenure. For example, I will talk vocally to ensure that personnel decisions at the department level are not discriminatory, I will call out colleagues and even superiors for sexism and implicit bias, and I will strongly advocate on behalf of female assistant professors, graduate students, undergraduate students, and postdoctoral fellows. It is my responsibility

Fig. 1 Mindy with her daughter "helping" teach chemistry

now and, in my opinion, a moral imperative when we are successful to help others who come after us to be successful as well.

One important note is related to the service requests that come in post-tenure. I had heard prior to receiving tenure that I was being effectively "shielded" from the majority of time-consuming service requests to serve on committees, to review manuscripts or proposals, or to do any other of the important, necessary, but nonetheless time-consuming tasks that fall under the broad umbrella of "service." I had some difficulty understanding that, because, in my opinion, I was spending significant amounts of time on service-related tasks. Once my tenure decision was official, however, the service-related requests skyrocketed almost instantaneously. Because I was still adjusting to being tenured, I continued to say yes to almost every request that came through, figuring that it was important to do so for continued professional development. This behavior pattern quickly became unsustainable, and my research productivity in the summer immediately after tenure (summer 2016) took somewhat of a hit as I struggled to figure out (or refigure out) the workload balance. The takeaway message from my cautionary tale: It is OK to say no to things. It is OK to say "no" pre-tenure, carefully and appropriately, but it is certainly OK to say "no" post-tenure to requests that you have decided are not a good use of your time and skills at that moment.

On Sexism/Bias/Discrimination

I wish I could say that this doesn't exist anymore, that it isn't a problem, and that future generations of girls and women interested in a career in academic chemistry wouldn't need to worry. I wish I could say that, but I can't. Most of the bias and sexism is subtler now, I have to believe that most of it is inadvertent and not malicious, but it is still there, and we still have to talk about it. Mostly, we have to

On Breastfeeding, Supramolecular Chemistry, and Long Commutes: Life as... 269

keep calling it out *every time* it happens, professionally and calmly, and make it clear that we will continue to do that every time in the future. Is this going to be tiring? Sure. But we owe it ourselves, to our gender, and to the future of female scientists to keep fighting this. Mostly, the best counterpoint to sexism in academic chemistry is to keep getting more girls and women interested in chemistry and to keep them interested and involved.

I'll give a few examples of some of the sexism/implicit bias I have encountered over my time in academia, with some of the details blurred to protect the less-than-perfectly-innocent, as well as my responses to those events:

1. A visiting seminar speaker sitting in my office (which is full of pictures of my children) told me, "I really think children belong at home with their mother." I wish I had responded better at the time, but for the next time this happens (and I can be sure there will be a next time), I will say, "I couldn't agree more. That's why my children are home with their other mother," calling attention both to his sexism and heteronormative biases simultaneously.
2. Somebody once suggested I bring a particular female student to an event. Because that student is attractive, he said, many more "horny old men" will be likely to stop by our outreach table. Again, I didn't say anything at the time but was sure to call out the bias in writing after I had time to process the information. I didn't bring that student to the event, nor did I attend myself.
3. A former colleague who saw me when I was pregnant with my first child told me, "That's so nice that you are having a child. Be sure you don't have any more children if you ever want to be successful in academia." I hope I have proven him wrong.
4. After a virtual grant panel review that had occurred completely during working hours, a grant manager told me, "Thank you for your work on this. I hope it didn't take away too much time from your children." I called out the manager and told him that while I appreciate the sentiment, I questioned the appropriateness of bringing my children into the discussion for a work event that had occurred completely during working hours.
5. Right around when I first started, one of my colleagues told me, "We used to have a woman in this department who was very pushy and aggressive. But don't worry, it didn't make us think that all women would be that way. We're glad to have you." When I recounted this anecdote to my husband, he wasn't sure why I characterized it as sexist. In response, I asked him to flip the genders for a minute and see how that would sound if a woman told a man the same thing. He thought for a minute, and then quickly agreed.

Science Outreach

The best counterpoint to continued sexism in the sciences in academia is to get more women and girls interested in STEM, sustain and nurture that interest, and support that interest if it turns into the desire to pursue a STEM-related academic career. To that end, I have devoted significant amounts of time and energy focused on conducting science outreach to girls at all educational levels, from elementary school through undergraduate and graduate students. The flagship outreach program, run by my group for the past 4 years, is Chemistry Camp for Girls, a full-time, week-long program for middle school girls that occurs over the April school vacation week. We accept 40 students into the program each year and are consistently delighted by the positive reception and engagement by all participants.

We have a newer high school girls program, Sugar Science Day for Girls, which is part of the broader impacts of my current NSF CAREER grant. In this program, up to 50 high school girls come to the University of Rhode Island for 1 day over their February vacation week and use the opportunity to study the science of sugar, make and eat rock candy, and launch rockets that run on a sugar-derived fuel. Two participants of that program are also welcomed for a paid, hands-on internship in our research laboratory over the summer.

There are a number of other outreach initiatives that I involved in, including running polymer workshops at a Girls Reaching Remarkable Levels (GRRL) Tech Initiative, planning and executing an elementary school science Sunday, and visiting my children's classrooms on multiple occasions. These are all time-consuming, but I have been fortunate to receive significant assistance from my graduate students to enable seamless execution. I have been more fortunate to have benefited from outreach (both formal and informal) and encouragement to get involved in science and stay involved and am happy to "pay it forward" by supporting the next generation of female scientists as well.

Reflections from Mindy's Children

One of the most interesting things for me is to hear how my children perceive my career, and this continues to get more interesting for me as they grow older. My children are now 8.5 years old, 6 years old, and 3.5 years old, and my oldest has been so remarkably articulate at so many stages. Some examples are discussed below:

1. I walked into the kitchen one day and saw that he had dumped a pile of coins and was counting them. I asked him, "What are you doing?" His response, "I am trying to see if I have enough money to give the parents who have to travel for work extra money to take their au pairs with them." "You need at least $300," I told him. "How much do you have?" "I'm getting there Mommy," he said, "I have at least 300 cents."
2. He also gave me advice on how to write a grant proposal, telling me to write, "Dear Misters, I am doing a lot of concentrating work. Please give me money."
3. He also filled in his most recent Mother's Day card by writing, "My Mom's favorite thing to do is: Go to work," but then later said, "I know your real favorite thing to do is to spend time with your two beautiful boys and one beautiful baby. But that was too long to write on the card. Also, I forgot."
4. One night he was in the bathroom supposedly brushing his teeth but spending quite a bit of time in there. I came in to ask him what he was

(continued)

doing and saw a tremendous mess. I started to yell, but he immediately said, "Mommy, I am doing an experiment. I am trying to make a soap that does not fall off your skin, so I added some sunscreen because I know that sunscreen stays on your skin. I also wanted it to smell nice so I put in extra toothpaste since I know that toothpaste actually makes your breath smell nice. Is that good experiment, Mommy?"

5. Most recently, our department was in the middle of two faculty searches, and I was discussing the search process with my 7-year-old. He suggested that instead of having to read all the applications, I should instead give the applicants a test about chemistry. He then proceeded to design the test, using his knowledge of chemistry, with the following questions:

   - How do you get a vaccine? Which type of chemical do you get when you get a vaccine?
   - What is an acid and what is a base? Give examples of each.
   - How does a diaper explode?

6. My oldest also recently decided to collect pebbles and try to sell them to me. "You can use them for science, Mommy," he said. "Can you pay me from your grant?"

7. We were recently in New York visiting my parents, when my middle child told them, "Mommy and I look a lot alike because we have a lot of the same genes. She doesn't have the gene for liking ketchup though. I think I must have gotten that from Daddy."

8. Recently when my boys were procrastinating going to sleep, they came up with a list of scientific questions that they wanted me to answer before they would go to sleep. The questions included: Why is glass so brittle? Why are people's skins different colors? Why do some children look a lot like one parent even if they are supposed to get half of their genes from each person? How do messages from your phone travel to Daddy's phone? What makes something reflective? How exactly does sun damage your eyes? Why do whales use sonar waves to communicate? What exactly is light anyway? What is an atom made of? How do rocks melt? Why does pressure help rocks melt?

9. My middle child was playing with a pretend gun recently and told me, "This is the biggest gun in the whole university." "I think you mean universe," I responded back. "No, university. The university is even bigger than the universe. Is that true, Mommy?"

10. My oldest son is fairly certain that when he grows up he is going to be a chemist and come to work with me, and my middle son is convinced he is going to be both a chemist and an engineer. I couldn't be prouder.

Reflections

In thinking about my career path, one question that comes to mind is to what extent my career path looks differently than it would than a parallel career path without children. To be honest, while I am sure that my life is busier because I made the decision to have children and a career, it is not clear to me that the career itself looks markedly different than it would if I didn't have children. The biggest career effect so far is probably that I restricted my job search to places that were commuting distance from my husband's workplace. Of note, work-related travel is something that is particularly important for professional development. While it is logistically challenging to accomplish such travel with multiple young children, with additional restrictions on the food I can eat (due to kosher), and on when I can travel (never between sundown on Friday and sundown on Saturday), in my experience, it is well worth it.

Most of how I approach my life, for better or worse (hopefully mostly better), can be traced in some part to my parents who raised me and the lessons they lived and imparted. My mother showed me the value of working hard and of making time for things that are important. I have so many memories of her being available for any and all family-related commitments and then completing her office work late into the night. She was juggling her career and her family responsibilities and showing me by direct example that it was possible to do both and to do both well. As my children get older and I see her relationship develop with them, I am seeing essentially that same kind of juggling act between the professional and family responsibilities. My father, a physician, recently switched jobs to work at a hospital that requires a substantial train and bicycle commute but has substantially more work-related responsibilities and hopefully concomitant job satisfaction. I give him a lot of credit and continue to be inspired by such career devotion.

Advice and Lessons Learned

Balancing academic chemistry life and motherhood is an endlessly fascinating juggling act. I'm full of stories about my life and about how I am caring for children, keeping and succeeding at my job, and appreciating my marriage to a man who has more patience than I ever would have imagined. Rather than this chapter being simply a collection of anecdotes, though, I would like to help people who are attempting similar balancing acts. To that end, I will also share some tips I have developed over the years that work for me:

1. *Make lists.* Make as many lists as you can and use them as often as you need to. In my case I maintain three active lists—a long range to do list that I make at the start of every semester which has a column for every grant, project, seminar, and other deadline, with a row designated for each week and what I have to accomplish for that particular week; a weekly to do list that I make from the long range list and

includes high priority tasks for each day of the week; and a daily to do list that includes all minutiae of this job, including things like "Upload solutions for problem set 3; format budget for new grant application; call vendor about broken rotavap; etc." I then make sure to devote the first part of every work day to the important work-related tasks and reserve the minutiae for the later part of the day when students start to interrupt my workflow.

I routinely put my students through a "paper blitz," where the goal is to submit one paper per week for as many weeks as we can. My papers get rejected sometimes, especially when we rush the submission process to meet this "paper blitz" goal, but overall the productivity of our research has been extremely high during these times, with 30 published papers and 2 submitted manuscripts over the past 6½ years. It's not a perfect system, but it seems to be working most of the time.

2. *Find extra time in your day.* This sounds ridiculous on some level, which is the level that understands that every day only has 24 hours, but recognizes that most people have time that is wasted and can be used instead for productive purposes. In my case this has been the early morning, and my definition of "morning" has been broadening throughout my years at this job. I now routinely wake up at 4:30 am to start my day and find that early time at work to be my most productive time throughout my day. I have tried to the extent possible to stop bringing home work at night and instead just keep getting up as early in the morning as necessary to get work done and to still be home in time for dinner with my family.

3. *Be efficient.* Find areas of your life where doing your best isn't necessary, but doing good enough will be fine. I will not delineate professional areas that may be satisfied with a "good enough" effort, but I can certainly give several examples of my personal life that are satisfied with "good enough"—the cleanliness of my house, the frequency of my children's haircuts, my children's cleanliness, and the quality of our weeknight dinners are some examples that come to mind.

4. *Recognize areas where good enough isn't going to work.* This area has been slower to develop for me and mostly occur as my children get older and need more. My oldest son is now in occupational therapy (OT) once a week for sensory processing disorder. This private OT is at a sensory gym that is 35 minutes away from my house in no traffic; at the time that we go, it routinely takes over an hour. The gym is also in the exact opposite direction from my office, which means that on the days that I take him to OT and then have to go into the office after, I usually get in around 11 am. This OT takes a tremendous amount of time that I certainly would not have said I had available in my schedule; however, the improvements in my son's behavior, in his social life, and in his school reports have been remarkable and have already demonstrated that the time investment is worthwhile. We stopped bringing him to private OT for 6 months when we decided this was too difficult for our schedule but soon learned that the time investment in the private OT pays us back in spades.

5. *Invest in high-quality childcare.* I am 100% sure that I have been saying this for as long as I have been a parent, but it wasn't until my first au pair joined our family in January 2015 that I understood why I should *really* mean it. My older

children were daycare kids, sometimes in the same daycare but often in two different locations. They were the ones who were dropped off at 7:29 am for a day that was supposed to start at 7:30 and picked up by a running, harried, and hassled mother at 5:55 when the center closed at 6 pm. In my case, these long days are partially caused by both my husband and I having extremely long commutes; as half of a dual-career couple, I consider us lucky that our jobs are only 100 miles apart and that we are still able to live under one roof. Days like those invariably required me to leave my work close to 2 hours before I got home, allowing for traffic, double pickups, negotiating children in and out of the car, and unloading everyone into the house. We all walked in at the same time around 6 p.m., with every member of the family tired, hungry, and stressed from a long day. The amount of time that I had to get food on the table from the moment we walked in the door, in order to avert a total and complete meltdown, was on the order of 30 seconds to a minute.

Since my first au pair joined the family in January 2015, there is someone else who does all drop offs and pickups of the older children from school. Someone else cleans up from breakfast, cooks dinner, takes care of my daughter, and generally makes sure the house is running smoothly. I am now able to leave work and go straight home and actually have enough energy left at the end of the that trip to enjoy spending time with my children once I get home. We are on our third au pair now, and I can say without qualification or reservation, **Get good help**. Put in the time and investment and whatever it takes to get yourself the help you need to stay sane; to keep your kids safe, healthy, and cared for; and to do the job you love (Fig. 2).

6. *Exercise*. I think that people who exercise are invariably happier, more productive, well-rounded individuals **on average** than those of us who don't. It is most

Fig. 2 Mindy with her three children and au pair

On Breastfeeding, Supramolecular Chemistry, and Long Commutes: Life as... 275

likely worth a little less time at work, or potentially a little less sleep, to be able to spend time exercising. In the past 5 years, I have completed two triathlons, and I am currently in the middle of training for my third triathlon. I am probably more tired from stretching my schedule even more to accommodate this training, but I am pretty sure it is worth it. Now that my children are a little bit older and much more aware of their surroundings, I think the exercise is worthwhile not only for my own health and sanity but also for the example that it sets for my children about healthy living and the importance of exercise.

7. *Find a partner.* Before I was married, I would routinely listen to talks about work-life balance, and invariably these talks suggested that the only way to "have it all" was to find a life partner who would be supportive of your career choices and value your career to the same extent that he/she values his/her own career. The problem always was I was never quite sure how one could tell whether a partner would actually be supportive in a future, hypothetical situation that may or may not arise and may or may not be what you are envisioning it to be.

I think it is more actionable advice to say that if and when you are lucky enough to find a life partner, make sure that you are explicit with him/her about what you need to be successful in your career and at home. Make sure that you decide together about important parenting decisions and about the division of labor with regard to housework and childcare. Once you have made those decisions, do not second guess or micromanage your spouse, as that is sure to undermine the main purpose of delegating and dividing the labor. In just one example of how this delegating works in my family, I am responsible for the food and food-related tasks, and my husband is responsible for the laundry. I do not know how to work the washer or dryer, I have almost never used the washer or dryer, and I have no interest in learning how to use them. My clean clothing appears in my drawers at random intervals, and that is how I know that the "laundry fairy," aka my husband, has been there. When my mother-in-law came to visit after the birth of my daughter, she asked me to show her how the washing machine worked so that she could run some laundry. "You need to ask your son," I told her. "I have never used that machine. I have no idea."

8. *Focus on your relationship with that partner.* I think there is probably a tendency to de-emphasize a relationship with a partner in favor of all the very demanding things you are doing with your life—your work, your children, the house, the bills, the laundry, etc. This is almost certainly a mistake. It is worth the time to invest in maintaining and growing your relationship with your partner. One day your children are going to grow up and move out (hopefully), and then where will you be in terms of your relationship with your partner? To that end, my husband and I have established "in-home" date nights once a week, where we commit to not working after the children go to sleep and instead watch a movie and eat ice cream straight from the carton (Fig. 3).

9. *Figure out your needs.* Figure out what you need and be assertive in stating those needs. If you need more (or less) time off to care for children, more or less resources, teaching accommodations, tenure clock extensions, or anything of that nature, be honest with yourself to decide what you need, and then keep

Fig. 3 Mindy enjoying time with her husband

asking for it, politely but firmly, until you get it. When I was a graduate student, I went to all kinds of events that were sponsored by the Columbia Chemistry Careers Committee, hoping to be talked out of a career in academia. I knew that the work would be stressful and that the salary was likely to be lower than what I could find elsewhere. However, I couldn't get past the fact that I could think of an idea today and then design an experiment and have someone in the lab tomorrow running an experiment to test whatever crazy idea I could come up with. That intellectual freedom is something that I knew I wanted and something that has been so fun and so exciting in this career. Finally, if you want assistance or advice or just someone to talk to, please reach out to me. I am always happy to encourage young women and to help them achieve their goals.

10. Believe you can do it. I think you have to be almost naïve in saying to yourself, "Nobody told me that I couldn't have children and still be a successful chemist at the same time, so I guess that means I can do it." I can work really hard during the hours I am at work, overall spend markedly less time on this job than many people without young children, and still be successful. I can delegate, prioritize, and accomplish goals. In doing this, I can be and am a role model for young women who may want academic careers but don't know how they can successfully manage that kind of career and a busy family life. I am a role model for my daughters and also for my sons. I hope that they know that when I am not with them, I am doing the work that I love, and that I am a better person and hopefully a better mother as a result. My life isn't clean, or perfect, or relaxing, but I am doing what I love and raising my family at the same time. We are doing it successfully. We can and will keep doing it successfully.

Acknowledgments   I have tremendous gratitude for my husband, Yedidya Hilewitz, who met me when I was a hard-working graduate student and has stuck with me as I transitioned to being a hard-working postdoctoral fellow, assistant professor, and now associate professor. The effects of this hard work and sometimes long hours on our house, children, and marriage are nontrivial and would only be possible with the full support, encouragement, and love of a life partner. Thank you.

For my children, Eli, Judah, and Dahlia: I love that you all have boundless energy, enthusiasm, and curiosity about the world. It makes it challenging and exhausting to parent you sometimes, but I wouldn't trade it for anything. I love that you are sure you will be scientists and that you already do your own experiments. Thank you for tolerating my work-related absences from the home and for learning to understand that it is because I am doing what I love. I hope that you see me as a positive role model of a working parent and that you grow up empowered to make your own decisions about marriage, parenting, and professional responsibilities. I love you.

To my parents: You raised me to care about my work, my family, and my religion. I hope I have learned the lessons successfully. Thank you for everything you have done for me, financially, physically, and emotionally. Special thank you to my father for providing me with so many internship opportunities in the biomedical fields and for helping me get excited about science and to my mother for being my emotional rock and for keeping me grounded.

To my friends: Part of being half of a dual-career couple and an academic chemist meant that Yedidya and I have chosen to settle in a community that is somewhat removed from our extended families. To our friends in that community who have become like family, thank you for everything. We are so very blessed.

To my PhD mentor, Professor Ronald Breslow: I was young and naïve and inexperienced in the spring of 2004 when you agreed to let me join your research group. I am inspired by your excitement for science, boundless energy and enthusiasm, and relentless drive. Thank you.

To my postdoctoral mentor, Professor Timothy Swager: I was only slightly less young and naïve when I started in your research group in the fall of 2008. Thank you for your support during my time at MIT and, even more so, for your professional support, encouragement, and advice since I left the group. I am deeply privileged to be part of such a successful group of Swager alumni and am thankful to continue to have that distinction throughout my professional career.

To my students: It is a deeply unfair fact of academia that the people who do all the hard work receive less of the credit and that I, sitting in my office, end up with most of it. I am so grateful for everything you have done over the past 6.5 years, for your hard work and deep dedication to science and to outreach and to the profession. Thank you for putting up with me as I continue to learn how to be an effective mentor, leader, and communicator. I would have nothing without you.

To my community of female chemists: I am deeply privileged to be part of such a highly accomplished, professional, and inspirational group of female chemists, most of whom are people I met through the American Chemical Society. I am thrilled to call you my colleagues, my inspirations, and my friends. Thank you for being on this journey with me.

I am devastated by the recent passing of my PhD advisor, Professor Ronald Breslow, in November of 2017. He had tremendous excitement for science, boundless energy and enthusiasm, and a relentless drive to always know more. The world is a less bright place without him in it, and I am grateful that I was able to benefit from his guidance and mentorship for as long as I did.

About the Author

Education and Professional Career

2003	BA Chemistry, Columbia University, New York, NY
2005	MA Chemistry, Columbia University, New York, NY
2008	PhD Chemistry, Columbia University, New York, NY

2008–2010	NIH Postdoctoral Fellow, Massachusetts Institute of Technology, Cambridge, MA
2010–2016	Assistant Professor of Chemistry, University of Rhode Island, Kingston, RI
2016–present	Associate Professor of Chemistry with Tenure, University of Rhode Island, Kingston, RI

Honors and Awards (Selected)

2016	Stanley C. Israel Award for Increasing Diversity in the Chemical Sciences, American Chemical Society
2016	Outstanding Graduate Student Mentor, University of Rhode Island
2016	Keynote Speaker, ACS Women Chemists Committee Regional Meeting
2016	ACS Women Chemists Committee Rising Star Award
2016	Northeast Section Younger Chemist Crossing Borders Awardee
2014	University of Rhode Island Early Career Research Excellence Award
2014	Featured Academic Young Investigator, Organic Division, ACS
2008	NIH Postdoctoral Research Fellowship
2008	Pegram Award for Excellence in Research
2007	ACS Women Chemists Committee/Eli Lilly Travel Award

Mindy has a deep and self-serving interest in recruiting more women to be interested in the STEM disciplines and keeping them interested in STEM so that they eventually pursue graduate studies and STEM-related careers. Her female-specific science outreach includes Chemistry Camp for Middle School Girls, Sugar Science Day for High School Girls, and polymer workshops for the GRRL Tech Initiative (Girls Reaching Remarkable Levels). She is also significantly involved with the Northeastern Section of the American Chemical Society as the 2017 Chair-Elect, believing that we need a robust professional society and that is in everyone's interest to support the society to whatever extent possible.

The Window of Opportunity

Nancy E. Levinger

Photo Credit: William A. Cotton, Colorado State University Photography

As a beginning graduate student, well before I met the man I would eventually marry, I recall considering my future. I was training to become a scientist, but would I find a partner? Would my career path allow me to have a family? Although these basic personal choices had always seemed inevitable to me as a child, in the frenetic schedule of a chemical physics graduate student, they were anything but given. At that point, studies had yet to appear showing the impact of advanced education on women's personal lives [1]. Still, it seemed clear; the likelihood of finding a partner while spending almost all my waking hours working on science was probably pretty small.

N. E. Levinger (✉)
Department of Chemistry, Colorado State University, Fort Collins, CO, USA
e-mail: Nancy.Levinger@colostate.edu

© Springer International Publishing AG, part of Springer Nature 2018
K. Woznack et al. (eds.), *Mom the Chemistry Professor*,
https://doi.org/10.1007/978-3-319-78972-9_20

After many musings, I made an active decision: the rich career afforded by my advanced degree would fulfill me whether I married or not or had a family. This decision played a role in my success as a graduate student. It allowed me to focus my energy on science and leave my personal life to chance. I thrived in graduate school, both academically and personally. About two years later, I met Pete, the man who would become my best friend, my husband, the father of my children, and the person who made it possible for me to succeed in my career as a professor and as a mother.

So what does success look like for Prof. Mother? What kinds of obstacles existed for me as a graduate student in the 1980s, an assistant professor in the 1990s, and now as a full professor [2, 3]? What extra challenges arose because of my personal choice to raise a family? Although I cannot claim to have universal answers to these questions, I believe that my life provides insight into some things that make it possible for women to thrive simultaneously as academicians and mothers and some of the significant issues that still remain. The individual path for each woman will vary, but some well-considered choices can increase the chances of reaching satisfaction in both career and personal life.

Early Career Decisions

I believe that the decision most critical to successfully balancing career and family rests on the choice of partner. An academic career places tremendous demand on individuals regardless of their personal choices. A supportive partner understands career demands. Children place tremendous demands on parents. A supportive partner understands and happily steps up to share the demands that come with raising children.

When Pete and I decided to marry, we discussed whether we wanted to have a family. Both of us hoped that our union would include kids. Our wedding vows included a line "I take you. . .to be the mother/father of my children" (which overjoyed my own father who until that point was not sure we wanted to have our own children).

When we would actually fit this into our research-full lives remained a mystery. Finishing my PhD and moving on to a very demanding postdoctoral fellowship, I wondered if we would find a way to start a family. Already in my thirties, we knew that risks associated with pregnancy would only increase and that our energy to keep up with kids would wane as we grew older.

When my search for an academic job loomed in the summer of 1991, Pete and I decided that we had a window of opportunity. We would stop using birth control for four months from June to September; if I became pregnant, we would have a baby in between my postdoc and professor jobs. If not, we would wait. Three months passed without a positive pregnancy test, but in September 1991, the last month in our window of opportunity, I became pregnant. We were elated and nervous. This meant that my academic job search would include balancing preparation and, hopefully, interviews while pregnant.

The Window of Opportunity 281

In 1991, just being a woman seeking a faculty position at a research-intensive institution placed me in a small minority. Adding pregnancy to that mix was not something I wanted to broadcast. So, except for a very small number of close friends and family, I worked to hide my pregnancy. I applied to jobs posted at 21 different institutions and was thrilled to receive invitations for several interviews. During January 1992, which was also the beginning of the second trimester of my pregnancy, I was interviewed at three different institutions. Though I had already gained weight, I believed that the pregnancy was not yet really showing. However, that did not deter a potential colleague at one of the interviews from stopping dinner conversation to ask me (in front of four other potential colleagues), "So, Nancy, it looks like you have gained some weight since we last met. Do you want to tell us why?" I was mortified and answered, "No." Years later the professors present at that fateful dinner told me that they knew I was pregnant and it did not matter. Whether it mattered or not, we'll never know. Nonetheless, I feared that confirming the pregnancy might jeopardize my possibility to receive an offer from that institution or anywhere else. When I finally revealed my pregnancy to my postdoctoral advisor, he was shocked but supportive.

In the end, I was thrilled to receive an offer to join the faculty at Colorado State University. I might have received other offers, but, having grown up in Fort Collins, CO, and with my parents and sibling still living there, the opportunity was a dream come true.

Incidentally after my interview, my soon-to-be Colorado State University colleagues also speculated that I was pregnant, a fact I did not confirm until I had accepted the job offered to me. Much later in my career, I learned that at least one of my colleagues openly questioned hiring a pregnant woman and made disparaging comments about my ability to succeed in the job suggesting that a pregnant woman could not possibly be serious about science. Lucky for me, I never knew of his views until after he retired and I had reached the rank of professor.

My first child, Ian, was born early in July 1992. Six weeks later, I started as a brand new assistant professor of chemistry at Colorado State University. Four years and two miscarriages later, my second son, Eric, was born in late July 1996. Both births and all pregnancies occurred prior to my earning tenure.

In 1995, Colorado State University began considering options for probationary faculty to extend the tenure clock for personal reasons such as childbirth. With a newly passed policy [4], I may have been the first faculty member at CSU to extend the tenure clock on the basis of childbirth. Exercising the option to delay my tenure decision helped to reduce my stress and gave me time to fill gaps in my academic vita. But the extra year of uncertainty about my ability to meet the expectations and earn tenure also added to personal stress. Through three pregnancies (two miscarriages and one viable), I had spent ~15 months pregnant, and the balancing act of work, managing exhaustion during pregnancy, and caring for infants left my curriculum vitae with obvious gaps. From frank discussions with more senior colleagues, I know that some questioned my productivity (or lack thereof) during the time of pregnancy and infant care. Although data suggest that extending the tenure clock can lead to salary inequities [5], the data also show positive impact on promotions. Evidence now also suggests that stopping the tenure clock can be more advantageous for men than women and can negatively affect women's ability to achieve tenure [6]. I continue to be grateful for the opportunity to postpone the tenure decision, an opportunity that is now nearly universal at academic institutions for women and men alike.

Not knowing the pressures presented by the simultaneous start of my independent academic career and family was probably good for me. Had I understood the challenges that both would place on me, I might never have chosen to have children. As it was, Pete and I managed. With the help of two spectacular nannies, I learned to balance demands of a career at a research-intensive university with the needs of infants, toddlers, and young children (Fig. 1). Initially, my husband worked in Boulder then in Denver, more than one-hour' drive away. So child emergencies were entirely my dominion. And there were plenty of them—the nanny calling in sick at 7 am when I had to teach my class at 9 am (Grandma to the rescue!), the fall that 4-month-old Ian took yielding a goose-egg-sized lump on his forehead and a skull X-ray (he was fine), and more. A few months before Eric was born, my husband achieved his career goal of gaining employment at the Hewlett-Packard site in Fort Collins. Trading his hour plus commute by car with a 10 minute by car, or 25 minute by bicycle, commute drastically improved our quality of life. He was much less tired and so much happier, which made him more able to support my frenetic schedule (Fig. 2).

Academic Success

When I entered graduate school after college, I was pretty sure that I wanted to pursue an academic career. The balance of research and teaching seemed like something I would really enjoy. My experiences as a teaching assistant assured me

Fig. 1 Nancy with her sons Eric (left) and Ian (right) on sabbatical in 2000

Fig. 2 Nancy's sons Ian (left) and Eric (right) in 2017

that I liked teaching; research, although demanding, was also stimulating and exciting. Shortly before earning the PhD, I discussed careers with my graduate advisor. When I expressed interest in an academic career, he probed me, asking what kind of academic job I felt would be best. At that point, I could not imagine what kind of research I could do at a primarily undergraduate institution, so I aimed for a position at a research-oriented university. In hindsight, this seems like a rather random way to pursue this demanding career. However, I feel that my career fits my talents and interests well; in hindsight, I would not change my choice.

During my postdoc, Dr. Tom Blackburn, then an ACS Petroleum Research Fund (PRF) program officer, led a workshop that provided critical guidance and gave me confidence that I could succeed in as a professor at a research-intensive university. Tom, assisted by University of Minnesota Prof. Larry Miller, led a group of chemistry postdocs and graduate students through a short exercise in which we came up with ideas for research proposals. Even though I had garnered a prestigious National Science Foundation (NSF) Postdoctoral Fellowship, I doubted my ability to generate fundable research ideas. This short exercise at a critical juncture in my life demonstrated to me that I had lots of fundable ideas. The confidence boost was enough to encourage me to apply for faculty jobs at research-intensive institutions. You might expect that gaining this confidence would be enough to allay my self-doubt from then on, but lack of confidence continues to follow me throughout my career. Despite many career accolades and successes, I still suffer from the imposter syndrome [7], feeling as if I have less talent than I actually have.

As an assistant professor, I applied broadly to granting agencies for funding. My applications yielded early fruit, netting funding from ACS PRF and then a prestigious NSF Young Investigator Award. I also teamed up with colleagues to write two instrumentation proposals, both of which received funding.

Although money came rather easily, papers were much more challenging. When my first full paper was rejected for publication, I did not know what to do. I had never experienced this as a graduate student or postdoc. Those results remain unpublished to this day because I did not realize that I could simply revise and resubmit the paper.

Earning tenure and promotion was probably the most gratifying result to a most terrifying process of my life. The period immediate before the tenure decision was particularly stressful for my family. As a young, assistant professor and parent, I was told by the chair of our promotion and tenure committee, "Nancy, your priorities are messed up. You need to place research way above everything else, significantly above your familial obligations and way above teaching and service." I responded that family and research could be equal, but I could not place research above my familial obligations. I hope that no junior faculty member of mine will receive the same advice. Still, I got the message and worked around the clock to meet my colleagues' expectations.

As the tenure decision neared, I believed that I needed to "spread the gospel of Nancy" so that potential tenure letter writers would know my work. I solicited and received many invitations to give seminars about my work at other institutions. In 1998, I visited 18 different institutions presenting talks at each, far too much travel for my family. Several multiday trips to speak at neighboring institutions kept me away from home for days. On one such trip, my younger son got sick and my husband spent most of the night up with him. To this day my husband reminds me of that night. Thus were born the "mama rules": (1) professional trips were limited to five consecutive nights away from home; (2) each professional trip required me to be home two weeks before the next could begin. After earning tenure, I attempted to honor these rules. As the children grew up, the rules relaxed to a certain degree. However, when possible, I still attempt to follow these rules as they demonstrate my commitment to family needs.

Five years after my promotion to associate professor, I requested a promotion to full professor and began preparing supporting materials. Then, the chair of the department's standing committee on promotion and tenure (SCPT) summoned me to meet with him. He told me that the committee had met and there was some sentiment that I ought to postpone the decision. Suspecting that one member of the committee opposed my promotion, I asked the SCPT chair if this meant I had to postpone the promotion. When he answered "No", I told him that I would go forward with the promotion. After all, what was the worst that could happen? That fall, the department voted unanimously in favor of my promotion. About a month later, I encountered the person on the SCPT whom I believe wanted to block my promotion. He greeted me with congratulations; more than a month had passed since the department vote, so his congratulations confused me. It was only when he mentioned how positive and effusive the external letters had described my contributions that I realized the congratulations were for my promotion.

Now, 25 years after embarking on this journey, I am proud of my accomplishments. Beyond my promotions, I have also earned several accolades at Colorado State University including being named one of 12 University Distinguished Teaching Scholars, one of two highest university honors bestowed on faculty. I have been elected to leadership positions in the American Physical Society (APS), American Chemical Society (ACS), American Association for the Advancement of Science

The Window of Opportunity   285

(AAAS), and as chair of a Gordon Research Conference. I earned fellowship in APS, ACS, and AAAS.

I try to use my own standing in these societies to further the careers of deserving colleagues, especially other women. One of the most gratifying honors came early in 2015. My former undergraduate researcher Dr. Kyle Kung, the first undergraduate to work in my lab, contacted me to tell me that he wanted to endow a scholarship in my name. Working together, we designed a fund that supports undergraduate students in research, which is a passion of mine. Sometimes I feel awkward to have the Nancy E. Levinger Undergraduate Research Fellowship named for me. But it is wonderful to be the source of funding for these talented students, to know that my role in Kyle's life translates into opportunity for them and to get to meet them!

Recognizing the challenges that I have faced, I work to connect with junior faculty as a mentor. I have served as a formal mentor that my department assigns to junior colleagues. In addition, I have volunteered as a mentor for junior faculty through the College of Natural Sciences Women in Natural Sciences (WINS) group. WINS pairs senior women faculty as mentors for junior women faculty across college departments. Thus, I have mentored junior colleagues from different departments. More often I end up in mentoring relationships because I notice the need. Through mentoring, I hope to help others avoid the mistakes I made.

Role Models, Encouragement, and Discouragement

Without doubt, my father had the most significant influence on my early scientific interest and success. I remember the excitement and fascination I had when my third grade class had a unit studying astronomy. Realizing my interest, my father, a mathematician, encouraged me to explore much further than my third grade class. He pulled a book of star maps from the shelf, and he and I poured over them to figure out what we could see in the dark night sky. Together we marveled at science in the first episodes of NOVA that began airing on public television around 1973. I was particularly enamored by the episode entitled "The First Signs of Washoe", reporting about a chimpanzee that learned sign language.

By the time I started junior high school, I sought and received the opportunity to take science instead of the required home economics course. I also participated in a program entitled, SCIP (Science Careers Investigation Program) that took girls and underrepresented minority students out of school for field trips to encounter science firsthand. My favorite class in the ninth grade was honors biology, from which I still remember a lot of information. In high school, I greatly enjoyed and excelled in my math and science classes. When it came time to apply to college, I knew I would pursue a science career.

Over my years in college, graduate school, and postdoctoral research, many people mentored me. My late CSU colleague, Branka Ladanyi, gave me my first opportunity to try research in the summer following my sophomore year in college. Although the Monte Carlo simulations I worked on that summer did not yield the results we hoped, Branka included me as a co-author on a paper she published, which undoubtedly impacted my applications to graduate school. Returning to college, I approached Rick Van Duyne who graciously offered me the opportunity to explore research in his lab. Though not very involved in my day-to-day activities, my graduate advisor Carl Lineberger's door was always open for discussion about science and careers. My postdoc advisor, the late Paul Barbara, also supported my career. Each of these official advisors provided a wonderful foundation for me, but I also found it useful to find a mentor who was not vested in my success, someone who could give me advice but for whom the advice had no impact on their own careers. Early in my career, Mark Ratner amply filled this role for me. As an undergraduate student, a graduate student, a postdoc, and then as a faculty member, I knew that I could always contact Mark for sage, unbiased advice.

As society started to accept women's abilities, most women of my generation, born toward the end of the "baby boom", did not encounter the enormous overt barriers to pursuing science that earlier generations faced [2, 8]. Even though I had wonderful academic mentors throughout my career, I did not initially have any female role models. In retrospect, I realize that role models existed but I did not connect with them. My rejection of female role models puzzles me now but fits a well-documented pattern. Raised in the same society, women are just as likely as men to demonstrate implicit bias toward women in male-dominated fields and roles [2, 9–11].

By now in 2017, overt gender discrimination is much less common than it was in previous generations, but sadly it is not completely gone. As an assistant professor, I received a prestigious NSF Young Investigator Award (the predecessor to NSF CAREER). A male colleague of mine, two years ahead of me on the tenure track, had applied for the same award but did not receive it. Rather than congratulate me, he told me that I had only received the award because I was a woman. Needless to say, his comment fed my insecurity making me doubt whether I really deserved the award. Although this overt sexism should not exist, 20 years later, a young female colleague of mine suffered the same response from our young male colleague when she received a prestigious award.

Unfortunately, implicit gender bias is still alive and well [11]. We all have biases. We use many of them automatically to make the decisions we constantly face in life [12]. The problem arises when bias limits our ability to pursue or achieve our goals. My challenge now is to recognize barriers when they arise. This may sound odd— we should recognize a barrier in the way of our progress. But often the discrimination can be difficult to identify. This is particularly true for the standards to which women are held compared to men [13, 14].

In my current position, I continue to work for gender equity. In the past few years, I have helped to initiate exploration in to gender bias at Colorado State University. Meeting regularly around my kitchen table informally with a group of full professor women faculty, many of whom have held administrative positions (e.g., department chair, associate dean), we precipitated a listening event with the university president, provost, and vice provost for faculty affairs (all men). This event led to an in-depth faculty salary survey that revealed significant salary gaps for women full professors

The Window of Opportunity 287

and to the creation of the Standing Committee on the Status of Women Faculty (SCSWF) [15]. Recently, the SCSWF released results from an extensive campus survey about the climate for women on campus [16]. Although the survey's methods revealed significant problems on our campus, it also provides recommendations to mitigate those problems. I feel proud that the safe, intimate setting of my home set in motion changes that will improve the climate for women at CSU and beyond. I am also indebted to my own family—all male—for supporting this group of strong women when we met over the course of nearly a year to set this ball rolling.

Impact of Career on Family and Family on Career

We usually think of family influencing career, but career can also influence family. Sometimes it is hard to figure out which way the arrow points, family $\rightarrow$ career or family $\leftarrow$ career? Surely those temporary issues like dealing with a sick child or being on time to pick kids up from day care fall under the category of family impacting career. Attending meetings, conferences, workshops, etc. in distant places or deadlines for work obligations fall under the category of career impacting family. But most of the time, a basic chemistry concept, that of thermodynamic equilibrium, describes the interaction precisely. We seek balance in equilibrium, *Family* $\rightleftharpoons$ *Career*.

So how have I found this elusive equilibrium? For the first 8 years of our children's lives, my husband and I chose to hire a full-time nanny, the most expensive, but also the most convenient, childcare. Initially, my husband's entire take-home pay barely covered the nanny's salary and the mortgage payments for our house, but it was worth every penny that we spent. We were blessed to have two exceptional nannies, Randi and Donna, for all but 4 months of those 8 years; in between these two, we had another nanny who did not work well with us. Our nannies did not live with us. They arrived at our home at around 7 am and stayed until 6 pm. They did so much more than tend to the children—all the laundry, much of the shopping, some food preparation—in addition to providing an attention-rich and loving environment for the boys. When both children were in school full time, we hired a series of wonderful after-school babysitters who picked the kids up from school and cared for them until we got home. Each had her own style and the kids loved them all. We continue to maintain contact with both our wonderful nannies and most of the after-school sitters for years after they stopped working for us. Our attention to our nannies' needs made it possible for them to stay with us for years. Indeed, we have continued to help both of them in times of need and they have returned the favors for us. Most of the weddings our kids have attended were of their former babysitters!

One way that my career has influenced our children is through their exposure to and reliance on lots of people other than their parents. Ian developed lasting ties to Randi, who cared for him from age 6 weeks to more than 4 years. Eric developed significant ties to Donna, who began caring for the children when he was 8 months old and only stopped caring for them when we left for my first sabbatical leave, a few weeks before Eric's fourth birthday. Both nannies worked as a team with my husband and me. They echoed our values; they read to our sons, engaged in enriching activities like swimming lessons, horseback riding lessons, visits to parks and playgrounds, regular visits to the public library for books galore, and so much more. Part of who my sons are today comes from their strong relationships with these wonderful nannies and after-school sitters.

When the children were young, I never felt as though I was doing enough, not as a faculty member, not as a mother. I remember telling this to a friend when Ian was about 4 years old. I intimated my concern that I was not spending enough time either at work or at home. Ian piped up and said, "but Mama, you spend lots of time with me." At that point, I knew that even if I did not spend all my time with him like many of my stay-at-home-mom friends, I spent *enough* time with him. I tried to stop berating myself about the amount of time I spent with my family and focused on making that time the best time possible.

My constant interaction with college students definitely affected the way that I treat my children. Working with college students taught me about students' behaviors; some I considered productive and others detrimental. I wanted my sons to treat their teachers as I like to be treated. As a professor, I imagined that students would take responsibility for their actions and I would not hear the "dog ate my homework" stories. I also thought that their parents would not intervene in their college students' academic struggles. Dealing with angry, accusatory parents of college students made me much less likely to intervene and advocate for my kids. I am much more likely to take a back seat and expect my child to solve his problem himself. Perhaps this put my kids at a disadvantage during elementary, middle, and high school. However, I believe that my expectation that they find solutions to their own problems has had long-lasting positive impact on my sons, developing them into independent, thoughtful, and engaged individuals. As Ian pursued his bachelor of music degree from CSU, I was able to help him navigate some of the CSU bureaucracy but mostly tried to keep out of his way. Getting to confer his degree when he graduated in 2015 was a real bonus (Fig. 3).

When asked how my career choice has impacted them, my sons responded predictably (Eric provides a few words of his own here). First, they have never experienced life with their mother staying at home. Even if other mothers stayed at home, they report that they do not feel my career choice negatively impacted their lives. Ian noted that with a Prof. Mother, a child is never on vacation. When the boys would pose a question, invariably they would stimulate the Socratic method in their mother, leading to many more questions than answers. Alternatively, a significant estimation would occur, for example, trying to figure out if there is a mole of grains of sand on Earth. Like many children of faculty members, my sons enjoyed interactions with the CSU Chemistry Club. On occasion, the kids would serve as guinea pigs for new outreach activities the club was trying out (Fig. 4).

My career choice enriched my family significantly through sabbatical leaves taken away from home. On three different occasions, the family (including a different cat on each stay) packed into the car and moved to the San Francisco

The Window of Opportunity

Fig. 3 Nancy with her spouse Pete (left) and son Ian (center) at Ian's graduation from Colorado State University in 2015

Fig. 4 Nancy's son Eric helping test a new chemistry club outreach experiment

Fig. 5 Family ski outing (from left to right: Ian, Pete, Nancy, and Eric) in 2016

Bay Area where I worked at Stanford University. Moving to the Bay Area also gave my husband work opportunities he never would have had in Colorado. These sojourns drew the family closer as we explored San Francisco, Half Moon Bay, Pigeon Point, and more. My sons became accustomed to riding public transportation, getting themselves to and from music lessons, and experiencing some less than comfortable San Francisco bus rides with nutty people on board. They learned that you can leave everything you know, arrive in a new place, and make new friends. I believe these experiences color their own lives now, helping them to be more open to diversity and change (Fig. 5).

Advice and Recommendations

Frequently throughout my career, I have entertained the question, "How do you do it? How do you balance career and family?" Early in my career, this question was hard to answer. "You just do it" is probably the best I could manage. At least once early in my academic career, I remember a woman graduate student who was taking the course I taught saying to me, "I don't want to be like you. I want to have time for my family!" This comment felt incredibly depressing. Instead of serving as a role model, I felt like an anti-role model. Indeed, balancing an academic career with spouse and kids has a different meaning for women academics than it does for men [17]. But this comment resonated with me and motivated me to speak about my experiences.

Advice is a dangerous thing to give and to take. Personal and professional situations vary so much! There is no "one size fits all". Still, I believe there are important lessons others can take from my career. Here are a few:

- Your choice of a partner is probably the most important variable you can control. Choose a partner who understands the demands of your academic career and wants to help you to succeed in it. Studies show that women tend to carry more than 50% of domestic responsibilities [18]. Choosing a partner who takes on significant domestic responsibility (cleaning, cooking, shopping, childcare, etc.) makes it possible to balance academic and personal tasks. Choosing a partner who is patient, supportive, and committed is the most important thing you can do to succeed in your position.
- When you are emotionally ready to start a family, stop using birth control. When you get pregnant, you will figure out how to work this into the equation. There are, of course, somewhat better and worse times to start or add to a family. But there is no real "right time"; so waiting to start can lead to problems associated with being pregnant later in life.
- Find out the policies which exist that can help you achieve your professional goals and exercise them. At this point (2017), many, if not most, institutions have family leave and extension to probationary periods (tenure clock) in place. Although stopping the tenure clock can seem dangerous, we must continue to educate colleagues that adding time to the probationary pre-tenure period should not raise expectations for productivity. This understanding seems to be taking root. Plan to stop the tenure clock to give yourself time to develop your relationship with your new baby. Do it and don't worry!
- If you are planning a family or expecting a baby, have a well-devised and comprehensive emergency plan. If your regular childcare falls through, have a backup plan. Whom will you call? Who can help you? Many people will want to help—let them!
- As soon as you have enough money, pay others to do the things you don't like to do or don't have time to do. Time is in short supply when you have a career, let alone a career and a family. It is worth spending money for someone to clean your house, do yard work, cook your meals, or whatever you would prefer not to do.
- Look for role models and mentors. Listen to their advice and work to incorporate it into your life. When you get through the most difficult and time-consuming stages, take on mentoring for others. You will find satisfaction in being able to help others navigate career and family.
- In his book, *The Four Agreements* [19], Don Miguel Ruiz lists four agreements to live by: (1) be impeccable with your word (don't gossip); (2) don't take things personally; (3) don't assume anything; and (4) always do your best. Of these, the first and last are pretty straightforward. Most of us do them without trying. The second agreement refers to praise and criticism, which often reflect issues that other people have with themselves. Not taking these personally allows us to analyze the situation without becoming hurt or glamorized. Likewise, assuming that others understand us can lead to significant problems. Much better to remove doubt about your words, actions, and intentions. These four agreements can be hard to follow, but if you can, your life will be easier.

Finally, find time for yourself. Life is short and you do not know where it will lead. Carpe diem!

> **Reflections from Nancy's Son**
> **Eric Levinger**
>
> The most important thing to know about having a chemistry professor as a mom is that for me, she's just my mom. And that has had a major impact on me, because she was there when I needed her. So what's it like to have a professor as a mom? It's no different than having a mom who has any other job. When you're a kid, they're just your mom; growing up, they're your mom; and after you move out, they're still your mom.
>
> One thing that I remember about from having a professor as a mom is all of her complaining about students. That's right; usually students complain about their professors but I heard the other side. Going into college I had inside information about how professors think about interactions with students. When I went to college, I knew how to talk to my professors and how to cultivate good relationships with them. Growing up with a mom in academia made many of the issues that most college students deal with seem trivial (Fig. 6).

Fig. 6 Nancy and her son Eric on Eric's 20th birthday in Washington, DC

The Window of Opportunity 293

Acknowledgments I am indebted to many people who made it possible for me to balance career and family. First, to my parents, who have believed in me throughout my life. Second, to my lovely sons, who inspire me to be the best I can be. Finally, to my husband, who has supported me, tolerated insane schedules, listened to my problems and helped me to solve them, raised two amazing children, and loved me through thick and thin.

About the Author

Education and Professional Career

1983	BA Integrated Science and Physics, Northwestern University, Evanston, IL
1990	PhD Chemical Physics, University of Colorado, Boulder, CO
1990–1992	NSF Postdoctoral Fellow, Department of Chemistry, University of Minnesota, Minneapolis, MN
1992–1999	Assistant Professor, Department of Chemistry, Colorado State University, Fort Collins, CO
1999–2005	Associate Professor, Department of Chemistry, Colorado State University, Fort Collins, CO
2005–present	Professor, Department of Chemistry, Colorado State University, Fort Collins, CO
2007–present	University Distinguished Teaching Scholar, Colorado State University, Fort Collins, CO
2009–present	Professor of Electrical and Computer Engineering (courtesy appointment), Colorado State University, Fort Collins, CO

Honors and Awards (Selected)

2016	Jack E. Cermak Advising Award, Colorado State University
2015	Nancy E. Levinger Undergraduate Research Fellowship, established
2014	Fellow, American Chemical Society
2010	Fellow, American Association for the Advancement of Science
2005	Fellow, American Physical Society
2004	Margaret Hazaleus Award, Women's Caucus, Colorado State University
1999	Colorado State University Undergraduate Research Mentoring Award
1994–1999	National Science Foundation Young Investigator Award

Nancy has published 90 peer-reviewed (including 4 invited reviews and 18 with 16 unique undergraduate co-authors) and presented at international, national, and regional meetings. She has mentored 10 PhD and 7 master's degree students and mentored over 40 undergraduate students and a high school student in research. She directed the NSF REU program at CSU for 7 years and founded and served on the NSF Chemistry REU Leadership Group, serving as its first chair. Nancy organized the CSU Celebrate Undergraduate Research and Creativity Poster Session for 10 years (campus-wide undergraduate research poster session) and continues to serve on its board. She has received > $5 million in grants as principal investigator and > $2.5 million as a co-principal investigator for multiuser, equipment, and collaborative grants. Nancy has served in leadership and governance of the American Chemical Society PHYS Division; American Physical Society, DCP Executive Committee; American Physical Society Council and Executive Committee; and Telluride Science Research Center Board of Directors.

References

1. Mason MA, Goulden M (2004) Marriage and baby blues: redefining gender equity in the academy. Ann Am Acad Polit S S 596:86–103
2. Valian V (1999) Why so slow? The advancement of women. MIT Press, Cambridge, MA
3. Minerick AR, Washburn MH, Young VL (2009) Mothers on the tenure track: what engineering and technology faculty still confront. Eng Stud 1:217–235
4. Colorado State University2012–2013 Academic Faculty and Administrative Professional Manual, section E.10.4.1.2 Extension of the Probationary Period. http://www.facultycouncil.colostate.edu/files/manual/sectione.htm#E.10.4.1.2. Accessed 7 Jan 2014
5. Manchester CF, Leslie LM, Kramer A (2013) Is the clock still ticking? An evaluation of the consequences of stopping the tenure clock. ILR Rev 66:3–31
6. Antecol H, Bedard K, Stearns J (2016) Equal but inequitable: who benefits from gender-neutral tenure clock stopping policies. IZA Discussion Paper, paper no. 9904
7. Clance PC, Imes SA (1978) The impostor phenomenon in high achieving women - dynamics and therapeutic intervention. Psychother Theor Res Pract 15:241–247
8. Monroe K, Ozyurt S, Wrigley T, Alexander A (2008) Gender equality in academia: bad news from the trenches, and some possible solutions. Persp Polit 6(2):215–233
9. Latu IM, Stewart TL, Myers AC, Lisco CG, Estes SB (2011) What we "say" and what we "think" about female managers: explicit versus implicit associations of women with success. Psychol Women Quart 35:252–266
10. Committee on Science, Engineering, and Public Policy (2007) Beyond bias and barriers: fulfilling the potential of women in academic science and engineering. National Academy of Sciences, National Academy of Engineering, and Institute of Medicine. National Academies Press, Washington, DC
11. Jackson SM, Hillard AL, Schneider TR (2014) Using implicit bias training to improve attitudes toward women in STEM. J Soc Psychol Educ 17:419–438
12. Kahneman D (2011) Thinking, fast and slow, 1st edn. Farrar, Straus and Giroux, New York
13. Kaatz A, Lee YG, Potvien A, Magua W, Filut A, Bhattacharya A, Leatherberry R, Zhu X, Carnes M (2016) Analysis of National Institutes of Health R01 application critiques, impact, and criteria scores: does the sex of the principal investigator make a difference? Acad Med 91:1080–1088
14. Magua W, Zhu X, Bhattacharya A, Filut A, Potvien A, Leatherberry R, Lee YG, Jens M, Malikireddy D, Carnes M, Kaatz A (2017) Are female applicants disadvantaged in National

Institutes of Health peer review? Combining algorithmic text mining and qualitative methods to detect evaluative differences in R01 reviewers' critiques. J Womens Health 26:560–570

15. Colorado State University Standing Committee on the Status of Women Faculty. http://cwge.colostate.edu/standing-committee-on-the-status-of-women-faculty/home/. Accessed 14 June 2017

16. Colorado State University Standing Committee on the Status of Women Faculty (2017) Experiences and perceptions of campus climate for women faculty at CSU. http://cwge.colostate.edu/standing-committee-on-the-status-of-women-faculty/scswf-report/. Accessed 14 June 2017

17. Park B, Smith JA, Correll J (2010) The persistence of implicit behavioral associations for moms and dads. J Exp Soc Pyschol 46:809–815

18. Misra J, Lundquist JH, Templer A (2012) Gender, work time, and care responsibilities among faculty. Sociol Forum 27:300–323

19. Ruiz DM (1997) The four agreements: a practical guide to personal freedom. Amber-Allen, San Rafael, CA

Pieces of a Puzzle

Sherri R. Lovelace-Cameron

Photo Credit: Youngstown State University

Start with the Corners

Who am I? I am a mom, wife, daughter, teacher, and chemist. These roles play a part in making up who I am. At certain times, some roles are more important than others, especially my family. So, I married a man who shared the same family values, as well as the importance of putting church and family first.

S. R. Lovelace-Cameron (✉)
Chemistry Department, Youngstown State University, Youngstown, OH, USA
e-mail: srlovelacecameron@ysu.edu

From the first time, I was introduced to science and math, I loved those courses. They were my best subjects during my school years. Growing up in the inner city of Pittsburgh, I personally did not know any women or African-American chemist. None of my family or neighbors had science or math careers. In elementary school, I constantly asked my parents for a chemistry set, but I never received one. I remember trying to do experiments with my "Easy Bake" oven. That was interesting. I would walk around the house, and whatever I thought would not be missed by my parents would somehow find its way to my room and make its way into the "Easy Bake" oven. Eventually, I asked my parents why I never got the chemistry set, and they simply responded, "we were afraid you would hurt yourself."

I attended The Ellis School for Girls, which was a private college preparatory school. The vast majority of students were Caucasian and from upper middle class and wealthy families. In an effort to diversify the population of The Ellis School, there were a few students from the economic middle and lower class that were awarded a scholarship or sponsorship. There was a local business man (we called him Mr. Cooper) that sponsored a couple of girls from a city track team. By being a sponsorship recipient, my parents could cover the remaining tuition that enabled me to attend this prestigious school. Every morning, I left my all black neighborhood and took a short bus ride on public transportation to my school. None of my classmates had any idea about my neighborhood, and I absolutely had no idea of theirs. Most of my classes had no more than 17 students, so you always had to be prepared to participate. There was no hiding or avoiding being noticed at The Ellis School for Girls. I tried. The Ellis school had virtually a 99% college attendance rate. I was told when I graduated in 1981 that my class of 33 was the largest graduating class.

I'm embarrassed to say that when asked the question of my career choice, my answers did not require an education beyond high school. I thank God for my parents who motivated me to go to college. Education was highly encouraged in the Lovelace household, and my parents explained that you are going to spend so much of your lifetime working; you should spend that time working in an occupation that you enjoy. My parents went on to explain that they didn't have that option. My mom would say "women have three jobs. Two of which you never have time off and they are mom and wife. And the job you have outside the home, you should really enjoy because you will spend a tremendous amount of time on that job." My parents did not want their children to be physical laborers (and none of us were), but I was the only one to attend and graduate from a 4-year college. My mom had some schooling beyond high school and worked briefly as a license practical nurse before moving on to office work. My father tested into an accelerated engineering program while he was in the military during WWII. When the war ended, so did the program, before he completed it, so he worked factory jobs.

I enrolled at the University of Pennsylvania as a chemistry major and never considered switching majors. By my second semester, I had to consider switching schools for financial reasons. During the second semester of my freshman year, I filled out the financial aid papers and received a notice that I was not going to receive any aid my second year. None, zero, goose egg, nothing. I went to my financial aid officer who told me to have my family re-mortgage their house. I left the office knowing I needed a plan B.

I applied to two schools in an effort to transfer at the end of the fall semester. I was accepted to both and decided to stay in Philadelphia where I transferred a few blocks down the street to Drexel University. Drexel is a co-op school, so the Bachelor of Science degree would take 5 years to complete. But for 3 years, I would take classes for two quarters and work in a paid internship for two quarters. I wanted to have as many different internship experiences as possible, so during the summer quarter, I would participate in Inroads Pittsburgh and work at companies at home, and during the fall quarter, I would work a co-op position in the Philadelphia region through Drexel University. My co-op experiences lead to my decision to consider a graduate degree. The money I earned through the co-ops helped with my rent and my other living expenses while I was enrolled in classes. My parents could focus on tuition.

Find All the Edge Pieces

I had never considered graduate school until after a few co-op experiences as an undergraduate. The co-op experiences and the curriculum at Drexel University which required STEM majors to take an economics class and my choice to take courses in finance and business law led me to consider the business path. During my senior year, I decided to apply to a graduate chemistry program and a MBA program.

My co-op experiences and summer jobs included:

- Cancer research with mice. I was around cages of mice all day and at times dreamt about little white mice.
- Chemical laboratory testing of paint and other coatings.
- Preparing blood products from samples collected at blood drives.
- Chemical sales to water treatment plants.
- Marketing with a chemical company.

The sales summer job took the luster off a traveling chemical sales career. No matter what city we were in, we spent long hours at wastewater treatment plants in a desolate part of town. I traveled with a Caucasian male, and some of the comments directed toward us by people who may have thought we were a couple when we went to eat at the end of a long day assured me that I would never want to constantly travel alone to different cities. Also, he shared that once he had been assaulted and robbed on a sales trip.

My cooperative experience helped me decide to go beyond the bachelor degree. I was not sure if I wanted to obtain a MBA or a PhD. On the morning, I was supposed to take the GMAT exam which is required for MBA programs; I started thinking that if I get an MBA degree, I will not have the opportunity to work in the lab, so I skipped the test and decided to only apply to PhD Chemistry and MS Materials Science programs. I returned to my hometown for graduate school at the University of Pittsburgh and moved back home with my parents. Pittsburgh has great public transportation, so I caught the bus to and from school. When I was required to work late nights in the lab, instead of catching the bus, I would drive to campus.

My research advisor, Dr. N. John Cooper, knew of my interest in teaching and suggested I apply for the Department of Education Fellowship. I used the fellowship that year to take two graduate courses in the Department of Education at the University of Pittsburgh. During that time, I spent a few months observing middle and high school classrooms in the City of Pittsburgh School District. It was an awesome experience. I thought it would be great to do K-12 curriculum development, but after talking to several individuals, I realized those jobs were coveted, and without K-12 teaching experience and at least one degree in education, that was unlikely.

My PhD research involved the study of the chemical and electrochemical reduction of organometallic compounds. Since I was part of a synthetic group, my research advisor suggested that I apply for a postdoc position with an electrochemist, Dr. William Geiger, at the University of Vermont. When I contacted Dr. Geiger, he encouraged me to apply for the Citibank Postdoctoral Fellowship. The purpose of the fellowship was to provide an avenue to increase diversity at the University of Vermont. I moved to Vermont in October of 1992, with an 18-month fellowship. In view of the fact that I was interested in academia, and that the fellowship ended in April 1994, Dr. Geiger provided funding for an additional 15 months which allowed me to stay until the end of July 1995. What appealed to me about an academic job is that I could teach, do research and community service. About 4 months after I moved to Burlington VT, I met my husband Daryl, who happened to be from my home town. Talk about small world.

Finishing the Border

When it was time to look for a job, I knew I wanted to go into academia and apply to schools where my main responsibility was teaching undergraduate and graduate courses. The position would require the involvement of undergraduate students in research. After three offers, I accepted a position at Youngstown State University (YSU). The deciding factor was the fact it was 90 minutes from my parent's home in Pittsburgh. The YSU is in the City of Youngstown in an economically depressed region of Northeast Ohio with a predominantly African-American school district. This meant there were opportunities for community service with a population similar to my childhood neighborhood. Although I was involved in a serious relationship, we were not engaged, and if the relationship did not work out, I figured that Youngstown was near three urban centers (Pittsburgh, Cleveland, and Akron) with black professionals. I moved to Youngstown at the beginning of August, but the fall quarter did not begin until late September. So, I went to visit Daryl in Vermont at the end of August, and at that time we got engaged.

Following the lead of my parents, I always volunteered or participated in some form of community service. In college, I joined the public service sorority Delta Sigma Theta Sorority, Inc., and when I moved to Youngstown, I joined the local chapter of the sorority and became involved with the community. A few weeks before the quarter started, I happened to be in my office and received a call from the dean asking would I serve on the Africana Studies Advisory Committee. Well of course I said yes. This would give me the opportunity to work with faculty from a variety of departments, and how do you tell the dean no. Early in my career, my desire and enjoyment of community and university service meant I was asked to do a lot of service. People knew I would be committed. As a result, I became overextended year after year. It took me years to learn that I could use my department chair as a buffer to avoid so many requests for engagements and service activities. I eventually learned to say no but would always suggest someone that had the same desire to serve on committees but was overlooked. After 20 years of seeing service involvement devalued, I now have no problem saying no. Now, I continue to do community service but mainly in partnership with non-university organizations.

In 1995 when I started at YSU, it was an open enrollment institution, which meant it offered admittance to all State of Ohio students if they had completed high school or passed their GED. The department offered courses for graduate students and classes for students who were not prepared for college-level work. Six years ago, a community college has opened approximately four blocks from campus. Also, the State of Ohio formula for funding changed from a focus on enrollment to completion rates, so there was a move to decrease the number of remedial classes offered . In 2007, YSU reorganized and the science departments were moved from the College of Arts and Sciences, to the new college of STEM, which had formerly been known as the College of Engineering. At that point and in my opinion, basic research began to be devalued, and there was a push to connect research projects with local companies. The new description of YSU was that of an institution of urban research.

Filling in the Center

At the end of my first year, I met with the chair of my department to discuss my annual evaluation. When he asked if I had any comments, I told him that when I start my family, I plan to take the full 6 weeks of maternity leave. Earlier that year, a nontenured faculty returned to work a few weeks after her child was born and brought her child to class. The chair assured me that it had been her choice and not a department expectation.

I met my husband during my postdoc in February of 1993, and we were married in August of 1996. I always said that I would wait 3 years after marriage to have a child to make sure there was a good foundation. That being said, I wasn't exactly a young chicken when I had my daughter. That, and the fact that I had a high-risk pregnancy and scary delivery meant as time went on, we decided to be one and done, when it came to family size. I never considered waiting until after tenure to start a family because I was not young and what if I was not granted tenure? When Khala was born in 1999, I recognized that I needed to be more efficient during my work hours on campus. I remember my departmental chair being very understanding when my daughter's school bus pickup time changed, and I asked him not to assign 8:00 am classes for 2 years.

Through a young lady at our church, we found a wonderful woman to provide childcare. And 18 years later, we still consider her a third grandmother to our daughter. An example of our childcare provider's devotion is when my daughter got chicken pox. I never had chicken pox, and she automatically offered to keep my daughter until she was no longer infectious.

In academia, your schedule changes every semester, and childcare can be challenging when you live in a city with no family. I would teach graduate courses in the evening and undergraduate courses mainly during the day. There were times when I did not get my final teaching schedule until a week before classes started on that Monday. My husband is a banker, and at times he would have to travel at least two and half hours away to visit a prospect or client. That combined with the weather uncertainty in northeast Ohio meant there was no guarantee when he would get home. Flexible reliable childcare was important.

My husband and I try never to miss one of Khala's activities, and there has always been at least one of us at every event. You can never get that time back to enjoy your children. Occasionally, now that my daughter is in high school and her sports events start later, or last longer, I may arrive late or must leave early to get back and teach a night class. I love hybrid courses which have face-to-face meetings and an online component. Since I get my daughter's schedule before my syllabus is given to the class, I try to give online assignments for my evening classes which conflict with her events (Fig. 1).

I must admit, I was hoping for an engineer or a scientist, but for as long as I can remember, my daughter's favorite subject has been social studies and history. She participated in many of my hands-on science demonstrations/activities with K-12 students, but when I asked why science wasn't her favorite subject, she always said "because they don't do science in school just worksheets." Through the fifth grade, I had to remind my daughter that she did not have a stay-home mom like many of the other children in her classroom. She had to understand I could not always volunteer to be at school on short notice or bring her something that she forgot at home. By the sixth grade, she had no problem telling teachers, friends, and coaches that both her parents work and cannot always adjust their schedules for drop-off or pickup times that conflict with their work schedule.

Fig. 1 Sherri's daughter Khala Lovelace-Cameron (age 10) celebrating a good outing at a karate tournament

Puzzle Complete (Reflections, Advice, and Lessons Learned)

I do not ever remember an obstacle that I fretted about overcoming. I have had a supportive family and department. Since my mother was retired and father working part-time, when my daughter was young, I never missed a meeting, conference, or professional development workshop, because one or both of my parents would always travel with me to help me take care of my daughter.

At three pivotal points in my life, I wish I had educated myself more before I made my decision even though I do not know if it would have made a difference. The first was choosing an undergraduate major. I liked chemistry and math but I never investigated what math majors do for careers. I never saw a job opening for a mathematician, but I knew chemist worked in laboratory or sales positions. Since I thought I knew what chemist did, I decided to major in chemistry. I am not sure it would have made a difference if I had known the careers possible for math majors. I may have still chosen to major in chemistry as an undergraduate.

When I reflect on my time in graduate school, what I would do differently is be more focused at the start of graduate school. Not having done undergraduate research, during my first 2 years of graduate school, I had to remind myself that me obtaining my PhD degree was not only based on the grades I earned but also how my research project was progressing. I would recommend that new graduate students learn from the seasoned group members. Don't wait for your research advisor to tell someone to show you a synthesis or assign the postdoc to work with you. It may never happen. If you are not the only one joining the group at that time, you are not the only one who needs attention. I would suggest asking one of the group members if they mind you shadow them and set up parallel experiments.

A third pivotal time was when I started the tenure track position at YSU. It took me years to learn that I could use my chair as a buffer to avoid so many requests on my time to engage in service activities. I had university and community activities I desired to embrace, and as soon as I did, someone would ask me to become involved with some other committee or activity. I can remember every year my annual evaluation said something like "Sherri should be careful not to become too involved in committee work/service which places too many demands on her time."

My advice for those going into academia is the advice someone gave me when I started working on my tenure packet, learn to market yourself. You are presenting your tenure and promotion application packages. Putting together a great package can help determine how your face-to-face meeting will go with the committee. In my department, there were three areas of evaluation which are teaching, scholarship, and service. By far teaching and scholarship were the most important in my department. Since the college committee also had a vote, I had to consider that individuals on the college promotion committee may have teaching and research in a different order of importance. I had a fourth section in my packet "cross categorical" which allowed me to show how my service activities crossed the boundaries of teaching or research. I believe this was an important section within my packet, because some of my service activities lead to grant collaborations with K-12 teachers and YSU education faculty, which resulted in publications and invited presentations. In addition, I incorporated a service learning assignment into a YSU science class, based on K-12 classroom observations I made and partnerships I developed with K-12 teachers.

When you start at your institution, learn the culture of your institution and the requirements expected by your department to achieve tenure. The University may have a bullet list but check your department's governance document to see what they emphasize. Post it at your desk as a constant reminder. Why, to make sure that the activities that you do move you toward the goal. To know what your department thinks is important since that group has the initial vote on whether you get tenure.

Mentoring can be important in choosing a career. I believe it's important for youth to see mentors that look like them. It can inspire them to see themselves in a chosen career path. I have always considered myself a positive role model for women and African-Americans. It makes me feel good when I know that I help to educate all students on campus and that they should not be surprised to have a woman or African-American in a decision-making/supervisory position when they become employed.

Fig. 2 Sherri (right) with her daughter Khala (left) and spouse Daryl Cameron, coming home from church

Reflections from Sherri's Daughter
Khala Cameron

Having a mom as a chemistry professor is awesome. When I was younger, one of my favorite experiments was making silly putty. I wasn't the only one who enjoyed those crazy experiments; my classmates loved when she came to our classroom to show off her talents. I can remember my fifth-grade science teacher asking me "Does your mom do your homework for you?" Of course, the answer was no, but I understood why she asked. One of the great things about having a parent in academia is that you have a "live-in" teacher at home. I am proud of my mom for all the accomplishments and goals she has achieved in her career (Fig. 2).

Acknowledgments I would like to thank the staff of Inroads Pittsburgh who provided a vehicle for internship opportunities that help me decide to attend graduate school, Professors N. John Cooper (University of Pittsburgh) and William E. Geiger (University of Vermont) who were my graduate school and postdoctoral research advisors, and finally my parents, siblings, and extended family who have regularly lifted me in prayer.

About the Author

Education and Professional Career

1981–1982	Chemistry student, University of Pennsylvania, PA
1986	BS Chemistry, Drexel University, PA
1992	PhD Inorganic Chemistry, University of Pittsburgh, PA

1992–1995	Postdoctoral Research Associate, Department of Chemistry, University of Vermont, Burlington, VT
1995–2002	Assistant Professor, Youngstown State University, Youngstown, OH
2002–2012	Associate Professor, Youngstown State University, Youngstown, OH
2012–present	Professor, Youngstown State University, Youngstown, OH

Honors and Awards (Selected)

2011	Distinguished Professor Service
2008	Ohio Student Achievement in Research and Scholarship Mentor of the Year
2002	Edna K. McDonald Cultural Awareness Award
2002	Libra Award, outstanding student organization advisor
2001	Distinguished Professor Community Service
2000	Master Teacher

Sherri is an active member in the American Chemical Society Penn-Ohio Border Section and the local American Association of University Women. She also serves as the faculty advisor for the student organization Youngstown State University Chapter of the National Society of Black Engineers.

Wanting It All

Cecilia H. Marzabadi

My Story

Growing up in the sixties and seventies when so many changes were happening in American society made me believe that my dreams were possible and that being a woman was no longer a barrier to the career that I wanted. I could be married with children and have a successful career. However, along with "wanting it all" is the reality that you have to prioritize and learn to balance what is most important to you in life. This is my story of how I have managed to wear the many hats that I do: as a

---

C. H. Marzabadi (✉)
Department of Chemistry & Biochemistry, Seton Hall University, South Orange, NJ, USA
e-mail: cecilia.marzabadi@shu.edu

chemistry professor at a midsized, private Catholic university, as a research mentor, as a wife, daughter, sister, friend, and as a mother.

As a child, I don't think I even knew any scientists. My mom was a commercial artist who drew department store ads for the newspaper. My dad was the owner of an industrial laundry. I am the only child from a second marriage; I have two half-brothers and two half-sisters. Growing up I had an interest in everything, including science. I loved reading and learning.

My parents divorced when I was five. I chose to live with my mom and stayed with my dad every weekend. The divorce was very disruptive to my childhood; my mother moved multiple times and every time I had to change schools. By the time I was a teenager, this instability led to behavioral issues. I started missing school and hanging out with friends who were older and not very motivated. I dropped out of high school.

Fortunately, I got help, and through this, education saved me. I passed my GED exam when I was 16 and started taking classes at a community college the same year. Initially, I had no idea what I wanted to do. I took an aptitude test that told me I would be best suited for a job both social- and science-based. The top two likely careers were a science professor or a physician. Initially, both involved taking similar courses.

After 2 years, I transferred to Saint Louis University. I took both biology and chemistry courses and, after about a year, decided that being a chemistry major was more suited to me. Biology seemed like too much memorization, whereas chemistry involved more problem solving. Organic chemistry in particular appealed to me; it wasn't as mathematical and had a large mechanistic component that I enjoyed a lot.

As an undergraduate student, I began doing research in organometallic chemistry. This enabled me to get to know the professors and other undergraduate and graduate chemistry students. It also gave me the feel of what it was like working on a research project in a lab; I loved it! During this time, I met my future husband, Mohammad, who was my teaching assistant in first semester organic lab. In my senior year, we began dating. Our relationship would significantly alter the subsequent career paths for both of us. He decided to stay local and pursue his PhD degree at Washington University. I decided to stay at Saint Louis University and to get my research MS degree in organic chemistry. I had the hope that by the time I finished my MS degree, he would be finished with his dissertation work and we could plan our next moves together, including my further pursuit of a PhD. Well, as they say, "the best laid plans." Unfortunately, research doesn't always follow a time clock, so I finished my studies before he did and looked for a job in the St. Louis area.

I was hired by Monsanto Agricultural Products Company to work in their metabolism group. Although my background was in synthesis, I readily learned how to do the analytical work required in metabolism. It was a very interesting job that taught me many new skills. In spite of this, I knew that eventually I would be limited in advancement in industry without a PhD degree. After 2 years at Monsanto, I left to pursue my doctoral degree. My husband was wrapping up things at Washington University. I had applied to and been accepted at several top chemistry departments. I had decided I would likely begin my studies at the University of

Illinois–Champaign Urbana. There was only one problem...what about Mohammad...what if he could not get a postdoc or a job in Champaign? In the end, I decided not to go. However, this was also influenced by another unfortunate event...the death of my father.

So my return to graduate school would ultimately be delayed three additional years as we cleaned out and sold my family home of almost three decades. Mohammad and I subsequently married, and after 2 years of marriage, our son John was born. I restarted my graduate studies at the University of Missouri–St. Louis when our son was 9 months old. I was 29.

Initially, I was very apprehensive about returning to school. I was not sure if I could keep up with my peers. I was concerned that I might have forgotten many things that I had learned both in my bachelor and masters programs. I was my doctoral advisor's second graduate student. He was a new, untenured assistant professor with a lot of enthusiasm for organic chemistry.

We hired a retired woman to come to our house and stay with John while I was at school and my husband was at work. Although it took most of my teaching assistant salary from the university, my son was well cared for, and we had greater flexibility than if we had to take him to day care. I had to get my work done within a normal 40-hour week. I learned to utilize my time at school effectively. Most of my studying was done at night after everyone else went to sleep, as I would spend the evenings with my family. Cumulative exams were held on Saturday mornings once a month. On the Fridays before, I would pull an all-nighter studying for my exams. Things worked out just fine, in spite of losing a little sleep. I excelled in the program, achieving straight A's in my course work and being the first student of my entering class to pass my cumes.

If I had to go to school on the weekend, I would try to save that time for running NMRs, so my son could come with me and draw his own NMRs with the colored pens in the NMR room. Otherwise, my husband would watch John. Unfortunately, we did not have family members locally who could help us out with babysitting.

At my graduate school, I was surprised that there was only one female faculty member in my doctoral department (out of 21 faculty). After all, it seemed like there were a fair number of female graduate and undergraduate students. In my entering graduate class, at least a third of us were women. The other graduate students said this female faculty member was very tough and that she did not like it when female graduate students were too passive. Eventually I had this professor for a course and got to know her by going to her office during office hours for questions about the course material. She wasn't as scary as people had made her out to be; in fact, I quite enjoyed talking to her. When I think back on things, I think she was just trying to prepare us for what lay ahead. I continued to interact with this female professor throughout my time in graduate school. She learned about my interest in a future career in academe and would often put articles in my mailbox about the glass ceiling for women in academic chemistry. I was a bit hurt at the time, when she told me I might do better in industry than in academe. Now I understand what she meant; as I am now, even in this day and time, the only female in my department (Fig. 1).

Fig. 1 Cecilia with her husband Mohammad, son John, and PhD advisor Chris Spilling at her PhD graduation from the University of Missouri-St. Louis, May 1995

In the meantime, my husband had done two postdocs in the area. He was offered a "real job" at a small start-up pharmaceutical company in NJ during my fourth year of graduate school. He moved to NJ and my son and I stayed in St. Louis. In my fifth year, when I was writing my dissertation, my son moved to NJ to live with my husband. It was difficult being apart from both of "my boys" for that year, but it forced me to focus on finishing my graduate studies.

The New York City area proved to offer a variety of postdoc positions for which I applied. I defended my dissertation and moved to New Jersey over Labor Day weekend. I started working the Tuesday after Labor Day at Hunter College–City University of New York (CUNY). My postdoc involved a commute into Manhattan every day.

I would catch the bus at 7 am every day and often would not get home until 9 or 10 pm. My husband's work, on the other hand, was only about 15 minutes away from home. My son attended kindergarten and elementary school near home, so my husband would pick him up every day from after-school care. I would do my best to make it home for school concerts, back to school nights, etc. My husband really held down most of the childcare responsibilities during this time; his help and support was critical.

I started looking for academic positions after about 2 years in my postdoc. Again, I was geographically limited in my choices, as I did not want to live apart from my family. I would apply for any openings I saw in New York and New Jersey. Sometimes I would even apply for positions in Connecticut and Pennsylvania, knowing that I would have to commute. I applied to all types of academic jobs, including those at predominately undergraduate institutions, though I knew I would not be happy at these types of schools. I had research in my blood, and having only part-time access to a fume hood in the physical chemistry teaching lab was not going to give me much opportunity for that. I wound up staying in my postdoc much longer than I had planned—almost 5 years. Now the question was . . . was I damaged goods, because of my age, for starting as an assistant professor in a doctoral program?

In December of 1998, I interviewed at Seton Hall University in South Orange, NJ. It is a private, Catholic university with both undergraduate and graduate programs (MS and PhD). The Catholic environment was very reminiscent of my undergraduate and MS days at Saint Louis University. However, I especially liked the fact that I would be able to mentor doctoral students. What I was not cognizant of, at the time, were the difficulties associated with doing graduate research in a small PhD-granting department, such as lack of resources and collaborators and difficulty getting grant funding.

In addition, the start-up money for setting up my lab was not great. However, Seton Hall is also one of the schools designated to receive money from the Clare Boothe Luce Program for promoting women in the sciences. I was hired and received an appointment as a Clare Boothe Luce Assistant Professor of Chemistry. The Henry Luce Foundation paid my salary for the first 5 years I was at the university and paid an additional 20% of my salary as a stipend that could be used in any way I deemed necessary (including for childcare). I used the money to help pay for lab supplies when I was first starting out. I am forever grateful to the Luce Foundation for this support; it really made a difference!

My new department had 13 other faculty members (12 men and 1 woman). The other female faculty member had serious health problems and died 2 years after I was hired. I was unfortunately unable to go to her for help and advice. Some of the senior male faculty in my department tried to help me along, though this was very informal and sometimes, quite frankly, wasn't even obvious. There was a Women's Faculty Association at the university, and I reached out to them for a female mentor. The mentor assigned to me was from the math department, which was an exclusively undergraduate department with different departmental requirements. I think we spoke briefly on only a couple of instances. I also became very active on different committees and on the Faculty Senate. This enabled me to meet women and other faculty in other departments.

I attended local Women Chemist Committee (WCC) meetings and met more female chemists at other universities and at local companies. At one of these meetings, I met Valerie Kuck, a retired industrial chemist and an active volunteer for the American Chemical Society (ACS). Valerie had an interest in doing research to elucidate the reasons for the underrepresentation of women on the faculty at top-ranked chemistry doctoral programs. This was a problem I had been keenly aware of since I was a doctoral student and in which I was also interested. We recruited two psychologists from Seton Hall, Susan Nolan and Janine Buckner, and began a collaboration that lasted almost a decade. To this day, I ascribe this collaboration as one of the major factors that helped me survive the isolation I felt in my department. Though the research we were doing was not chemistry, and probably wasn't respected by some of my departmental colleagues for this reason, we were able to get several grant applications funded and published and presented papers in multiple social science and chemical education venues. In fact, I remember an off-colored remark at a faculty meeting about how "maybe we all should apply for grants for basket weaving." In spite of this disdain for our work, I believe this collaboration helped strengthen my tenure application package and definitely gave me a support network.

I received tenure in 2005, the year my son turned 17. I went for my first sabbatical, the year my son left home for college. Looking back, I always thought that I would have more than one child, but I just couldn't find a way to manage it. About a month after I was offered the job at Seton Hall, I found out I was pregnant. I worried and lamented on how to handle the discussion with my new department, but I never had to, as I miscarried a month later. I am in awe of those people who have managed to have several children and to have productive and successful careers. There are different formulas for working out a successful professor–mother balance, and having other supportive people to help you out is key. You have to follow your instincts and not be afraid to do for yourself what you feel is necessary.

Since coming to Seton Hall, I have mentored more than 50 high school, undergraduate, and graduate students, as well as postdocs and visiting scholars. I have graduated eight PhD students.

I was promoted to full professor in 2012. I look back on the days when I was struggling to balance work and being a parent. How much I worried at the time about not being there for all of my son's school events and the guilt I felt. I am blessed to have a wonderful, intelligent young man for a son with a good head on his shoulders. After all, everything did work out (Fig. 2).

Fig. 2 Cecilia with her son John in 2013

Lessons Learned

You can't do it all. Marriage and child-rearing require a partnership. When both spouses work, there has to be a mechanism to share household and child-rearing responsibilities. In my case, my husband is not a good cleaner, and only in recent years has he begun helping with the cooking. He was extremely helpful to me in terms of taking care of our son so that I could stay late at work or get schoolwork done at home. For about the past 10 years or so, we have also hired outside help for cleaning the house and doing the yard work. The helpers only come every other week, but it makes such a huge difference. It is so nice to come home to a clean house and to have time to relax on the weekends and spend time with the family.

Pick your department wisely. Some departments do not have a children-friendly mentality. For example, at my university, all of our graduate courses and departmental seminars are held in the evening. This is done to accommodate our large number of part-time students. This makes it difficult when you have children and need to be at home in the evenings. Also, consider whether many of the faculty (male and female) in your department have children. This may affect their views on your needs to be available for your family.

Don't limit where you look for mentors and support. If there is no formal network of support in your department or university, do not hesitate to find your mentors elsewhere. Get involved with university/college committees. Go to the faculty lunchroom and meet people in other departments. Go to local and national ACS events and network.

Don't be afraid to say no. As a professor and a mother, you have plenty to do without being talked into all kinds of university, departmental, or community service. Pre-tenure, it may be difficult to refuse, but do not be afraid to step away from these jobs once you have done your service. Also, I think it is important that everyone does their fair share.

Don't beat yourself up. There will always be guilty feelings about why you were not there for this or that with your children. Simultaneously, we feel guilty for not getting work done: not getting the paper submitted when planned, not putting enough time on that grant application, etc. Just do the best you can and realize there will have to be give and take (Fig. 3).

Don't be afraid to do what is right for you. Listen to your inner self and do what you need to be happy. Don't worry about what others think, and remember *there is more than one possible formula to success and also more than one definition for it.*

Addendum

It has now been 18 years since I started my "professorial adventure" at a midsized R3 institution. I have enjoyed working with students, particularly the graduate students at my institution over the years. They have been the extended family I did not have.

Fig. 3 Cecilia with her son John (left) and husband Mohammad (right) in 2015

Working with my research students and with the many students in my organic lectures is what has sustained me. Because I have felt that I have made a difference in their lives, I have persisted. Don't get me wrong, I truly enjoy research and, in particular, the freedom of academic research. However, in retrospect, I realize that my institutional choice 18 years ago had a profound impact on my career. By staying at a smaller, private institution, I have affected my chances of getting big funding and major papers and even having access to supportive colleagues who value what I do for research and who want to collaborate with me. Life choices like these are difficult, and we have to make the best of what we have or, if we are so driven, make the changes we need. Overall, I am not dissatisfied. I have come a long way in my journey and have accomplished a lot.

The gender composition of my department has not changed much in the time I have been at Seton Hall. Though for a short while there were two tenure-track women faculty, we are now back to only one (out of 12 total). So for 14 of the 18 years that I have been at my university, I have been the only woman. What is worse is that none of the administrators or other faculty members find this low percentage concerning. I became department chair about a year and a half ago and have tried to increase the numbers of women faculty and the diversity overall in the department. The Clare Boothe Luce Professorship that was available to me has not been available to new female faculty in STEM for more than 10 years, due to a decision of the faculty committee that controls the spending of the endowment. I have tried to make them see why it is important to fix the pipeline at those places it is most leaky, but the politics of who gets the talented undergraduate students as researchers in their labs has prevailed. I don't understand the logic to it to all, as we already have more than 50% female majors across all STEM disciplines at the undergraduate level.

Initially I resisted the chairperson job. I thought I would not get any research done. Now I realize it was important for me to take the job on several levels. Most importantly has been on a personal level, because it increased my self-confidence and made me feel like my life was going somewhere. It has been very challenging at times, especially when dealing with difficult colleagues. My research has slowed. Somedays I find it impossible to get out of my office. I would not recommend that a colleague who is not a full professor take the position of chairperson.

I am gradually becoming more assertive and willing to fight for what I need. I look back over the years and at how I have changed. I also am impressed with the strong, young women I see becoming chemistry professors. They seem more willing to "toot their own horn" when necessary. They are more confident and assured in their own abilities...thankfully! There is hope.

My husband and I celebrated 30 years of marriage this past December. My son has grown up to be a handsome, intelligent, and caring adult, and I am very proud of him. He has assuaged all of my concerns about being a working mom.

Five years from now, I hope to be working as a Dean or a Director of STEM Diversity or something along those lines. I would not mind living in another part of the USA.

I want to continue to see young women succeeding and fulfilling their dreams. I also want to see more underserved minorities do better. No one's dreams are unimportant. We need more mentors who are like the students they teach. In the end...I may just be...wanting it all... (Fig. 4).

Fig. 4 Marzabadi Research Group (left to right): Gisela Diaz, Emi Hanawa-Romero, Sumiea ElTayeb, and Vikram Basava in 2015

Acknowledgments I would like to acknowledge my husband and my son for all the patience and support they have provided to me over the course of my training and career as a professor. I would also like to thank my mom for her constant encouragement, especially in the times when I felt most discouraged.

About the Author

Education and Professional Career

1982	AB Chemistry, Saint Louis University, St. Louis, MO
1984	MS Chemistry, Saint Louis University, St. Louis, MO
1984–1986	Research Chemist, Monsanto Agricultural Products Company, St. Louis, MO
1987–1989	Laboratory Technician, Washington University, St. Louis, MO
1989–1994	Teaching and Research Assistant, University of Missouri-St. Louis, MO
1994	PhD Chemistry, University of Missouri-St. Louis, MO
1994–1996	Postdoctoral Research Associate, Hunter College-City University of New York (CUNY), New York, NY
1996–1999	Adjunct Professor, Hunter College-CUNY, New York, NY
1999–2004	Clare Boothe Luce Assistant Professor of Chemistry, Seton Hall University, South Orange, NJ
2005–2012	Associate Professor, Seton Hall University, South Orange, NJ
2006–2007	Visiting Associate Professor, Harvard University Medical School, Boston, MA
2012–present	Professor, Seton Hall University, South Orange, NJ
2013–2014	Visiting Researcher, Sloan Kettering Cancer Center, New York, NY
2016–present	Chair, Department of Chemistry and Biochemistry, Seton Hall University, South Orange, NJ

Honors and Awards (Selected)

2004	Manchester's Who's Who Among Executive and Professional Women
2003/2008	American Chemical Society Project SEED Service Awards
1993	Dissertation Fellowship, University of Missouri-St. Louis

Since 2001, Cecilia has been affiliated with Seton Hall University's Women's Studies Program, for which she has an Adjunct Appointment and was Acting Director (2005–2006). She also serves on the Women Chemists Committee of the American Chemical Society.

On Our Own Terms

Sara E. Mason

An Outspoken Overachiever

I do not recall exactly how old I was when I came down the stairs of my family's home one morning to declare that I was never going to get married or have children. I was grappling with career choices; the possibilities at that time included concert French hornist, hard news journalist, or justice of the Supreme Court. It was at least a decade before I would have any thoughts about attending graduate school, and several years before I would discover my queer identity. While many of my thoughts

S. E. Mason (✉)
Department of Chemistry, University of Iowa, Iowa City, IA, USA
e-mail: Sara-mason@uiowa.edu

and opinions would shift and change as I grew up, it would be a long time before I would reconsider my stance on starting a family.

I was a voracious learner throughout grade school. My teachers gave up trying to keep me occupied in class and would regularly dismiss me to go to the library by myself to read. I was confident and proud. I did not dwell much on my lack of popularity or athletic ability, as I valued my intelligence and believed in the potential that my future held. Sadly, my sense of self-worth would abruptly leave me for a time at the onset of adolescence. I will save the details of my life between the ages of 13 and 17 for another chapter of another book. What is important to know for this story is how my choices and circumstances during that time affected my education and career choices. And it was not good. The net result was negative in that I barely graduated from high school and thus got little more value out of my secondary education than the paper on which my diploma was printed.

Future Chemist, Interrupted

After high school I spent a misguided year at the University at Albany, State University of New York. I never declared a major and rarely attended a class. By the middle of the spring semester, I recognized that I was wasting time and resources and decided to withdraw at the end of the year. I moved back into my parent's house and started working in food service. I truly enjoyed restaurant work, especially in the back of the house in the kitchen or on the grill. I loved the adrenaline when there was a rush of customers on Friday night, and my co-workers appreciated how efficiently I could handle big jobs during shifts with staffing shortages. It was a long way from the Supreme Court bench, but by now I had dismissed those high childhood aspirations as nonsense. I was just starting to enjoy life again. I was not ready to challenge myself, and I did not have the confidence to do so.

However, all was not well in my culinary fairyland. I was living off of my parents and could not make enough money to move out on my own. Furthermore, I was not blind to the fact that even my highest-ranking managers were clearly struggling financially, despite working 50–70 hours a week. I have never aspired to be rich. What I wanted was to live comfortably and on my own terms. To me that meant not relying on someone else for my essentials and not needing to compromise on my convictions. After a year or so in food service, I confronted myself with the need to form a new strategy. And what I came up with was brilliant: I decided to enroll at community college.

Back to School

Admittedly, the decision to attend community college was an accidentally brilliant plan as opposed to one that was brilliant by design. The intentional aspect of my plan was that I wanted to obtain a degree that would give me more earning power. This limited my choice of major, as some majors are more suited for academic pursuits than for financial success. I knew that anything in the medical field was out of the realm of possibility, as I am not comfortable around blood. Pouring through the college catalog (this was still in a largely pre-internet world), I came across the program in chemical technology. While my high school experiences were largely negative, chemistry was actually a class in which I did well. I attribute this fact largely to the amazing chemistry teacher I had who treated me with compassion, seeing past my circumstances at the time and through to the voracious learner still tucked inside my then-struggling psyche. (Thanks, Mr. D!) What I did not know in making these plans was that the community college environment was in fact much better suited for someone such as myself who did not come out of the gate equipped with previously honed study skills.

My experiences at Monroe Community College (MCC) laid the foundation for my future career in chemistry. I decided that it was not good enough to be back in school. I wanted to also be a straight A student. However, I didn't know how to be a straight A student. First off, I found a bunch of straight A students and befriended them. My new chemistry clique was comprised of mostly nontraditional students. They were older than I was and focused on success, but they were also supportive and nurturing toward this black sheep that had wandered into their pastures. With their help, and with the countless instructor office hours that I consumed, I hit the honor roll with a perfect 4.0 my second semester at MCC. I still did not think I was much of a scholar, but the relatively short time line to completing an associate's degree gave me confidence that my goal was attainable.

My time at MCC was about rebuilding myself. Therefore I was not actively thinking about starting a family. I needed to make sure that I could afford to live my life on my own terms. Graduation was a light at the end of the tunnel that represented freedom to me. However, it was undeniable that I was also starting to develop genuine scholastic interest in the sciences. I would confront these opposing forces between independence and education while taking analytical chemistry during my last semester at MCC. The class had only two other students in it, all 20 years apart in age, with me being the youngest. The instructor was tough. In fact, there were seven of us in the first half of the course in the fall, but only us three passed to continue the spring semester. The three of us could not have less in common, but we bonded over the challenges of the class and were all holding on firmly to A grades. We were doing so well that later in the semester my two classmates chose not to turn in an optional lab report. I was enthused regardless to complete the assignment. I recall taking great pride in a figure that I photocopied from the manual and modified for use in the methods section of my report.

Completing that extra lab report was one of the best decisions I had ever made in my life. Or at least it was one of my best up until that time. It was April, and graduation was looming. I was drooling in anticipation of my degree, which I viewed as my golden ticket to freedom. I was certain in my path. That is, until my analytical instructor posted the grades for the lab assignment. The class (all three of us) was seated when he went to the board and wrote all of our names. Next to the other two names, he drew big, round goose-egg zeroes. Next to my name, he wrote my now-favorite letter: A. The dramatic display was punctuated by one of my classmates sarcastically snarking, "Well good for you, Dr. Mason." Without taking an instant to pause, and in his steady, serious tone, my instructor responded simply: "I can see that happening." This was the nudge that I needed, and I immediately started to look into options for transferring into a 4-year school to continue my education.

The new goal of becoming Dr. Mason meant that my wings would stay clipped for a couple more years while I completed undergraduate studies under the support of my parents. Because it was April when I decided to continue my education, I had limited choices in what school I could attend. I found a small local private school, St. John Fisher College, and enrolling there was my next bit of accidental brilliance. Fisher, much like MCC, was a nurturing environment where I would not fall between the cracks. I was nervous to continue my education, but Fisher was the right place for me. Some of my nerves stemmed from the fact I would be taking advanced classes in chemistry. This meant that I finally had to take the big one. The one we only whispered about at community college. The one known to kill the GPA's of even the strongest students: P-chem. I took it, I aced it, and I loved it. When I learned about quantum mechanics in the second semester of P-chem, I loved it even more. The fact that one could do chemistry through equations instead of beakers or instruments fascinated me like nothing before. I found a faculty member who was willing to supervise me in independent study (in which we did more quantum mechanics) and research (in which I first ventured into computational chemistry). In my senior year, I started looking into graduate programs that had faculty using quantum mechanics in their research.

Flying the Coop

In addition to the excitement of research possibilities, the choice of where to go for graduate studies also presented my first real chance at independence. The offers I had included stipends that were sufficient for me to take care of myself while pursuing my doctoral degree. This was no small perk in my eyes as my entire reentry into academia was based on my desire to live on my own and on my own terms. Now, I just needed to decide where. And again, my choice would be accidentally brilliant. I say this because while Fisher was a great environment for me, it was a bit sheltered, and I did not really know what considerations to weigh in selecting a graduate program. My decision came down to indecision. The deadline for accepting offers was fast approaching when one of the chemistry faculty members at Fisher inquired about my status. I told him I was on the fence between the University of Pennsylvania and one other school. While visiting Penn, I had seen Prof. Andrew M. Rappe discuss his research using a quantum method called density functional theory to study materials, and I was starstruck. However, my visit to the other school had also been impressive to me. At the time, I did not really know about or understand the implications of school rankings. The professor listening to my top two choices clearly did and was willing to clue me in without subtlety. Throwing his hands in the air, he all but commanded me "Go to Penn!" And in the end, I did, joining Prof. Rappe's group in 2001.

Deciding to pursue graduate studies at Penn was undoubtedly an excellent move for me career-wise. It was also a decision that I made as a single person with no children and with no immediate plans to change either status. And while I did end up meeting the person whom I would ultimately marry late in my first year of study, I never had even fleeting thoughts of starting a family during my PhD training. I am not equipped to decipher if my lack of interest in having a child during graduate school came from my own innate wishes at the time. I also felt that I was in an environment that was incompatible with that path. On the one hand, I was enjoying the brutally challenging academic and fast-paced research aspects of the program, and I typically worked long hours and over the weekends. On the other hand, I did not see other students, even the married ones, having kids while in grad school. Worse yet, those that I knew of that did have kids were mostly men and ended up leaving without completing their degrees so that they could earn more money to support their families. There were a couple of cases I knew of in which extended family was able to move in and take over child care. Looking beyond my classmates, I noticed assistant professors who were in their offices on the weekends and late into the evenings. Some of them had small children who would occasionally be brought to visit them while at work. Some professors had no children, and some male professors had children cared for by a stay-at-home wife. Some professors, regardless of whether male or female, appeared to wait until they were professionally well-established, as indicated by tenure or other major success, prior to starting a family.

In addition to the focus on my own personal growth and the culture of graduate school and academia, a third reason not to ponder motherhood was my queer identity. Even if I overcame the obvious question of how to get pregnant, what sort of family could I build with my partner? Marriage was not available, and the university offered no domestic partnership benefits. I had no idea where to even begin trying to unravel if or how my medical benefits could be used to help me to conceive. I had no strong role models of successful gay parents. I was brought up to feel ashamed of who I was, and I had no support network. Any emerging personal desire to have kids was completely drowned out by the complete medical, financial, and practical impossibilities that I felt existed.

Of all of my perceived obstacles to starting a family, money and job security were perhaps the most significant, as many other problems can be solved with sufficient resources. It was explained to me that a postdoctoral position was all but required for most careers in chemistry, and it seemed by definition this would extend the period of time during which salary and security remained concerns. By securing a nationally competitive fellowship, I was able to obtain a relatively high salary but only on a 2-year term. That fellowship is what brought me to the National Institute of Standards and Technology (NIST) in 2007.

Ready, Set, Setback

My time as a postdoc is when I started to verbally express interest in having a baby to my partner. The sheer act of saying the words "I want to have a baby" felt momentous. But I still had no roadmap for how to move forward. My postdoc advisor did not have children, and I never discussed with her my growing desire to become a mother. I was mostly around other postdocs, and all of us were on term appointments and thus quite consumed with securing our next respective positions. Then, before I could act any further on navigating how to start a family in my situation, my situation changed. I got sick. Not the kind of sick that ruins weekend plans, but a life-altering kind of sick that turns your world upside down. I will save the details of my illness for another chapter in another book and only discuss the impacts this had on my career and detoured road to motherhood.

My illness never jeopardized my life, but it did threaten my career. Postdoctoral positions are not permanent, and my funding had an expiration date. I had to be thinking about my next job even before I was properly diagnosed or under effective treatment. This resulted in me doing all that I could to keep up appearances by still attending professional workshops and conferences. Adrenaline is a fantastic weapon, and I used it to power through traveling and presenting as needed. But in between the draining fits and bursts of appearing fine, I was not. There was a real possibility that all that I had worked for in obtaining my PhD would go to waste and that I would be forced into a medical retirement before I turned 30. Needless to say, those nascent feelings of wanting a baby went to the wayside. Thankfully, once I was properly diagnosed and under treatment, I started to improve. However it was not a fast process, and there were bumps in the road. When I started to search for my next position, I placed as few restrictions as possible on my job search and did not even consider work-life balance or cultural climate as criteria. Ultimately, a few radically different options emerged, and the one I chose was to accept a tenure-track assistant professorship at the University of Iowa, starting in summer 2010.

Tenure Clock, Meet Biological Clock

My position at Iowa came with improved health and the ability to marry my partner of 8 years. We wed on a Monday morning, in a lawyer's office, with strangers as witnesses. It was during my second interview (though I had already accepted the position so it was really a house-hunting trip), and within an hour of getting married, I was meeting with future colleagues, administrators, and staff on campus as part of my visit schedule. I did not mention my marriage to anyone, nor had I mentioned my partner at all during any of my interviews. From my personal experiences, I did not expect broad acceptance. I was also coming from one of the most liberal and diverse parts of the country to a rural city and needed time to adjust to such a drastic change.

Over time, I discovered the LGBTQ Safe Zone Project on campus and later joined the effort as a facilitator. Helping to spread a visible message of inclusion, affirmation, and support of LGBTQ people in the university community has been a positive experience for me. I often get to hear stories of allies going out of their way to be open and supportive, and I am particularly impressed by the compassion and acceptance expressed by students. While my role as facilitator is to be there to guide them through the program training, the experience is also healing for me as I did not always have a supportive environment.

As I settled into my new role as assistant professor, my feelings of wanting a baby resurfaced. By now, years had passed since I first verbally put my wishes into the universe. I was in my early thirties, which is getting late in the window of ideal child-bearing years, and I also questioned if my illness could have affected my fertility. Before I even saw a doctor to discuss my options, my doubts already had me consoling myself by considering my role as an instructor and mentor as alternatives to motherhood. My intuitions were shown to be correct when, at a doctor's appointment for another matter, it was discovered that I needed surgery for a condition that was not only affecting my general health but which would also have made it impossible for me to carry a child.

The surprise surgery, along with my ever-advancing age, catalyzed discussions and planning to get me pregnant, if possible. My wife and I were thrilled to learn that my medical benefits could help financially, and the doctors and nurses at the university hospitals and clinics were experienced in working with female couples trying to conceive. Everyone, including myself, was full of optimism that I would be pregnant within a few tries. Or maybe I'd be pregnant within several tries. Or no, something must be wrong.

And something was wrong, and because of that, we had to resort to in vitro fertilization, or IVF. IVF is not a simple process. It involves taking lots of medications, many by shots that are required multiple times a day. It requires very close monitoring, meaning several visits to the clinic every week. It requires a minor surgery to retrieve the eggs, if any can be retrieved. It requires waiting several days after retrieval to learn if any embryos were successfully created, and it involves decisions about how many embryos to transfer back to attempt a pregnancy. Then comes the dreaded 2-week window before you learn if, after all of that, you are actually pregnant. Tallying up all of the time spent related to my surgery and attempts to get pregnant and preparing for the IVF, I was up to at least 2 years. That was 2 years of enduring the high emotional and physical tax of trying to conceive while also on my tenure track. I recall commenting to my wife during my 2-week window that if I failed to get pregnant, I was probably going to commit myself. And I was only half joking.

I had to teach the day I got the blood test that would reveal whether or not I was pregnant. I was able to get the blood drawn prior to my lecture, and after my lecture, I went home to wait for the results. The clinic used an online system to share test results with patients, and I was reloading it every few minutes. I had never felt this level of anticipation, though I have been known to obsessively check on journal and grant submissions in a similar fashion. When the result came in indicating that I was pregnant, I was overcome with joy. I had not planned on how to share the news with my wife, and I ended up grabbing a hot dog roll, putting it in the oven, and sending her a photo from my phone. In case that was not sufficient, I also took a picture of a jar of Prego sauce and sent that along to her. It was goofy, but it felt good to be playful amidst such a harrowing and stressful time.

While getting pregnant was difficult for me, pregnancy turned out to be a breeze. Other than the nerves of getting through the first trimester safely, I had very little discomfort the entire time. Professionally speaking, my pregnancy seemed very well-timed: I got pregnant about 6 months after my first major research grant was funded, and my due date was in early June, which is convenient in terms of the academic calendar. I liked that the outward appearance of my pregnancy seemed so perfect, despite the struggles behind the scenes.

Practically speaking, I thought it would be a good idea to discuss my pregnancy with my department chair. We agreed to have me team-teach in the spring, which was a normal teaching load and also offered flexibility in case I had to go out early on leave. We also had a meeting with the dean to discuss my plans for leave. I will say that I did not go into that meeting with the best information or much of a plan, and if I had to do it over, I would have been more proactive in discussing and negotiating the terms of my maternity leave. However, since I was due in June, it was fairly straightforward to plan that I would be off for the summer and would return for the start of fall classes. In addition to my duties to the university, I also have obligations to funding agencies. At the time of my pregnancy, I had one federally funded grant to support my research program. I looked into how to take my maternity leave into account in terms of the progress on the grant projects, but I did not find any accommodations that I could use.

My daughter was born June 5, 2014, at around 12:30 pm. I had completed my spring teaching assignment without incident. I even gave a research talk at a neighboring Midwest university while 36 weeks pregnant and was at work the day before I went into labor. The ease of events leading up to the birth left me with the expectation that I would be back on conference calls and working remotely from home within a week or so. This is where my lack of a support network came to a head: I was not ready to jump back into work mode within days of the birth. While my wife was able to take several weeks off through FMLA (which had only become available to same sex married couples when the Defense of Marriage Act was overturned on March 27. 2013), we had no family in the area. Both of us were the youngest in our families, and neither of us had ever taken care of a baby. I was unprepared for the impacts of postpartum hormonal changes on my well-being. We were in unchartered waters, and it took time for our little family to get into a groove.

The fall semester started when my daughter was about 11 weeks old. Her first day of daycare coincided with the first day of class. We were all tired, but I was excited to get back into a classroom and excited to have more freedom to work while my daughter was in good hands. That fall was also an active grant writing season for me, and I submitted two proposals on which I was a lead PI and was also part of a large collaborative grant effort with a team comprised of professors from many universities. I was ecstatic when my teaching evaluations for my first semester as a mother came back with high scores from my students. On my CV, that fall looked successful.

However, I was also still figuring out how to juggle my work and mothering duties, and to be honest, I dropped a few balls. I was not the only one who noticed, and there were conversations in which I was confronted over issues such as my lack of timely responses to email requests and absences from meetings and calls (which were largely due to illness, as all three of us were sick the entire semester due to our lack of immunity to the bugs going through the daycare center). It was quite humbling, if not worse, to need to be told that my performance was having a negative impact on those around me. I did my best to not take the feedback too personally, as intellectually I was certain that I was not the first new parent to struggle to keep up with work and a newborn.

I also did my best to adapt new practices, such as flagging emails or setting reminders up in my calendar. There was no secret or trick or quotable wisdom behind the adjustments I had to make in order to be both a professor and a mother. As with many other challenges in life, a lot of it came down to just rolling with the punches, learning from mistakes, and celebrating successes.

Catching up to present day, my daughter (Fig. 1) just turned 2 years old, and I am up for tenure in the fall. Over time, the problem of how to juggle my responsibilities

Fig. 1 Sara's daughter

has not simplified. If anything, as my research group and daughter both grow, it seems there are always more and more balls to keep in the air with just my two hands. Having more students and more active projects, many of which are collaborative, means more research meetings, more collaborators to keep updated, and more journals to read. Service roles are also expected to expand with experience, and I constantly challenge myself to generate new and innovative curricula and course materials.

As my daughter grows, I get to know her better each day. She is loud and funny, strong-willed and spirited. She wags her finger in my face to tell me that no, it is not time for a bath (or bed or whatever else I may have planned). She often shows no interest in the breakfast I have prepared for her and instead demands mashed potatoes at 6 am. When we take her to the playground, she runs right past the preschool jungle gym, opting instead for the tower of terror designed for kids ages 10 and older. I look forward to hearing many headstrong declarations from her. And I will keep working my hardest to bring the best for my family, on *our* own terms.

Lessons Learned

The Philosophical: No two families (or families in the making) are the same, and there is no silver bullet for how to pull off a gratifying career in academic science while creating, maintaining, and growing a family. Sure things can fall through, and long shots can pan out. Keep in mind there are many ways to be successful, and many paths can lead to happiness. And just as some successes have an element of good luck associated with them, some failures are simply a case of bad luck. Not all outcomes in life are a reflection on your character.

The Practical: Reproductive technology is at an amazing stage, but there is currently no way to overcome a shortage of healthy eggs. Looking into having eggs retrieved for storage is an option that I would suggest to young women embarking on long programs of study. You do not need to wait to start your family, but in case that is what you decide, having healthy eggs in storage will give you more options in the future. In my situation, where ICSI/ICF was used, I was fortunate to live and work so close to a hospital with facilities that offer such services. As I wrote here, I often needed to be at the hospital 2–3 times per week for the procedures undertaken to get me pregnant. Therefore, depending on your situation or plans, factoring in access to medical resources when job hunting may be appropriate.

Whether you are a student, postdoc, or faculty member, knowing your rights before disclosing a pregnancy will put you in a better position to negotiate your plans, if any, for leave.

A lesson that was difficult for me to learn was that I was not in full control of my level of ability shortly after giving birth. I put undue pressure on myself to act as though giving birth was "no big deal." For me it was a big deal, and making professional commitments before I was truly ready backfired. If at all possible, leave in flexibility, such as returning to work gradually or working more remotely.

Epilogue, July 2017

I have been promoted to Associate Professor with tenure and am expecting a second child literally any day now. My wife and I have been together for 15 years (married for 7), and we look forward to living life on our own terms.

Acknowledgments I am eternally grateful to my wife, best friend, and partner in crime, Rebecca D. Dudley.

While all of my instructors and professors played an important role in directing and defining my scientific career, I would like to call special attention to my high school chemistry teacher, Mr. Tom Di Gaetano, who probably has no idea that I so adored his class since I was always wearing black and crying during those days. Mr. Sherman Henzel (Monroe Community College) is why I pursued my education past my associate's degree. The MCC Chemistry Club is how I learned to be a good student and kept me on my toes getting those straight A's. From St. John Fisher College, Profs. Thomas Douglas, Leslie Schwartz, and Dan Piccolo all played distinct but important roles in teaching and mentoring me. I am still amazed that Prof. Andrew M. Rappe (University of Pennsylvania) accepted me into his research group. By pushing me beyond what I thought my scientific limits were, Andrew made sure I was ready for everything I would face professionally after graduate school. My postdoc mentor, Dr. Anne Chaka, taught me a great deal about how to be versatile and how to find connections between theory and experiment. At University of Iowa, I was extremely fortunate to have Prof. Vicki Grassian (now at University of California, San Diego) as a mentor early in my professorship and continue to look up to her as a role model and accomplished scientist. I am also grateful to all of my colleagues and the students in the Center for Sustainable Nanotechnology for helping to advance and extend the research opportunities in my group. Finally, to all of my research and classroom students, for giving me such a sense of purpose. The success of those I've had the privilege of teaching and mentoring provides a uniquely exhilarating sense of accomplishment and desire to do all I can to lift others up to toward their dreams.

About the Author

Education and Professional Career

1999	AAS Chemical Technology, Monroe Community College, Rochester, NY
2001	BS Chemistry (minor in Mathematics) *summa cum laude*, St. John Fisher College, Rochester, NY
2007	PhD Chemistry, University of Pennsylvania, Philadelphia, PA
2007–2010	National Institute of Standards and Technology-National Research Council Postdoctoral Fellow, National Institute of Standards and Technology, Gaithersburg, MD
2010–2017	Assistant Professor, Department of Chemistry, University of Iowa, Iowa City, IA
2017–present	Associate Professor, Department of Chemistry, University of Iowa, Iowa City, IA

Honors and Awards (Selected)

2014	Science and Technology Rising Star, St. John Fisher College
2013	National Science Foundation CAREER Award
2012	Emerging Investigator, Journal of Environmental Monitoring
2007	National Research Council Postdoctoral Research Associateship
2001	Ahmed Zewail Fellowship, University of Pennsylvania

Sara's research group applies and develops theory and modeling to study nanomaterials in the environment and energy applications, using quantum mechanical methods to derive molecular-level understanding of systems such as complex metal oxides, Al-nanoclusters, and mineral-water interfaces. The over-arching goal is to use the chemical understanding obtained from our modeling work to guide the rational design of more sustainable materials for energy and water remediation processes.

Putting Family First Led to My Extraordinary Career

Saundra Yancy McGuire

Photo Credit: Eddy Perez, Louisiana State University Communications

Recently a friend of mine, a successful provost at a large public university, said to me, "You know, I never set out to be an administrator. I just tried to do the best job that I could in every position I held." I told her that I also had never aspired to a career in administration, but I rose through the ranks from instructor to professor to administrator because I loved every position that I ever had, did the best job I could,

---

S. Y. McGuire (✉)
Department of Chemistry, Center for Academic Success, Louisiana State University, Baton Rouge, LA, USA
e-mail: smcgui1@lsu.edu

and was repeatedly offered higher-level positions for which I would never have applied.

When I retired as assistant vice chancellor and professor of chemistry from Louisiana State University on June 30, 2013, I ended a 43-year career as a chemical educator, learning specialist, and administrator. The path my career took was not the result of carefully crafted goals, objectives, or timelines that I had established. Rather it was the result of flexibility, family priorities, and serendipity. To understand the ending, we must start at the beginning.

Growing Up in a Highly Educated Family

I grew up in the 50s and 60s in Baton Rouge, Louisiana, the deeply segregated capital of the state. I was blessed to be born into a highly educated family, and my parents and other relatives demonstrated by example the importance of higher education. My grandmother, the daughter of a freed slave, had attended Leland College in Baker, Louisiana, and was a teacher in rural Greensburg, Louisiana, where she, my grandfather, and their nine children lived on a large farm. She decided that all of her children would attend college (a revolutionary idea in a rural community where very few students—Black or White—even graduated from high school). And although my father and his siblings had to walk 2 hours each way to attend a segregated high school (passing by a White school right there in their hometown), they graduated and attended college. Seven of my grandparents' nine children went on to obtain master's degrees or even education specialist degrees, attending some of the nation's best graduate schools in their respective areas. Because the state of Louisiana preferred at that time to pay other schools to educate African-Americans rather than desegregate LSU, four of my grandparents' children earned their degrees at Iowa State University (my dad's alma mater), Kansas State University, Michigan State University, and the University of California, Los Angeles. The other three earned their degrees at LSU, which had finally become integrated by the time they graduated from college.

The sacrifices my dad and his siblings made to travel to far away schools to obtain graduate degrees made an impression on me, my three siblings, and nine cousins, who collectively hold two PhDs, four MDs, and six master's degrees. I decided by middle school that I would pursue a PhD someday, even though I had no idea what that might entail. However, I was also strongly influenced by my mother's decision to be a stay-at-home mom, even though she had a bachelor's degree in elementary education and was credentialed to be an elementary school teacher. I loved coming home from school to find my mother welcoming me with tuna fish sandwiches, juice, and rich conversation about our respective days' activities. I found myself feeling sorry for my classmates who went home to an empty house each day and had to fend for themselves. (I found out years later that they had enjoyed having the house all to themselves and felt sorry for me because I never had that luxury!)

Inspired by my father to obtain a PhD and inspired by my mother to be at stay-at-home mom, I declared that I was going to get a PhD but be a stay-at-home mom until my youngest child graduated from high school. At the time I saw no conflict between career and parenthood, but I later realized how unlikely it was that my plan would pan out. I did, however, get to combine having a PhD and being at home with my kids by working part-time until our younger daughter started high school. Working part-time allowed me to volunteer at our daughters' Montessori preschools, conduct science demonstrations and experiments with their elementary and middle school classes, and participate in the Huntsville High School Band Parents Association. As president of the association, I led the fundraising effort that made it possible for the band to appear at the 1988 International Youth and Music Festival in Vienna, Austria. (And as a chaperone, I heartily enjoyed my authentic wiener schnitzel.) My part-time work life was also very rewarding. I taught general chemistry, organic chemistry, and biochemistry in addition to providing learning assistance to STEM students. It was the perfect combination for me. But I'm getting ahead of myself. Let me return to my childhood to illustrate how I ended up with a PhD in chemical education and an active family life with a husband and two daughters.

I was the second oldest child of four children, and our father was a high school agriculture teacher who also taught general science. He would often come home from a day of teaching science and enthrall us kids by asking questions like "Did you know that the melting point of ice and the freezing point of water are exactly the same temperature?" We would quickly respond "Noooooo, that's impossible. How would it know whether to be water or ice???" He would laugh and then explain the phenomenon to us. I'll never forget the nights he took us out into the front yard to gaze into the heavens to see Sputnik. Science fascinated us, and we all wanted to become scientists.

My Decision to Major in Chemistry

But pursuing an undergraduate degree specifically in chemistry was not always a given. One afternoon during the spring semester of my freshman year at Southern University, Dr. Vandon White, chair of the department, came out of his office and happened to start walking in my direction. When we met he stopped me and said, "Miss Yancy, what are you going to major in"? I replied, "I don't know, Dr. White; it'll be one of the sciences or mathematics, but I haven't decided which." He then said, "Why don't you major in chemistry?" And I said, "OK." That succinct conversation was the beginning of my pursuit of a career in chemistry.

I had a wonderful time in undergraduate school. Because of efforts to increase the number of African-American students who would successfully pursue STEM careers, there were a number of special programs that provided opportunities to study at prestigious universities. I was selected to participate in the Harvard-Yale-Columbia Intensive Summer Studies Program, funded by the Carnegie Corporation and the Ford Foundation. Consequently, I attended Columbia University during the summer after my sophomore year and Harvard University after my junior year. I was also afforded the opportunity to spend my entire junior year at the University of California, Berkeley, funded by the Crown Zellerbach Corporation. Two students from Southern were selected to attend—one young man and one young woman. We could choose to attend either UC Berkeley, Stanford University, or UCLA. The young man selected was Stephen C. McGuire, a physics major from New Orleans, LA—the wonderful guy I had begun dating my freshman year!

Steve and I had an exciting time in California, at UCLA and Berkeley, respectively. However, the 1968–1969 academic year at Berkeley was culture shock for me. There was a different demonstration each quarter, the first of which was a call to boycott classes until the university agreed to offer a program in Black Studies. I was very conflicted about what to do. I strongly supported the efforts of the Afro-American Student Union to demand a Black Studies Program at Berkeley, but I knew that boycotting my chemistry classes would be very detrimental to my success. Besides, I was *loving* my classes, especially my organic chemistry class that was being taught by Professor Melvin Calvin, an entrancing lecturer who had wonderful models to demonstrate the reactions he was discussing. However, when I saw the peer pressure applied to "scabs" who went to class, I absolutely did not want to be seen ignoring the boycott. So I stayed out of classes for an entire week. But then something dawned on me. My friends who were boycotting were social science majors who were Berkeley students year round. They could get incompletes in their courses and make them up after the boycott was over. I, on the other hand, was a chemistry major who would be at Berkeley for one year only. I couldn't afford to get incompletes or, even worse, fail courses due to absences. So after being out of class for a week, I found the back route to the chemistry building and resumed classes. Thankfully, my professors were sympathetic to my plight, and I suffered no ill consequences of the boycott. The experience further taught me that I did not ever have to choose between apparently competing goals. Yes, I *could* be an academic and a mother. Yes, I *could* work in the sciences and promote social justice. With flexibility and open-mindedness, there is always a solution.

Although I found my time in California invigorating, I was quite happy to return to the nurturing, but challenging, environment of the Southern University chemistry department. It really was the ideal place to pursue my undergraduate chemistry degree. Dr. White, whose PhD was from Purdue University, had amassed a group of faculty of 15 African-American PhDs (the most of any institution in the nation), so seeing Black role models in science was business as usual for me. The example of these scientists had strengthened my belief at Berkeley that I need not choose between career and community, and they would continue to inspire me as I furthered my academic career. After applying for graduate school admission and supporting fellowships in my senior year, I was thrilled to be awarded a Danforth Foundation Fellowship and a place in the Cornell University chemistry department. Serendipitously, Steve would remain close and attend the University of Rochester to pursue graduate studies in physics.

Bitten by the Teaching Bug

My first semester in the graduate program at Cornell was quite interesting. Although my chemistry courses were going okay, I found that I poured more of my time and energy into my teaching assistant duties, which included conducting recitations and labs and holding office hours to help students sort out challenging material. The high point of that academic year was receiving the Department of Chemistry Outstanding Graduate Teaching Assistant Award.

I loved teaching and seeing the students excel, and I decided that I would pursue a teaching career rather than a career as a research chemist. So after my first semester, I decided to obtain a master's degree in science education. As fate would have it, when I changed career paths, my advisor turned out to be renowned science educator and concept-mapping guru Professor Joseph Novak. Professor Novak's ideas resonated with my experience in helping students learn. Conceptual understanding and connecting new ideas to what the learner already knows are essential to learning. His seminal book, *Learning How To Learn* (Novak and Gowin 1984), describes his approach, and I have been fortunate to be able to apply it with great success over the years. When I began to pursue studies in chemical education, a whole new world opened up. I started taking courses in educational psychology, the history of education, educational research, and other areas that shed light on teaching and learning. I knew this was my calling, as opposed to basic chemical research, and I embraced it enthusiastically.

I was also very happy with my new direction because I felt that it fit much better with my future family plans. Steve was studying to become a research physicist, and I knew that would require long hours in the laboratory. I figured that a teaching career would be much more conducive to being able to spend time with kids. Steve and I were married after our first year in graduate school, and I moved to Rochester, New York, with freshly minted master's degree in hand.

My time in Rochester was fulfilling. I taught general chemistry, algebra, and chemistry for cosmetologists at the Educational Opportunity Center of the State University of New York at Brockport. My students excelled. Our first daughter, Carla Abena, was born during this time, and I was very happy to spend my days with her and my evenings teaching chemistry.

One Foot in the Classroom, the Other in the Campus Learning Center

When Steve completed his master's degree, we returned to Ithaca where he pursued his PhD in nuclear science and engineering at Cornell and I joined the faculty of the Learning Strategies Center as a lecturer in chemistry and a learning support specialist. Our second daughter, Stephanie Niyonu, was born during my second stint in Ithaca. We had planned the pregnancy perfectly, and she was born on June 13, giving

Fig. 1 The McGuire family (left to right), Steve, Stephanie, Saundra, and Carla, in 1978

me the entire summer before I would resume my teaching. And when the fall came, I arranged my teaching schedule so that I could teach when Carla was at the Montessori school and Stephanie was in daycare part-time. Whenever I had a meeting at Cornell when Carla was not in school, I took her along with me. I remember how excited she would get whenever I was going to meet with Professor Roald Hoffmann because he always gave her his molecular models to play with while we were meeting. Each time she would ask me "Can I play with Dr. Hoffmann's molecules?," and I loved seeing her little face light up when I answered "Yes!" I was both professionally and personally fulfilled during that time, spending quality time with my family while still enjoying working with students and colleagues (Fig. 1).

After completing his PhD, Steve took a position as a staff scientist at the Oak Ridge National Laboratory, and I became a part-time instructor at the University of Tennessee at Knoxville. Shortly after classes started at UTK, I received a call from a Cornell faculty member, Joanne Widom, who was writing a new general, organic, biochemistry textbook with Stuart Edelstein, a biochemistry professor. The book would have an accompanying student study guide, and they had decided that—based on my success helping chemistry students at Cornell—I was the only person they wanted to write it. That was the first of several ancillaries that I would be asked to write for chemistry textbooks over the next 20 years.

As if teaching and writing weren't enough to keep me busy, I began my doctoral studies at UTK during my second year there. I encountered concepts about learning even more fascinating than the ones I'd already been exposed to, concepts like cognitivism, constructivism, social constructivism, and the like. Because I was intrigued by the different approaches students took to problem-solving, I was curious to see if the newly emerging ideas about right- and left-brained information processing were relevant, and I did research in this area. My dissertation was entitled "An Exploratory Investigation of the Relationship Between Cerebral Dominance and Problem Solving Strategies Used by Selected High School Chemistry Students." Little did I know that this research would be relevant to my position as director of the Center for Academic Success at Louisiana State University (LSU), a position that I would accept almost 20 years later.

By the time I completed my doctoral research, Steve had taken a position as a professor at Alabama A & M University in Huntsville, AL, beginning in the fall of 1982. We moved the family to Huntsville with only my dissertation remaining to be written. I completed the writing in March 1983 and was offered a job teaching biochemistry at Alabama A & M when the professor who previously held the position did not return after spring break. I participated in the spring graduation exercises at UTK and continued teaching part-time until I was offered a tenure-track faculty position in the fall of 1983.

My tenure-track position at Alabama A & M involved teaching, research, and service, and I was actively engaged in all three. I was providing workshops for elementary and middle school teachers demonstrating hands-on activities they could do in their classrooms with everyday materials, and I was publishing my work. Steve was teaching physics and conducting research at both the university and at NASA Marshall Space Flight Center, where he was honored with NASA's Office of Technology Utilization Research Citation Award for his work in cosmic-ray physics. And our daughters loved Huntsville and thrived. Their dad played a very active role in the girls' activities. He would borrow the school's computer on the weekends and teach them how to program their own video games. Although the games were very unsophisticated, compared to the commercially produced ones their friends owned, Carla and Stephanie were very proud of their creations and enjoyed playing with them.

It was also during our time in Huntsville that Steve spent several summers conducting research at the Lawrence Livermore National Laboratory in Livermore, California. The entire family went for two summers, and I worked with the scientists there to conduct workshops for California teachers. I even got a commitment for funding and materials to conduct more workshops for teachers in Huntsville. The girls loved our time in California because we were able to take them on weekend excursions to places like the Grand Canyon, Lake Tahoe, Big Sur, and Sequoia National Park where they saw the giant California redwood trees. On the return trip, we stopped at Los Alamos National Laboratory to visit friends and to tour the lab. It was a productive and enjoyable period in our lives.

Returning to Familiar Territory

In 1988, after spending 6 years in Huntsville, Steve was invited to join the faculty of the Department of Nuclear Science and Engineering in the College of Engineering at Cornell University, and the girls and I accompanied him. This would be my third stint in Ithaca, and I was looking forward to rekindling past friendships and professional acquaintances. Although I was due to come up for tenure the year that we left, and it seemed certain that it would be granted, I had no qualms about leaving that behind; my primary interest was in the welfare of my family. Plus, my job prospects in Ithaca were excellent. I was welcomed back to the Learning Strategies Center and picked up where I had left off some 10 years previously. I was again lecturing in chemistry and helping students learn.

The 12 years that I spent at Cornell were very productive both professionally and personally. I was promoted to senior lecturer and won the coveted Clark Distinguished Teaching Award based on the nomination of my students and colleagues, and I became the Director of the Center for Learning and Teaching. During those years our older daughter, Carla, matriculated at Howard University and graduated summa cum laude with a full scholarship to attend Duke Medical School. After graduating from Duke, she did her residency at Baylor College of Medicine and married her college sweetheart, Eric. Those years saw our younger daughter, Stephanie, graduate from Ithaca High School and MIT as a Marshall Scholar and attend the University of Oxford where she obtained a D.Phil. in auditory neuroscience. Again, my professional life and personal life were fulfilling.

In 1999, Steve was called to return to Southern University in Baton Rouge, our undergraduate alma mater, to chair the physics department from which he had graduated 29 years before. He accepted the call, and I was again on the move to accompany my husband to his next assignment. As was the case in the past, I had no idea what I would do in Baton Rouge, but I was confident that I would land something that I would enjoy and that would be professionally rewarding. I had no idea about the gold mine that was waiting for me.

Teaching Students *How* to Learn

As fate would have it, just as we were returning to Baton Rouge, the director of the learning center at LSU was leaving because her husband had completed his graduate studies. When I learned about this from my sister, an advisor at LSU, I immediately thought "That's my job!" I applied for the position posthaste, was granted an interview, and received an offer. I readily accepted and would start work within 2 weeks.

The learning center at LSU proved to be the most challenging and productive environment of my entire professional life (Fig. 2). This was because, unlike at Cornell, faculty readily referred lots of students to the Center for Academic Success. Also, unlike at Cornell, where I primarily taught chemistry and had excellent colleagues who taught students general learning strategies while I helped them with chemistry, I was expected to also teach general learning strategies, about which I was clueless! But with excellent mentoring from Sarah Baird, a great learning strategist who literally took me under her wing to introduce me to effective strategies, and after reading everything I could get my hands-on about helping

Fig. 2 Saundra teaching learning strategies to an organic chemistry student. Photo Credit: Jim Zietz, Louisiana State University

students become independent, self-directed learners, I learned how to help students develop metacognitive learning strategies that resulted in immediate and dramatic improvement in student performance. In fact, the improvements were so dramatic (e.g., students' increasing their scores by 50 points or more after *one* strategies session) that I could not believe what I was seeing. But over the past 15 years, I have seen these improvements over and over again. Moreover, not only have I personally witnessed the transformative power of learning strategies, but I have also traveled the world, giving presentations about how learning works to students, faculty, staff, and administrators, and I have received feedback from hundreds of learners and instructors about the life-changing potential of the strategies. As a result, I have become utterly convinced that it is not difficult to teach students how to significantly improve their learning. My recently published book, *Teach Students How to Learn: Strategies You Can Incorporate Into Any Course to Improve Students Metacognition, Study Skills, and Motivation*, was co-authored by our younger daughter Stephanie and published by Stylus, LLC in October 2015.

My work over the past 15 years has resulted in my being elected a Fellow of the American Chemical Society (2010), a Fellow of the American Association for the Advancement of Science (2011), and a Fellow of the Council of Learning Assistance and Developmental Education Associations (2012). Additionally, I received the Presidential Award for Excellence in Science, Mathematics, and Engineering Mentoring in a White House Oval Office Ceremony (2007, Fig. 3), the National Organization for the Professional Advancement of Black Chemists and Chemical Engineers (NOBCChE) Lifetime Achievement Award (2014), and the 2015 Lifetime Mentor Award from the American Association for the Advancement of Science. I also received promotions to the position of assistant vice chancellor and to the rank of full professor at LSU. Upon my retirement from LSU in June of 2013, I was granted Emerita status, and a scholarship was established to honor the work I did in learning assistance. I continue to work with students, and I travel the country to teach other faculty and administrators how to teach students strategies for academic success. I am still surprised by the interest that my work has generated among educators. I've delivered keynote addresses or presented well-received student and

Fig. 3 Saundra receiving the 2007 Presidential Award for Excellence in Science, Mathematics, and Engineering Mentoring from President George W. Bush in the White House Oval Office. Photo Credit: Rodney Choice, Choice Photography

faculty development workshops at over 400 institutions in 46 states and 9 countries, most recently South Africa, Nicaragua, and Japan.

Following My Husband, Charting My Own Path

It is probably apparent to you by now that my professional moves were dictated by the moves that my husband made. As you will recall, I decided as a child that I would work part-time (or not at all) until my youngest child graduated from high school. And as a young adult, I understood that as a research scientist, Steve needed to have the freedom to determine which career moves were best for him, without having to deal with a "two-body problem." I often joke that we really have a one-career family. My husband has the career, and I get a job wherever he ends up! Jokes aside, this strategy has resulted in an astonishingly satisfying, well-rewarded, long, and celebrated career in chemical education. Every day I am grateful for the respect and prominence that my work has earned from my peers. And I have been blessed to have phenomenal opportunities to work in fulfilling, stimulating positions while still being able to devote a significant amount of my time and attention to my family. I have always believed that, of all of the responsibilities I assume at any given point in time, the only ones that cannot be immediately assumed by someone else, should I suddenly cease to exist, are being a mother to my children and a wife to my husband. My choice to put family first and career second might sound like a set-back for women's equality, but I believe that true equality means having the ability to choose which path to take. This path has worked for me, and I would like others to know that

Fig. 4 Saundra with her daughters Carla (left) and Stephanie (right) in 2007

Fig. 5 Professors Stephen and Saundra McGuire in 2009

it is possible to have a successful professional life that takes a back seat to one's family life (Figs. 4, 5, and 6).

The recurring theme in all of my activities, professional and personal, is that I pursued what I was passionate about. My family always came first, but I pursued professional opportunities and accepted positions that were in line with my passion—helping students succeed. And I'm very proud of the accomplishments of our family. Steve is an elected Fellow of the American Physical Society and a member of the LIGO Scientific Collaboration, the multinational team responsible for proving the existence of gravitational waves, a result that earned the project's three head

Fig. 6 Saundra and her grandchildren (from left to right) Joshua, Daniel, Ruth, and Joseph in 2016

scientists the 2017 Nobel Prize in Physics. Carla is Professor and Chief, Section of Immunology, Allergy, Rheumatology and Retrovirology, Department of Pediatrics, Baylor College of Medicine, and Stephanie holds a D Phil in Neuroscience from the University of Oxford and is an accomplished opera singer who has performed at Lincoln Center. She is also the co-author of my recently published books, *Teach Students How to Learn* and *Teach Yourself How to Learn: Strategies You Can Use to Ace Any Course at Any Level*, published by Stylus, LLC. I certainly agree that it's not possible to have it all, but I think that my experience proves that it *is* possible to have it the way you want it.

Reflections

I have been truly blessed to have had a fulfilling career and family life. I am often asked about advice I would give to younger women who want to combine an academic career with family. After thinking about it over the years, I would offer the following advice. Pursue your passion, even though this may not be what your peers, your advisors, or your parents think you should do. Keep an open mind and surround yourself with positive people who respect your choices.

Putting Family First Led to My Extraordinary Career

I would also advise women at the beginning of their careers to seek out mentors or, where none are available, develop the ability to actively mentor *yourself* through self-assessment and reflection. Those of us interested in mentoring ourselves need only to think carefully about those things that we want from a mentor and then systematically devise a plan to provide those things for ourselves. For example, if we need information about the culture of an organization or its unwritten rules, we can use our well-developed powers of observation and critical thinking to discern these rules. If it's strategies for success and advancement we need, then we can talk with others who have already advanced to find out how *they* did it and use similar strategies. If it's encouragement we'd seek from a mentor, we can find encouragement in our peer networks and develop the skill of encouraging ourselves, constantly reminding ourselves that the skills and gifts that have brought us to this point will lead to future success. Self-mentoring is a powerful way for young women to take responsibility for their own professional welfare and advancement when the circumstances they find themselves in are not ideal.

> **Reflections from Saundra's Daughter**
> **Dr. Stephanie N. McGuire**
> My mother has written that her career path "was the result of flexibility, family priorities, and serendipity," but she conveniently left out her Herculean work ethic, penetrating intelligence, and boundless optimism and energy. (Seriously, the woman can function at near optimal capacity on frighteningly little sleep.)
>
> My mother is simultaneously flexible and unyielding. Her strong faith informs her that whatever unfolds in her life will always be the best for her, but she never uses that belief as any kind of cop-out; on the contrary, she always goes above and beyond her human responsibility to make the most of every opportunity given to her. As part of this book project, my mother was asked to list significant obstacles in her career. Knowing what she endured growing up in the segregated South, I was astonished to discover that she believed she had encountered none. It was then that I realized my mother doesn't even *see* obstacles, only opportunities.
>
> I cannot overstate the impact of growing up with a working mother who also passionately loves her work. Seeing my mother's name on the spines of books mightily impressed me as soon as I was old enough to read. I loved it when my mom picked me up from school, and I also loved seeing her in the dining room, poring over texts and preparing lectures, books scattered as far as my child's eye could see over the lacquered surface upon which, on other nights, sumptuous meals would be served at family gatherings. I remember my mother full of happiness, always generous with her time, and endlessly interested in what we were doing even though her life was clearly full to the brim.

(continued)

Sometimes I would see her face bearing a scowl of concentration, familiar to anyone who loves someone who thinks for a living, but soon enough it would be replaced with the broad grin of satisfaction she often wore when she returned home late after a lecture. Sometimes she returned with a wearier expression, evidently exhausted, but she never, ever complained.

I was a taxing child and an impossible-to-please teenager. But my mother never tried to tone down my intensity or strip me of my intellectual independence, even when we vehemently disagreed. Of course, she was still a parent, so she did want me to wear patent leather shoes for Easter and behave demurely in front of her parents, which I tried my best to do, even as I was full of righteous indignation.

My mom's passion for and enjoyment of her own work have enabled her to relate to and support the different twists and turns I have taken in my life. Music? *Great!* Organic farming? *Why not?* Editing? *You'd be so wonderful at that!* In fact the only thing she ever outright vetoed was my plan to do stand-up comedy. Wise woman.

The older I get, the more "siblings" I discover I have, people who call my mother "mom" because of the outsized impact she made on their lives. She is so much more than an educator and mentor. Empowerment is a word that is often thrown around, but my mother is a vessel for empowerment. You know that proverb, *Give a man a fish. . .Teach a man to fish. . .?* My mother is like a master fisherman who knows deep in her bones that everyone else can become one too. She has made it her life's mission to spread that message of justice and equality.

Acknowledgments There are so many people to thank for the successes that I have experienced throughout my life.

First I want to acknowledge and thank the wonderful family of educators into which I was born. My grandmother, Mrs. Effie Jane Gordon Yancy, my parents Mr. and Mrs. Robert (Delsie) Yancy, Jr., and most of my aunts and uncles were outstanding educators who inspired their students to excel. Special thanks also go to my brothers, Robert Yancy III and Dr. Eric A. Yancy, and to my sister, Annette L. Yancy, for their part in the wonderful childhood (and adult) experiences that continue to inspire me.

I am extremely grateful for the unwavering support, encouragement, guidance, and inspiration that my husband Steve, the love of my life, has provided throughout the four and a half decades of our marriage. And I offer sincere thanks to our daughters, Dr. Carla McGuire Davis and Dr. Stephanie McGuire and to our grandchildren, Joshua, Ruth, Daniel, and Joseph Davis, all of whom have been willing subjects as I used them to test the effectiveness of the learning strategies that are the foundation of my current work.

Sincere thanks go to the colleagues with whom I have worked at various institutions and organizations over the years—especially colleagues at the LSU Center for Academic Success and the Department of Chemistry, the National Organization for the Professional Advancement of Chemists and Chemical Engineers (NOBCChE), and the American Chemical Society (ACS). I especially thank the ACS Women's Chemists Committee for including my story in this book. And last but certainly not least, I am truly grateful for the thousands of students who I taught over the years and from whom I learned even more!

About the Author

Education and Professional Career

1970	BS Chemistry, Southern University and A & M College, Baton Rouge, LA
1971	MAT Chemical Education, Cornell University, Ithaca, NY
1983	PhD Chemical Education, University of Tennessee, Knoxville, TN
1983–1988	Assistant Professor of Chemistry, Alabama A & M University, Huntsville, AL
1988–1991	Assistant Director and Lecturer, Learning Strategies Center and Department of Chemistry, Cornell University, Ithaca, NY
1991–1997	Associate Director and Senior Lecturer, Learning Strategies Center and Department of Chemistry, Cornell University, Ithaca, NY
1997–1999	Acting Director and Senior Lecturer, Center for Learning and Teaching and Department of Chemistry, Cornell University, Ithaca, NY
1999–2009	Director, Center for Academic Success and Adjunct Professor of Chemistry, Louisiana State University, Baton Rouge, LA
2009–2013	Assistant Vice Chancellor and Professor of Chemistry, Louisiana State University, Baton Rouge, LA
2013–present	Retired Assistant Vice Chancellor, Professor of Chemistry, and Director Emerita, Center for Academic Success, Louisiana State University, Baton Rouge, LA

Honors and Awards (Selected)

2017	American Chemistry Society Dreyfus Award for Encouraging Disadvantaged Students to Pursue Careers in the Chemical Sciences
2017	Induction into the Louisiana State University College of Science Hall of Distinction
2015	American Association for the Advancement of Science Lifetime Mentor Award
2014	National Organization for the Professional Advancement of Black Chemists and Chemical Engineers Lifetime Achievement Award
2012	Council of Learning Assistance and Developmental Education Associations Fellow
2011	American Association for the Advancement of Science Fellow
2010	American Chemical Society Fellow

Saundra remains actively engaged with speaking and writing about her research on metacognitive strategies for increasing student learning. She has presented her work in 46 states and 9 countries on 5 continents. Her latest book, *Teach Yourself How to Learn: Strategies You Can Use to Ace Any Course at Any Level*, was released by Stylus Publications, LLC in January 2018.

Family Matters: How Family Influenced My Career

Jin Kim Montclare

Photo Credit: Jin Kim Montclare

A Love for Science and Importance of Family

My family emigrated from Korea, and I was born and raised in the Bronx. I was taught to value education, and my parents worked several jobs just so my brother and I could go to good schools. I loved math and science all throughout my life and got to experience research during my high school career. My love was fostered by my teachers, and, while I never considered an academic scientist as a career, it was my teachers and mentors who led me on my path.

During high school, many things happened that really changed my life. First, my younger brother was diagnosed with cancer (osteogenic sarcoma, his right femur), which resulted in years spent in the pediatric cancer ward at the hospital. Fortunately, his life and leg were spared. Second, I met a boy who would later become my husband. After school, he would drive me to the hospital to see my brother.

Marriage and Pursuit of PhD

By the time I enrolled in college, my brother was thankfully cancer-free and my boyfriend was already attending Fordham University as an English major. I majored in chemistry and minored in philosophy. I made two crucial decisions during college, neither of which made my parents happy. I decided to pursue a PhD in chemistry, instead of medicine, and I asked my high school sweetheart to marry me (we have now been married for 20 years)—all of this happening in my senior year of college.

An important thing to note here is that having a supportive significant other has been crucial to me being able to balance my career and family. One of the best decisions I made was to marry my husband, Brandon. He was the one to encourage me. He said that since I loved chemistry, I should consider a PhD. He was also always there to help me cope with family and professional challenges along the way. As you will read later, he also made some significant choices so I could pursue science.

My PhD Advisor and Female Role Model

I decided to work in the burgeoning field of bioorganic chemistry/chemical biology under Alanna Schepartz at Yale University. What inspired me was her research on DNA-binding proteins and creating artificial mimics. She was the first and only female chemistry faculty at Yale when I was a PhD student. As such, I was able to have a direct role model of a successful female academic.

As a PhD student working long hours and experiencing a multitude of failed experiments, I was not sure that I wanted to pursue an academic career. In fact, one day I was expressing my doubts about it very loudly in the lab that shared her office wall. Alanna came out of her office, walked to our lab door opening, and told me that I should not doubt going into academia and that I was capable. Her words of encouragement were very important to me, and, to this day, I am grateful for them, because they have served me well.

Importantly, I also saw how Alanna was able to manage her growing family while being a professor. When I started in the lab, she already had one daughter. During my graduate work, she had her son. In fact, I would babysit her eldest when she brought her to the lab and at her home, occasionally. I recognized the flexibility of an academic position, especially when having a family.

Family Matters: How Family Influenced My Career 347

Comic Books and Research

While I worked toward my PhD at Yale, my husband owned a comic book shop. He had co-owned this shop with his business partners when we were both in college and had continued on with it. I would spend some of my weekends helping him sell overstock at comic book conventions.

After completing my PhD degree, I was offered a postdoc position at Caltech in Dave Tirrell's group. As it would be impossible for him to continue on and be in the same state as me, Brandon made the decision to sell his portion of his shop. We packed our Toyota and drove to Pasadena, where I began my postdoc.

Job Search Distraction: Cancer Again

In California, my husband enrolled in a master's program in English and worked on adaptations of Manga/graphic novel into English as a freelance editor for TokyoPop. I started to seek faculty positions and was beginning the interview process when my father was diagnosed with stage 4 colon cancer. I flew back home to New York in the midst of job searching, and, in between interviews, I would visit him in the hospital. Somehow during this blur, I managed to secure offers from a couple of institutions.

While this would normally be a time for me to be excited and select an institution to start my academic career, I could not. I was focused on my father's chemotherapy treatment and desperate for his survival in his battle with cancer. My family had fought this disease before, and I was not going to give up. As I was dealing with this, I realized that I couldn't start a faculty position. I declined all the offers, choosing to focus on my father. Miraculously, my father responded to the treatments and surgeries and, against terrible odds, was cancer-free.

During this time, my postdoc advisor, Dave, was supportive and provided me the time I needed to address all of this. He also did not pressure me in any way to make a specific decision, but rather helped me assess what my priorities were at the time.

Priorities Adjustment: Comic Books and Faculty Position

With my family back on track toward recovery and my head back in science, I decided to carry out another faculty search the following year. This time, I was focused without distractions and began the interview process.

Midway through this, my husband landed a position as an assistant editor at his dream company—DC Comics in NYC! For those of you unaware, DC Comics is the publisher of Superman, Batman, and many well-known characters. After letting go of his ownership of his comic shop and being offered a position where the whole industry of comics began, it was my time to shift my priorities for my job search—I had to find something close to NY. I had offers from various schools, and fortunately one of them was in NY!

I started as an assistant professor in a joint department of Chemical and Biomolecular Sciences and Engineering at Polytechnic University. As soon as I started, the department split, and I got shuffled into the Chemical and Biological Sciences.

Several things that were promised in my offer did not materialize, including the key thing of my lab being set up. An important thing I started doing was to request meetings with my department head and facilities, getting them on schedule to get my labs completed. It was a tumultuous time since the departments were reorganizing. Rather than simply wait for things to settle, I learned early on that it helped to initiate coordination while keeping a record of meetings and emails so that things would not be forgotten. At the same time, I experienced several department head, dean, and provost changes, which also put me in the habit of keeping detailed records—something I encourage everyone to do.

Family Planning and Academic Mergers

As I began building my research profile and securing grants, my husband and I decided to try starting a family. At first we thought this would be simple. However, it took us one full year of trying in order for me to get pregnant.

Because it took so long and we were afraid of potential miscarriage, we didn't tell anyone until I was through my first trimester. I also was extra careful about being protected when in the lab with my students. I then notified my department head and set up maternity leave as well as my tenure clock pause.

This was also an interesting time, as my institution was being integrated with my new institution, New York University. The year 2008 was the start of a transition to become Polytechnic Institute of NYU. This was accompanied by another set of new administrators and departmental changes that would ultimately lead to the merger of the split department and the formation of the Department of Chemical and Biomolecular Engineering.

As I prepared for both the transition of being a parent and being in a new department, I realized that I couldn't oversee every little thing in the lab and department, especially when I would be on maternity leave. I began slowly delegating responsibility to the graduate students in my lab.

One of the first things I set up as my lab grew in numbers was a subgroup leader for each project. Any issues that arose with students were supposed to be dealt through the subgroup leader and then reported to me.

Another thing I did was set up an online system to monitor projects so I could manage things from home. Finally, I asked my group to have group meeting at my apartment so that I would be able to nurse and minimize time away from the baby.

Thankfully I had set this all up ahead of time as my daughter, Violet, arrived two weeks earlier than expected via an emergency C-section. While I was spending time recovering, I felt that my lab was in good shape to continue to function.

Fig. 1 Montclare group photo New York University in 2012: (left to right) Haresh More, Ching-Yao Yang, Jasmin Hume, Carlo Yuvienco, Jennifer Sun, Andrew Olsen, Jin, Violet (Jin's daughter), Rudy Jacquet, Ekta Sharma, Joseph Frezzo, Liming Yin, and Jennifer Haghpanah

Family and Lab Gratitude

Once I recovered, we began group meetings at the apartment with the baby in attendance. To this day, I am grateful to my students during that time. They continued to do awesome research and helped me manage things with the baby. As Violet grew older, she continued to be part of the lab, and, as Alanna's daughter did with me, my daughter bonded with many of my students (Fig. 1).

While I've always been thankful to my group, I have learned to express it and make a point of it after having my daughter. I think the most important thing I learned was to trust others and give them opportunity to lead. Every day, I am amazed by how my students, technicians, and postdocs rise to the occasion and come up with interesting experiments and conduct the very best science.

Child Rearing 50:50

While I was coordinating my return from maternity leave, my husband and I were going over daycare options. While there were plenty of options, Brandon decided to quit his job as an editor at DC Comics and work as a freelance writer in order to spend more time with the baby. This was something I did not suggest, and I tried to convince him otherwise—especially since he loved his job. However, he was adamant and took over the responsibilities of child rearing while I went back to work.

During the days he would take care of Violet, and, as soon as I got home from work, it was my turn so that he could write. We split the responsibilities 50:50, and, to this day, our daughter has an especially strong bond with her dad.

What I find quite interesting is how raising Violet has impacted Brandon in his writing. As a writer, he has written more and more comic books with young, female protagonists—many of them strong-willed and smart and several of whom love science! In fact, some of the storylines suggested by Violet made it into books. I get to participate by making sure the science is accurate and feasible. Most recently, I was profiled in Marvel's Unstoppable Wasp to inspire comic book readers, specifically girls in STEM (https://news.marvel.com/comics/59499/unstoppable-wasp-meeting-minds-pt-2/) (Fig. 2).

Branching Out and Expanding the Family Again

After securing tenure (a separate story for another time), I have branched out into other interests including entrepreneurship. Because my group continues to self-lead, I have been able to work on start-ups and educate others within NYU about entrepreneurship.

Fig. 2 Jin as illustrated by Elsa Charretier for issue 2 of Marvel's Unstoppable Wasp Agent of G.I.R.L

Fig. 3 At the 65th ACS Undergraduate Research Symposium at Fordham University in 2017: (left to right) Jin's mother Clara Kim, daughter Zinnia, Jin, undergraduate advisor James Ciaccio, and undergraduate department head Diana Bray

Just recently, we had another baby girl, Zinnia. This time around, I embraced technology as a means of communication. While on maternity leave, I would google hangout into group meetings with the baby on a sling and coordinate projects from home. Again, I was grateful that my group was accommodating. As with my first, my second continues to be part of the lab, occasionally interrupting with babbles. Since my oldest is seven years older than the baby, she has been a great help with taking care of her baby sister.

Most recently, I was honored to present the keynote lecture at the 65th ACS Undergraduate Research Symposium at my undergraduate alma mater, Fordham University. My family came with me to the event, and I was able to introduce my children to my chemistry professors who inspired me at the beginning of my journey into academia (Fig. 3).

Reflections and Advice

As I reflect on my path in academia and being a mom, there are several important things that helped me.

1. *Prioritize!* I had learned this early with my family battles with cancer. Identify what key things are most important to you. While these may change, it is important to evaluate and identify what your priorities are.
2. *Surround yourself with support.* My husband has been my number one supporter throughout my career, cheering me on while also equally sharing the

responsibility of child rearing. This has been key for juggling the demands of academic and family life.

3. *Delegate.* While delegating responsibilities may feel like one is losing control over something, it is not. By giving others responsibility, they are able to exercise leadership and rise to the occasion. My students and postdocs participate in grant writing with me and are able to meet seminar speakers when I cannot. This provides them with extraordinary exposure and writing skills. At the same time, this frees up my time to attend certain conferences, work on a collaborative projects, and spend it with my family.

4. *Document everything.* As a professor, you will have to deal with coordinating people, and while this is not a skill that is taught as a PhD student or postdoc, it is something that one can learn. To help yourself in management, document everything. It is good to provide follow-ups of meetings so there is a record and mutual understanding.

5. *Embrace technology.* With Skype, Google Hangout, Dropbox, Evernote, Slack, etc., there are a multitude of applications that you can use to make your life easier in terms of sharing data and interacting virtually. Use it to help you balance work with family.

6. *Thankfulness.* Always be thankful and provide appreciation to those who have helped you. Pay it forward and help others. This also helps you develop a network of support.

For me, my family has directly influenced my career. Since time is precious, I choose to do things that I believe are worthwhile. While this is hard to do, it becomes a necessity for me, especially since I want to spend time with my family. Because I am passionate about both my research and family, I have found that my family and laboratory work have somehow permeated each other.

Since it was important for me as a PhD student to see my advisor balance her life and career, I do the same for my students and postdocs. They are exposed to my children when I bring the kids to work, invite my lab to my home, and coordinate various celebratory parties for my students. My group understands when interruptions happen due to a sick kid, as I understand when things come up in their lives.

Acknowledgments I acknowledge my extraordinary husband (Brandon Montclare), two daughters (Violet and Zinnia Montclare), parents (Clara and John Kim), and brother (Peter Kim) for keeping me level headed. I thank my teachers throughout K-12, mentors from Fordham University (Diana Bray, James Ciaccio, Shahrokh Saba, Robert Beer, Father Cloney, Father Heggie), undergraduate research mentors from Albert Einstein College of Medicine (Raju Kucherlapati, Kate Montgomery), PhD advisor (Alanna Schepartz), and postdoctoral advisor (David Tirrell). I am grateful to *all* the former and current students, postdocs, and technicians who have worked in my lab. Finally, I thank my Executive Leadership in Academic Technology and Engineering (ELATE) cohort and countless colleagues who have provided me sound advice and sanity, when needed.

About the Author

Education and Professional Career

1997	BS Chemistry major, Philosophy minor, Fordham University, Bronx, NY
2003	PhD Bioorganic Chemistry, National Science Foundation Predoctoral Fellow, Yale University, New Haven, CT
2003–2005	National Institutes of Health Postdoctoral Fellow, California Institute of Technology, Pasadena, CA
2005–2012	Assistant Professor, Department of Chemical and Biomolecular Engineering (CBE), New York University (NYU) Tandon School of Engineering, Brooklyn, NY
2012–present	Associate Professor, CBE, NYU Tandon School of Engineering, Brooklyn, NY
2014–present	Associate Director for Technology Advancement for the NYU Materials Research Science and Engineering Center, Brooklyn, NY
2015–2017	Graduate Studies Director for CBE, NYU Tandon School of Engineering, Brooklyn, NY
2017–present	Director, Convergence for Innovation and Entrepreneurship Institute, NYU Tandon School of Engineering, Brooklyn, NY

Honors and Awards (Selected)

2016	Rising Star Award, Women Chemists Committee, American Chemical Society
2015	Agnes Faye Morgan Research Award, Iota Sigma Pi
2014	Executive Leadership in Academic Technology and Engineering Fellowship
2014	Distinguished Award for Excellence, Dedication to Invention, Innovation and Entrepreneurship
1997–2000	National Science Foundation Graduate Research Fellowship
1997–2000	Pfizer Fellowship
1996–1997	Barry M. Goldwater Scholarship
1995–1997	Clare Boothe Luce Scholarship, Henry Luce Foundation

Jin is performing groundbreaking research in engineering proteins to mimic nature and, in some cases, work better than nature.

Mother and Chemist: Every Pitfall is an Opportunity to Rise with a New Beginning

Ingrid Montes-González

Photo Credit: Peter Cutts

My goal in this chapter is to share with you what has been my trajectory throughout life. As for many, indeed most women, but particularly those with the dual roles of mother and professional, it has been full of challenges, both as a woman (daughter, sibling, wife, mother) and as a professor of chemistry. It has certainly been one filled with blessings, satisfactions, and achievements. In addition, I have had the opportunity to fulfill my mission of service for the benefit of our profession and the community in general.

I. Montes-González (✉)
Department of Chemistry, College of Natural Sciences, University of Puerto Rico at Río Piedras, San Juan, Puerto Rico
e-mail: ingrid.montes2@upr.edu

Fig. 1 Ingrid at her Bachelor's degree graduation in 1980

Let's Get to Know More About Ingrid!

I am the youngest of three children and have one brother and one sister. My mother worked for a while as a secretary, and she always took care of us. My father was a veterinarian and, from a young age, encouraged my scientific curiosity. Since I was a little girl, I liked science, especially observing my father when he was working with animals, both in his clinic and on farms. I always loved playing with and caressing all animals, as they have always been my passion.

Undoubtedly, my father and my science teachers, especially my high school chemistry teacher, were key role models in motivating me to pursue a degree in chemistry. Although my initial goal was to become a veterinarian, organic chemistry courses led me to becoming passionate about research, and I chose to seek my doctorate in organic chemistry. Certainly, an excellent research mentor and role model, Dr. Gerald Larson, inspired that decision.

At the age of 21 at the end of my bachelor's degree in chemistry (Fig. 1), I made the decision to marry. In making that decision, I rejected the options of pursuing my graduate studies in the United States. For many years, it did not seem like a good decision; however, I can now say that it was. This decision was greatly influenced by my Latino culture, which in those years had as a rule that women should marry very young and, once married, take on the primary role of being a wife and having children. That is, the professional aspect should be considered afterward, if at all.

My father influenced and motivated me to pursue graduate studies at the University of Puerto Rico, Río Piedras Campus. I will never forget his words: "The only thing that really belongs to us in life and nobody can take away is our education. That is why we must aspire to the highest degree, that will assure you more and better opportunities in life ...." These words marked my life from that moment, and my intention is to encourage every reader to sustain them forever.

I undertook my doctoral studies in organic chemistry under the supervision of Dr. Larson, who was also my research advisor as an undergraduate student. Being married and a graduate student at the same time involved having a lot of organization, rigorous time management, and extensive efforts devoted to my studies as well as quality time for fulfilling my personal responsibilities.

After I completed my first year and qualified successfully for the doctorate, I became pregnant with my son Gerardo José. Certainly the blessing of becoming a mother added other responsibilities. Initially, a lot of stress was added because his health was very delicate for his first 3 months. At that time I thought that I could not continue my studies; however, thanks to the support I received from the whole family and mentors, I overcame that belief and was able to finish my PhD (Fig. 2). Of course, it was a great challenge, full of many sacrifices and responsibilities, to be able to fulfill the roles of mother and wife and to continue with my research and my commitment to complete my doctoral degree. Every time I looked at my son growing up and smiling, I was inspired and got from him all the strength required to continue.

Fig. 2 Ingrid at her PhD graduation in 1985

Fig. 3 Ingrid with her daughter Mariana and son Gerardo in 1986

After I graduated, I had several job offers from industry, all of which were excellent. However, I felt strongly that I could contribute more in academia. I got my first job in the School of Pharmacy at the University of Puerto Rico, Medical Sciences Campus. I raised funds for my research, recruited students, and successfully became a researcher and professor.

That was when I became pregnant with my daughter, Mariana del Carmen. She was another blessing! At 2 months of age, she also had health complications, so I had to give up my newly started career. It was a very difficult time in my life because it was also complicated with my divorce; no doubt my children gave me all the strength and courage to move forward (Fig. 3).

The Sun Always Shines

After my daughter overcame her health crisis, I set out to find a new course in my professional life. It was precisely at that moment that I implemented the words of my father. I promptly began as a part-time lecturer in the Chemistry Department of the University of Puerto, Río Piedras Campus, the same institution where I completed my doctorate. Initially, it was somewhat difficult because my teachers were now my

Fig. 4 Ingrid with her daughter Mariana and son Gerardo in 1987

colleagues. I had the challenge of opening my way professionally and at the same time taking care of my two children: Gerardo (five years old) and Mariana (1 year old) (Fig. 4).

During the early stages, I taught several courses and their respective laboratories. I understood that many changes were needed in the teaching-learning process of organic chemistry, and I decided that this was my niche to begin with. In the teaching position that I held, I was not allowed to do research. Nevertheless, I committed myself to study education. Thanks to many mentors, I was able to start doing research in chemistry education and get my first publications in that area. I devoted myself to the development of new experiments for the organic chemistry course. Also, I sought to better understand the problems of the learning process and to do research in the area. Within the experiments that I was developing, I was immersed in what is now one of my research areas: Synthesis of derivatives of ferrocenyl chalcones.

My research was evolving as my children grew and became more independent, allowing a more flexible schedule for me to engage in research. I also obtained a tenure-track position and quickly moved through the ranks, becoming an associate professor in 1992 and full professor in 1998.

I admit that it was not easy to fulfill my professional development and my role as a mother. I remember how I had to organize myself to be able to attend my children's sports events, help them with their homework, and drive them to their extracurricular activities, especially during the summer time.

My two kids were always leaders and very cooperative with their classmates. They volunteered to bring all the goodies (sandwiches, brownies, etc.) or meals to cook for their different activities. According to their teachers, they always claimed, "My Mom knows how to do the best one." One day, I decided to ask them why they always volunteered for those things, and they explained that the other kids always volunteered for things that their moms could buy directly from the supermarket, so someone needed to bring the "other things." At that time, it was stressful and not funny at all. Now I always remember this with nostalgia and with a big smile.

Currently I have a research group that involves two areas: (1) synthesis and characterization of ferrocene derivatives, mainly based on ferrocenyl chalcones, with potential applications in molecular materials, such as redox sensors, polymers, and drug design, and (2) chemical education research on the instruction, the practice, and the assessment of organic chemistry, based on the theoretical perspective building on constructivist and meaningful learning theory. There are some graduate students working simultaneously in both areas.

Something very special to me is to see my research students growing intellectually and professionally. It is a satisfaction that could be compared to a mother seeing her kids growing up. I have supervised 8 graduate students and over 80 undergraduate research students, and I am very proud that 75% of them have continued graduate studies (chemistry, PharmD, or medicine).

A New Horizon Appears

Since my beginnings in the Department of Chemistry at University of Puerto, Río Piedras Campus, I committed myself to helping the members of the American Chemical Society (ACS) Student Chapter. Since 1989, I have been their faculty advisor, a role that I am proud to be in. It has been one of the most rewarding tasks that I have undertaken. For the past 24 years, the chapter has been recognized with the ACS Outstanding Chapter Award. The students' energy, commitment, professionalism, and dedication have always given me strength and enthusiasm.

It was as the chapter faculty advisor that I started to become involved with ACS. My first step was serving as an evaluator of student chapter annual reports. On that occasion I met many people who became my mentors and encouraged me to continue my voluntary service.

I began to work on activities as part of National Chemistry Week (NCW), not only with my chapter but also with other ACS Student Chapters in Puerto Rico. My interest grew and evolved much more than I could imagine. Currently, I continue to coordinate activities that involve all the ACS Student Chapters of Puerto Rico and recently also the ACS High School ChemClubs. I am still the coordinator for NCW and Chemists Celebrate Earth Day (CCED), and I can count on over 500 volunteers! These future leaders and their outreach activities will positively influence the image

Mother and Chemist: Every Pitfall is an Opportunity to Rise with a New Beginning 361

and understanding of chemistry's role in daily life of all citizens they reach. The thousands of elementary, middle, and high school students who have been reached through different events of the NCW and CCED activities in Puerto Rico will become better citizens by being exposed and informed about chemistry and its importance.

My volunteer engagement with the ACS gradually led me to participate in tasks related to education at the national level. I worked as a member of the task force for undergraduate programming, then coordinating the First Latin American Congress for undergraduate students, and as a member of the Society Committee on Chemical Education. I also became involved in the Division on Chemical Education Program Committee and Long Range Planning Committee, which I chaired for several years. I was also a member of the advisory board for the magazine *ChemMatters*, which I also chaired for 3 years. My biggest contribution to this wonderful magazine was to translate one of the articles to Spanish for the benefit of Hispanic teachers. I am currently a member of the Editorial Advisory Board of the *Journal of Chemical Education*.

Since my first steps, I was also part of the NCW Task Force, which evolved and is now the Committee on Community Activities. I was privileged to chair this committee for 3 years. While on the committee, I also was involved in the initiative to translate the publication *Celebrating Chemistry* into Spanish, an activity that continues today.

All the aforementioned experiences helped me in developing leadership skills, confidence, and the ability to work closely with a host of wonderful people. They expanded my networking and, most importantly, gave me volunteer service opportunities outside the boundaries of Puerto Rico.

In 2005, I proposed the establishment of *Festival de Química* to the current ACS President Dr. Bill Carroll, suggesting that this outreach event focused on the Hispanic population be held as part of the ACS National Meeting held that year in San Diego, California. He embraced the idea. Thanks to the many volunteers and to the contribution and support of all faculty advisors and the Puerto Rico ACS Student Chapters that eagerly worked for the event, it was very successful. After the success in San Diego, we decided to adopt that event in Puerto Rico. Since then, it is celebrated twice a year as part of NCW and CCED.

The *Festival de Química* is a community outreach event that, through demonstrations, emphasizes the importance of chemistry in daily life. Initially designed, developed, and implemented in Puerto Rico, this model was adopted by ACS in 2010 to broaden its impact internationally. Having become the ACS Chemistry Festival program in 2015, events have been successfully implemented in 15 countries around the world, positively impacting thousands of people. This program gave me the opportunity to continue expanding my impact and volunteer service internationally and to train students and volunteers in many countries, including Colombia, México, Chile, Panama, China, and Malaysia, among others. Serving as co-founder and coordinator for the Spanish webinar series of the American Chemical Society and Mexican Chemical Society provides additional opportunities for international contributions.

In 2013 I was elected to become Director-at-Large, being the first Hispanic serving on the ACS Board of Directors. In 2016, I started my second term as Director-at-Large. Being a board member is one of the greatest and most important responsibilities that I ever dreamed to have. It is a different way to contribute to ACS, to our profession, and to continue my voluntarism.

In addition to serving at the national level, I have also been very active at the local level. I have occupied various leadership positions on the ACS Puerto Rico Board of Directors, including being a Councilor and Chair of the Local Section three times. I am still the outreach coordinator and am also the Project SEED Program coordinator, a very successful program in Puerto Rico.

When I started my voluntarism in ACS, I never imagined how it would change my life. As I explained in the first section of this chapter, my education, professional development, and personal life were limited to Puerto Rico. With this limited outside exposure and also the influence of my culture, I felt insecure when speaking and expressing my opinions. However, ACS helped me develop leadership skills and provided me the networking to find wonderful and inspiring mentors. Moreover, they have enabled me to mentor and inspire thousands of students nationally and internationally, including many females and Hispanic women.

Every Pitfall Is an Opportunity to Rise with a New Beginning

Reflecting on my life, I can only say that I have been blessed every day. When we are young and things are not going perfectly or according to what is our plan, the tendency is to think that it is something negative. However, after my many challenging experiences, I have to be grateful that I had those situations in my life. In every challenge, I grew and became stronger as a woman, mother, and professional.

As a Hispanic woman, I had faced discrimination, the "machismo" aspect of my culture, and diverse circumstances. Without any doubt, that was important for becoming more sensible and aware of people's needs and brought to my life new perspectives, mentors, and opportunities to mentor others.

All my experiences have enriched me and have increased my commitment to motivate the younger generation, especially females. My message to you all is to always stay positive, believe in yourself, keep focused on your goals, and never expect anything from others. You cannot control others—you only control what you give to others, and that will be your greatest satisfaction!

I have been blessed with my career as a professor. I have found wonderful colleagues and mentors who always inspire me. Moreover, I feel privileged for being part of the formation of the future leaders and professionals. I am extremely proud of the accomplishments of each one of my students, especially from my research students, and those students have sparked and marked my pathway. My philosophy has been and will continue to be to assist students in their career development by encouraging networking, teamwork, effective communication, and critical thinking. In addition to this role, I strongly believe in the development of students' character and leadership by encouraging strong work ethics, service awareness, and building self-esteem and self-insight. Together, this combination and foundation help to instill and inspire integrity and a better overall professional.

Being a professor has been very rewarding and an extension of my motherhood. I have a very unique relationship with my students, sometimes a little bit maternal, that enables me to share my experiences in life. That is probably the reason that they use to call me "Mommy Ingrid"! Because they are so important in my life, I asked them to contribute their reflections to this chapter.

> **Reflections from Ingrid's Students**
> Even when you have the full support from your family and friends, sometimes you need that extra push to make you keep moving forward and reach for higher goals in your life. That is the value of having Dr. Ingrid Montes.
>
> For many students, the word motherhood for her is not a compliment, rather a statement: being a person who dedicates her life to motivating and guiding in countless aspects, with a spark of inspiration to serve as your role model. From the person who once was only your organic chemistry professor arises a true scholar, your research advisor, your mentor, and your friend.
>
> Chemistry is not only a research theme, as chemistry in a student-advisor relationship is crucial for a fructiferous future in your career, in your decisions, in opportunities, in achieving your goals, and in advancing in your life. Sharing her experiences and thoughts, she makes you feel you are in a world where the unachievable is achievable, where every pitfall is an opportunity to rise with a new beginning. She provides a different point of view, looking at the horizon as a north of your personal and professional goals.
>
> Thank you, Dr. Montes, for all your support but more than that, for opening new pathways into a life fulfilled with dedication, passion, and love to what we believe in.

Finally, at a personal level, I have also been blessed with my lovely and supportive family: my parents; siblings; my wonderful husband, Tony, who is also a chemistry professor; and my two kids. They have greatly inspired each day of my life and fill me with love, motivation, and happiness to deal with any challenge. Years have run away really fast! Gerardo is now 35 and became a doctor. Mariana is now 31, a nutritionist, and soon-to-be mother of a baby girl, Sofia Isabel, who will arrive very soon. Now perhaps I will become "Grandmommy Ingrid." That will be the starting of my role as a grandma! Another blessing to add and a new motivation to face any new challenges on my life! Please stay tuned because my next chapter will be "Grandma. . .the chemistry professor"!! (Fig. 5)

Fig. 5 From left to right, Ingrid's husband Tony, daughter Mariana, Ingrid and son Gerardo in 2017

Reflections from Ingrid's Daughter
Mariana del Carmen

Ingrid Montes Gonzalez—everyone knows her as a professor, professional, chemist, member of Board of Director of ACS, but many don't know her as mother and soon-to-be grandmother of a girl. Every moment that I think of my mother, I think of sacrifices of pure love. She raised us (Gerardo Jose and me) in a home full of love with the best education and family union. Every day she showed us her love since we woke up in the morning. I am 31 years old, and every day mom reminds me how important I am for her, no matter how far away she is with work travels.

Every day Mom taught me how to be a better human being, a strong woman, and a successful professional, sister, and wife. Shortly I will apply all the love she provided us, as mother of my first baby, always with her full-time support. I am sure that I can't do neither without her support and encouragement to be better person every day.

Mom is what her name "Ingrid" means—beautiful and beloved—from the inside to the outside of her soul. Mom is the type of person who maybe you think is always working. But I will tell you a secret of her days. She wake up very early, cooks, does the garden and laundry, organizes and cleans the house, answers every email, and then goes to work to fulfill all her responsibilities with her students and university. I do not mention all the things she is preparing for our little blessing who will come in September—her first

(continued)

granddaughter, who will be the one who changes our lives with illusion and happiness and more blessings.

She is a multitasking woman. I don't know how she does everything and never stops being the best one. I admire her as a mother, as a woman, and as a professional; she is my role model and my example. Every time that she receives recognition, my heart beats full of pride and happiness. I am proud of all the things she does, not only as a professional but also in her personal life. She deserves the best mother chemist award! She is my mom and best friend; my life does not make any sense without her.

Acknowledgments I want to thank my family—my parents, for all their love, patience, and support, and my sister and brother, my husband Tony, my daughter Mariana, and my son Gerardo for their support and for always being there for me, allowing me to reach my goals. My gratitude goes to all my undergraduate students, student chapter members, research students, and, in a very special way, to my graduate students (especially to Juan Carlos, Sara, and Johanna for their suggestions and feedback while preparing this chapter), who always gave me energy, enthusiasm, and much love. I also want to thank my mentors and, in a very special way, my research advisor, Dr. Gerald Larson. Finally, I thank the American Chemical Society for giving me many opportunities to serve and take my footprint far beyond what I ever imagined. All of them have been an integral part of the person, the educator, and the professional who I have turned out to be.

About the Author

Education and Professional Career

1980	BS Chemistry, University of Puerto Rico, Río Piedras Campus
1885	PhD Chemistry (Organic), University of Puerto Rico, Río Piedras Campus
1985–1986	Lecturer, School of Pharmacy, University of Puerto Rico, Medical Sciences Campus
1987–1992	Assistant Professor, Department of Chemistry, University of Puerto Rico, Río Piedras Campus
1992–1998	Associate Professor, Department of Chemistry, University of Puerto Rico, Río Piedras Campus
1998–present	Professor, Department of Chemistry, University of Puerto Rico, Río Piedras Campus
2015–present	Assistant Dean of Graduate Studies and Research, College of Natural Sciences, University of Puerto Rico, Río Piedras Campus

Honors and Awards (Selected)

2017   Award for Distinguished Women in Chemistry or Chemical Engineering, International Union of Pure and Applied Chemistry
2016   Special recognition by the Puerto Rico Senate
2013   Honorary Member, Golden Key International Honour Society
2012   Volunteer Service Award, American Chemical Society
2010   Fellow, American Chemical Society Fellow
2006   Leonardo Igaravidez Award, American Chemical Society, Puerto Rico Section
2006   Fellow, International Union of Pure and Applied Chemistry
1999   Chemical Education Award, American Chemical Society, Puerto Rico Section
1997   Excellence and Productivity Award, University of Puerto Rico, Río Piedras Campus

Ingrid has devoted her career to research in organic chemistry and chemical education and incredible extensive public outreach in Puerto Rico, Latin America, and the world. She has been an extremely active chemistry leader at the national, international, and local level through the American Chemical Society and the International Union of Pure and Applied Chemistry. Elected in 2013, she continues to serve as Director-at-Large on the ACS Board of Directors. She is the co-founder and coordinator of the Spanish webinar series for the ACS and Mexican Chemical Society and the founder of the Chemistry Festival, recently adopted as an ACS program. The Chemistry Festival is a major community outreach event that, through demonstrations, emphasizes the importance of chemistry in daily life. Since 2010, it has been successfully implemented in 14 countries around the world impacting thousands of people.

Taking an Unconventional Route?

Janet R. Morrow

The path to my current position as professor of chemistry at a large public university could be considered to be a conventional one. I became interested in biological and chemical research as an undergraduate college student. I pursued that interest in graduate school and received my PhD in chemistry. I was a postdoctoral fellow in two laboratories (one international and one in the USA) prior to landing my current job. Along the way I got married and had two children. When I embarked on my career 28 years ago, it was less common for women to major in chemistry, and very few moved into an academic position. This was especially true for women with children. When I look at my current working life and my path to it, I feel very fortunate to have landed in such a challenging and interesting job. There is nothing

J. R. Morrow (✉)
Department of Chemistry, University of Buffalo, State University of New York, Buffalo, NY, USA
e-mail: jmorrow@buffalo.edu

more rewarding than having a family, and there is nothing more enthralling than doing research.

Remarkably though, being a female faculty member in chemistry at a research institution in the year 2017 is still not as common as one might expect. Women are underrepresented as faculty members, especially given the large number of women majoring in chemistry. As a mentor, I have seen many talented women decide not to pursue an academic position even though I thought that they might have been happy and successful in taking this route. I write this brief account with the hope that it will guide junior scientists in their efforts to balance work and family. I hope my story will encourage women to pursue a career in academics.

Growing Up in LA-LA Land

I had a wonderful childhood in Southern California. My father was an electrical engineer who worked in industry. My mother was a nurse, although after the birth of four kids in 4 years, she practiced nursing only upon request by friends. She still has an encyclopedic knowledge of medicine and physiology and obviously takes much enjoyment in it. My father was a traditional working father, but he encouraged me and my three brothers to study science or engineering. In fact, he was pretty insistent that we should be scientists or engineers. Two of my brothers and I took his advice. Over dinner and on vacations, my dad would question us about how we thought the physical world worked. He would ask us how we thought electrons might travel in different media and discuss what he knew about atoms and molecules.

My three brothers and I bodysurfed, skied, climbed, and hiked our way through childhood. We spent hours in the backyard swimming pool playing Marco Polo and soccer. My family traveled throughout the Western USA for vacation, camping and backpacking along the way. My father noticed my interest in ocean life and encouraged me to study marine biology, a trendy topic at the time because it was forecast that humanity would need to cultivate the oceans for food.

College in Santa Barbara

I went to the University of California, Santa Barbara (UCSB), to study marine biology. The only universities I considered attending were different campuses in the University of California system. The UC campuses are renowned for science, and the tuition was a bargain at that time. My parents had to plan for having four children in college at nearly the same time, and this was clearly a factor in the decision. It turned out to be a good choice for me because I was able to take a range of courses to try out different science and engineering disciplines as well as to start undergraduate research projects early on in my studies.

The introductory science courses at UCSB were large (my biology class was over 400 students), and I wanted a more individualized experience. On the recommendation of a friend, I started research in the laboratory of Professor Barbara Prezelin in the biology department during my sophomore year at UCSB. I loved the research, and I enjoyed being part of a larger team of people with a focus on research. I spent a summer and a couple of academic years culturing dinoflagellates and attempting to isolate a type of chromoprotein that was involved in light gathering for photosynthesis. On one memorable field trip, we rented a boat from the Scripps Institution of Oceanography and collected samples of dinoflagellates along the coast of Santa Barbara. Another memorable experience was trying to get my dinoflagellates to grow in seawater media. At one low point in my project, I paddled out into the lagoon in the center of UCSB and collected samples of bird guano to determine if that would help the dinoflagellates grow. I finally found that the missing ingredient was EDTA, a metal chelator. That experience launched my interest in metals in biology.

My protein isolations and purifications were difficult, and I was frustrated by my lack of background on how to improve separations. I enjoyed thinking about science in terms of molecules and liked the way that chemists could propose solutions to my questions on protein purification and chromatography. I transferred to the chemistry department in my junior year and started two new undergraduate research projects, one in ultrahigh vacuum studies of platinum surfaces with Professor Arthur Hubbard. Then I moved on to Professor Peter Ford's group to study the photochemistry of platinum organometallic complexes. After I graduated with my BS in chemistry and was waiting to attend graduate school in the fall, I worked on the preparation and characterization of micelles under the direction of Professor Henry Offen.

UCSB was a great place to do undergraduate work. Not only were the research opportunities fascinating to me, the ocean was right there lapping at the edges of campus. My boyfriend Boyd (later to become my husband) and I had a boat, and we took it to go scuba diving out at the Channel Islands to collect abalone and swim with the sea lions.

The experience of working in four different research laboratories in different areas of chemistry/biochemistry influenced my approach to science in later years. Chemical research is becoming increasingly interdisciplinary, and this early experience encouraged me to dive into new collaborative research projects later in my career. I planned to go to graduate school because I so enjoyed these research experiences. I toyed with working for a year at a job in the food industry but in the end decided to go right to graduate school on recommendation of my advisors. Spending a year in industry might have been beneficial, but clearly given the relatively long period required for graduate school, it was best to get through with little delay given that I was sure I wanted to pursue a graduate degree.

Graduate School: Moving East

I chose to attend graduate school at the University of North Carolina, Chapel Hill, based on the strength of the department in analytical and inorganic chemistry. At the time of sending out applications, I couldn't decide which of these topics to pursue. Once I got there, I decided to study organometallic chemistry in the research group of Professor Joe Templeton. I liked the combination of synthesizing molecules, coupled with spectroscopic and structural characterization of the molecules as well as theoretical calculations.

The big change in scenery and culture from Southern California was a maturing influence. For fun, I learned to white-water kayak and to rock climb (Fig. 1). Chapel Hill was rich in music, especially Blue Grass and Irish music. I remember that time fondly.

Research went well and the bulk of my work was published prior to graduation. This was a tribute to my graduate advisor who submitted manuscripts in a timely way. I wasn't savvy enough at the time to appreciate it, but graduate student publication rate and quality is one very important metric to look for in a PhD advisor.

Fig. 1 Janet in her climbing days at Chapel Hill

Postdoctoral Studies: Moving Across the Atlantic

Before I knew it, it was time to move on. I started searching for a postdoctoral position to continue my love of research, travel, and the outdoors. I chose to apply for international postdoctoral experiences and was awarded a fellowship from the National Science Foundation to study at the University of Bordeaux for 15 months with Professor Didier Astruc, an organometallic iron chemist. I defended my thesis and got married in the same week. A few weeks later, we left for France to start my postdoctoral position. Subsequently, we would honeymoon in the French Pyrénées and climb with members of the French Alpine Club.

Working in a French laboratory took some adaptation on my part. I had to change my research plan according to the available resources which were, at the time, different than in laboratories in the USA. For example, the NMR spectrometers were limited to hands on for staff only, but there was an EPR spectrometer and instrumentation for electrochemical analysis of iron complexes that I could use freely. I also struggled to adapt to a different culture and language.

All in all, it was wonderful adventure. My husband and I traveled widely throughout Continental Europe during the long summer, Christmas, and spring breaks. During one climbing escapade in the Wilder Kaiser mountain range that took place after I had attended a conference on organometallic chemistry in Vienna, I fell 150 feet down a cliff face and broke several ribs and punctured my lung. I was rescued by helicopter and spent a couple of weeks in the hospital followed by a painful train trip back to Bordeaux. That took a few weeks' time out of my research projects and convinced me to move on to tamer sports.

This period in Europe was full of formative experiences that would build my confidence and courage to tackle challenging projects and to work with people from different backgrounds and nationalities. One of the most challenging was a presentation at the end of my stay that I gave to a committee made up of French and German scientists. This was more stressful for me than my PhD thesis defense because it had to be given partially in French.

After 15 months of adventure and hard work, I was ready to return to the USA. I wrote to several faculty members and had three positions to choose from. I chose to work with Professor Bill Troger at the University of California, San Diego (UCSD), based on research interests and his reputation as a good mentor. I also considered the university setting. UCSD is a large university that I thought would have good professional development and job placement opportunities. Bill was a wonderful mentor and allowed me to sit in on courses and to initiate a new project in the area of bioinorganic chemistry. After 11 short months, I was on the job market. I did apply for a couple of industrial jobs but then put most of my efforts into searching for academic positions. I wanted a position in a large university with an emphasis on research, similar to what I had experienced.

Life as a Faculty Member

I accepted a position at the University at Buffalo, State University of New York (UB). UB is a large public university with a comprehensive set of colleges/schools including a college of engineering and a medical school. I moved there with my husband who soon started his own small business in violin making, sales, and repair. One of the things I did not appreciate at the time was that Western New York was also an excellent choice for raising a family. The commuting time was short, the public schools were excellent, and the family-friendly atmosphere made it relatively easy to bring up a child.

Nothing quite prepared me for the challenge of my first faculty position. It always is a busy lifestyle, but the learning curve at the start is steep. I juggled teaching new courses, leading students in research, being in the lab myself, writing manuscripts and proposals, and giving presentations as a faculty member. I made the error of starting too many projects and spreading myself too thin. I naively expected that my students would be as motivated as I was. Management skills were not something I learned very much about in graduate school and had to develop as I went along.

The birth of my first child, Erin, at the start of my second year as a faculty member forced me to narrow my choices of research projects. I chose the most promising ones and focused on them more effectively than I had before her birth. At the time, there was no maternity leave policy, and I took off 2 weeks from teaching after she was born. As the only young female faculty member in the department, I received some negative comments on the pregnancy. One of my colleagues told me flat out that "some things are more important than tenure." A graduate student asked me: "Are you going to quit?" Other faculty members were more supportive and even offered to babysit when I had meetings.

I was given the semester off the following spring. This turned out to be of key importance for my research. I could spend more time in the lab and I traveled to conferences as well. This time for travel enabled me to develop key collaborations, one in academics with Professor Bill Horrocks of the Pennsylvania State University and one in the biotechnology industry with Dr. Brenda Baker of Isis Pharmaceuticals. These collaborations enhanced the scope of my research and garnered the attention that was needed for a successful tenure case.

Post-tenure

Our son Garrett was born a year after I got tenure. This time, I took off 6 weeks from teaching. Like his sister, he was an easy-going and happy kid, and this made it feasible to travel with him and to bring him to work as needed. However with two small children at home, I traveled less during this period. Despite the reduced stress from having attained tenure, I remember this period as being especially tough to balance work and home life.

Still, we managed to balance work and family, travel, and fun. I would often bring one of my kids with me on a work-related trip. Erin came along on trips to New York

Fig. 2 Janet (center) taking a break with her daughter Erin (left) and son Garrett (right) in Quebec City

City, to Washington DC, to Mexico City, to Quebec, and to an extended stay in France as part of a sabbatical. Garrett went along for a 10-day trip to Switzerland, and we traveled by train to hop between cities for university lectures (Figs. 2 and 3).

One of the ways that I tried to influence the departmental atmosphere is through administration. I was associate chair for 6 years. In this position, I participated in the administration of the department and interacted with university colleagues in other departments. During this period, I observed that the attitude toward junior faculty had improved. We made the effort to hire more women and now have 5 female faculty members out of 30 total.

I have resisted taking on more administrative duties, because research is still one of the most rewarding activities for me. Interdisciplinary and collaborative research requires a large time commitment in order to stay abreast of new fields and evolving research projects. Currently my research has a heavy component in magnetic resonance imaging contrast agents that contain iron and cobalt and in the design of metal ion complexes that interact with unusual nucleic acid structures.

Advice and Lessons Learned

If there is anything I would emphasize, it is to be determined and committed to your science but be flexible and keep a sense of humor. Doing scientific research takes a combination of focus and direction but also inspiration and creativity. Read widely,

Fig. 3 Helmut Sigel, Janet, her husband Boyd and son Garrett in Switzerland taking a break from lectures

attend conferences to network, and seek out collaborations. Tell people about your research projects to get feedback and suggestions.

Maintaining a balance between family and career is challenging. In my situation, neither set of grandparents lived close-by, nor could we afford a nanny. So I combined everyday activities such as grocery shopping with fun when the kids were small. We would get helium-filled balloons and cookies at the store while we shopped at Wegmans grocery store. I would make up exams for them to work on while I was grading mine. I took them on business trips and combined these trips with vacation activities when I could. Erin, my pre-tenure child, traveled with me to conferences, university lectures, study sections, and international trips right through her teens. We joke that this travel and continually new and changing experiences contributed to her restlessness as an adult. At 27 years of age, she has already traveled to more countries than I have, including a recent 2-month stay in Antarctica. She graduated with a BS in engineering from the University at Buffalo and a master's degree in electrical engineering from Columbia University and is currently taking courses for a master's degree in computer science at Johns Hopkins University where she works at the Advanced Physics Laboratory. My son Garrett is an undergraduate in mechanical engineering at the University at Buffalo.

Both of my children did experiments with me in the laboratory on weekends. We put together chemistry demonstrations for their classes. During one of them, an unexpectedly large chunk of flaming sodium metal shot up right into the middle school classroom ceiling. The ceiling was fortunately fireproof. The neighborhood kids remembered me for that one. Several years later, my daughter published a research paper with me while she was an undergraduate engineering student. She came to appreciate my career from a different perspective when she entered college herself.

Did my career influence my family time? Of course, but this is likely the case for any career woman. I generally work on the weekends, especially to catch up on reading. During certain periods of her life, my daughter was vocal about my lifestyle and commented that I was not a girly girl. When I asked her to explain, she said that I wasn't an avid shopper and didn't fixate on decorating the house. I didn't rush out to the furniture stores to get the best deals. I didn't fuss over my kid's clothes and appearance as her friends' moms did. In my defense, I remember doing many girly things such as taking long shopping trips in search of the perfect dress for middle school and for high school dances. In any case, your life will inevitably be different than most people's lives because you are a scientist. Your kids will let you know that.

Finally, it is of utmost importance to find a good balance between work, family, and other interests. There have been times in my life when things seemed especially

Fig. 4 Janet and her family on vacation in the Cayman Islands

challenging, and it was important that I could fall back on friends and family or an activity outside of science for a brief respite. As your children grow, you will develop new interests and activities to share with them. I feel very lucky to have balanced raising children with an academic career. I hope that my route becomes a conventional one for women (Fig. 4).

Acknowledgments I would like to acknowledge many mentors, both personal and professional, for their support. First, I would like to thank my parents for encouraging me to study chemistry and to earn a PhD. I thank my three brothers for their help in maintaining a good life balance through shared adventures in climbing, hiking, canoeing, traveling, and skiing. I am grateful to my late husband, Boyd, for his support of my career and his willingness to move across the country and around the world with me. My children, Erin and Garrett, have been supportive of my career, and it has been fun to see them choose to become engineers. Finally, I would like to thank my professional mentors who encouraged me as an undergraduate, graduate student, or postdoctoral fellow. My academic life has been made livelier by my interactions with my 42 past and present graduate students, 94 undergraduate students, and with many visiting scholars, postdoctoral fellows, and faculty collaborators.

About the Author

Education and Professional Career

1980	BS Chemistry, University of California, Santa Barbara, CA
1985	PhD Inorganic Chemistry, University of North Carolina at Chapel Hill, Chapel Hill, NC
1985–1986	Postdoctoral Fellow, University of Bordeaux, Bordeaux, France
1986–1988	Postdoctoral Fellow, University of California, San Diego, CA
1988–1994	Assistant Professor, University at Buffalo, State University of New York (SUNY), Buffalo, NY
1994–2003	Associate Professor, University at Buffalo, SUNY, Buffalo, NY
1996–1997	Visiting Professor, University of Rochester, Rochester, NY
2003–present	Professor, University at Buffalo, SUNY, Buffalo, NY
2005–2015	Director, research experiences for undergraduates (NSF-REU), Chemistry Department, University of Buffalo
2016–present	University at Buffalo Distinguished Professor, University at Buffalo, SUNY, Buffalo, NY

Honors and Awards (Selected)

2015	University at Buffalo Exceptional Scholar
2014	Schoellkopf Medal, American Chemical Society, Western New York Section

2007–2009   National Science Foundation Special Award for Creativity
1994–1996   Alfred P. Sloan Fellow
1985–1986   National Science Foundation Postdoctoral Fellowship

Janet's research focuses on the synthesis of inorganic complexes for biomedical diagnostics, sensing, or catalytic applications. She is an associate editor of the ACS journal, *Inorganic Chemistry*.

The Path to Academia and Motherhood: It Takes a Village

Emily Niemeyer

Photo Credit: Andy Sharp, Williamson County Sun

Growing up in rural Ohio in the 1980s, I never met a university professor. In fact, no one in my immediate family had even attended a 4-year college. And although I clearly remember almost everyone within our small community, I can't recall a single child who had two mothers. Yet today I enjoy a successful career as a chemistry professor at a primarily undergraduate institution *and* am happily raising a 7-year-old daughter who thinks nothing of the fact that she has two moms. How did I arrive at this point? Hard work, persistence, and, of course, a little bit of luck helped. But just as important were the encouraging mentors and a supportive group of family and friends who helped me successfully navigate the path to become a scientist and educator as well as a partner and parent.

E. Niemeyer (✉)
Department of Chemistry and Biochemistry, Southwestern University, Georgetown, TX, USA
e-mail: niemeyee@southwestern.edu

Becoming a Chemist

Mr. Art Holman was an amazing high school science teacher. I eagerly anticipated his physics and chemistry classes: they were hands-on, interactive, and completely engaging. He skillfully made his students partners in the learning process, whether we were involved in class-wide competitions to test the strength of hand-built balsa wood bridges or learning about phase changes by making ice cream in soda cans. Those high school courses—and Mr. Holman—not only inspired me but also motivated me for the first time in my life to seriously consider going to college. My parents played a large role in this process as well, continually emphasizing the importance of furthering my education. In 1989, I enrolled at Ohio Northern University (ONU), unsure of my major, but, largely due to my experiences in Mr. Holman's classes, I registered for a slate of science courses: introductory chemistry, biology, and calculus.

Within my first few weeks of college, I realized that I was in way over my head, a feeling that I attribute, in part, to being a first-generation college student. Even with a solid academic foundation from a small public high school, I struggled with the rigor and expectations of college-level coursework. I credit much of my ability to survive that difficult transition during my first year to my academic advisor and mentor, Dr. Alan Sadurski. Dr. Sadurski had an unwavering confidence in my abilities, even when I did not. After my first quarter in college, I was surprised to find that I had received a B in Dr. Sadurski's general chemistry course when I was relatively certain I had earned a C. When I finally mustered the courage to ask him about the grade, he explained that if he had given me a C (the grade I had earned), he knew I would have quit studying chemistry, and he believed I was too good to do that. He also informed me that he would only give me such a chance once, and, after that, the rest was up to me.

Dr. Sadurski's faith in me as a student—and a chemist—boosted my confidence, and I worked diligently to live up to his high expectations. In the process, I found my academic home in ONU's Chemistry Department, a program with a supportive and close-knit group of students and faculty. I quickly joined Dr. Sadurski's lab, a transformative experience that helped me truly understand the joys and frustrations of doing scientific research. Because of ONU's comprehensive curriculum, I also gained an appreciation for a wide variety of disciplines and began to see how my growing passion for chemistry related to other fields outside the sciences. The summer before my senior year, I participated in an internship at a petrochemical company that proved pivotal in my career. Not surprisingly, the pay was fantastic, but the job turned out to be tedious and mind numbing. Lesson learned. When Dr. Sadurski invited me to work as a student teaching assistant for an upper-level laboratory course the next fall, I jumped at the chance. Compared to my summer internship, the pay was paltry, but I loved it. Teaching was exciting, fun, and extremely rewarding—I even discovered that I was good at it! That experience pushed me to apply to graduate school, a possibility I never once entertained before arriving at ONU. With the encouragement of Dr. Sadurski and the Chemistry

The Path to Academia and Motherhood: It Takes a Village 381

Department faculty, I entered the graduate program at the University of Buffalo (UB), State University of New York, in 1993 with the goal of ultimately becoming a chemistry professor at a predominantly undergraduate institution like my alma mater—a place where I could teach and do research.

But what kind of chemist? The answer to that question was unclear until my second semester at UB when I took a graduate-level analytical course with Dr. Frank Bright. Frank had a huge personality, high expectations of his students, and a reputation for being tough and more than a little intimidating. He was also an incredible teacher, and although I knew virtually nothing about his research, I wanted to work with him. As intimidating as Frank could be, he was funny, smart, and an internationally recognized chemist—plus I just liked and respected him. Although I had heard that his lab was full, I still found the courage to ask him if I could join his research group. During that short 15-minute meeting, Frank initially said he didn't have room for another student. But as we continued to talk, he changed his mind, assuring me that he would find a way to stretch the lab budget to accommodate another person even if we were all "eating beans and weenies" that summer.

As a research advisor, Frank ran a tight ship. At times, I felt his expectations of me were so high that I could never realistically hope to achieve them. But I often did, and when I didn't do so initially, I tried again until I succeeded. I never once regretted my decision to ask Frank to join his research group. Frank was an outstanding mentor and a strong advocate for my future success. With his guidance, I became a better chemist while gaining the confidence and skills necessary to eventually lead my own research group. Frank wholeheartedly supported my aspirations to teach at an undergraduate institution, encouraging me to continually pursue new opportunities to help attain that goal. In fact, with Frank's recommendation, I spent a summer teaching an introductory chemistry course within UB's Educational Opportunity Program. That course—in which I was completely responsible for designing and teaching the class—further solidified my commitment to becoming a faculty member. The additional teaching experience, along with the numerous research articles I was able to publish with Frank and my fellow group members, also provided an important foundation for my success in ultimately obtaining a tenure-track position.

Becoming an Educator

Southwestern University was a lot like ONU, but it was also decidedly different. Southwestern is in Texas—a long way geographically and culturally from Ohio—and I was now a faculty member, not a student. Arriving at Southwestern in 1998, 27 years old, and straight out of graduate school, however, I actually *was* closer in age to my students than I was to most of my colleagues. What I lacked in experience, however, I made up for in enthusiasm. Like my alma mater, Southwestern was a selective residential undergraduate college and exactly the type of institution where I wanted to be. I quickly dove into creating a research program with my students and developing a repertoire of courses in general and analytical chemistry as well as classes for non-science majors.

As one of the six new tenure-track hires that year, I was fortunate to join the Southwestern faculty during a time of rapid growth and expansion. In fact, my cohort—containing faculty in the sciences, social sciences, and humanities—was crucial to my development as a liberal arts professor. These faculty members opened my eyes to teaching and research possibilities beyond my own discipline. Along with two other members of my cohort, I became deeply involved in issues of campus sustainability and was appointed as the chair of Southwestern's new Environmental Studies Program. My collaborative research program with students underwent a fundamental shift, as we began to explore questions of ecological importance using the tools of analytical chemistry. With the support of my chemistry department chair, my teaching also changed, and I created new courses in environmental chemistry for both science majors and nonmajors.

In hindsight, although the years leading up to tenure were exhausting and at times incredibly stressful, they were also filled with creativity and excitement. I was—and still am—fortunate to work at an institution that values teaching, scholarship, and university service at a level commensurate with my own professional expectations. Additionally, Southwestern provided the resources, support, and freedom for my research and teaching to grow and thrive, both before tenure and ever since.

Partner, Parent, and Professor

My junior faculty cohort also provided an important social support network. We went to concerts, picnics, and hiking trips together, often inviting other faculty that we knew from across the university. One person I met was Maria, a faculty member in Southwestern's sociology and anthropology department. We soon started dating and eventually married. Working at the same institution as your significant other occasionally may present challenges, but for me, the benefits have been innumerable. Academic careers can be all consuming: waking up to a day with 5 hours of scheduled meetings, an exam that needs to be written by noon, and a to-do list of tasks that never seems to grow any shorter. Having a partner who truly understands such demands—and the nuances and intricacies associated with them—has been invaluable.

And working with my partner has offered some amazing opportunities for personal and professional growth as well. For example, in 2005 we were both selected as the faculty members to lead Southwestern's London Semester program. Having never previously lived abroad, I relished immersing myself in such a vibrant and diverse city. Teaching in London had a profound impact on my pedagogy too, helping me understand the power and importance of experiential learning. It also felt a bit like "Introduction to Parenting 101," as Maria and I collaboratively dealt with

The Path to Academia and Motherhood: It Takes a Village

383

the myriad academic and interpersonal issues that arose among the large group of Southwestern students participating in the program.

Upon our return to Texas in 2006, we began to talk seriously about starting a family. The timing seemed ideal: we were both tenured and established in our careers. However, we had unique planning issues relative to many other couples: Which of us would actually have the baby? How would we find a donor dad? Like true academics, we completed a lot of research to answer those questions, debating the many pros and cons. We ultimately decided that I would be the one to carry the baby and a dear family friend graciously agreed to be our donor and bio-dad.

What we didn't anticipate were the difficulties I would experience during pregnancy. After being diagnosed with preeclampsia (a maternal disorder characterized by high blood pressure) in my eighth month, I was placed on bed rest. Thankfully, this situation arose during the summer, at a time when I was not formally teaching and was lucky to have two independent and very talented research students. Everyone was incredibly understanding, yet I worried that I was letting my research students down by missing several weeks of our 8-week summer program. My preeclampsia continued to worsen until eventually my doctor believed that the condition was becoming dangerous to the baby. In July 2008, I was medically induced and our daughter, Lucy, was born: tiny at just over 5 pounds but otherwise healthy. Typically, preeclampsia resolves immediately after the baby's birth, but mine did not, and I was forced to remain on bed rest for an additional 3 weeks after Lucy was born. It was a challenging time, and Maria and I were grateful for the continuous support of friends and family, particularly my mother who flew down from Ohio to help us.

Five weeks after giving birth, in a sleep-deprived state and still suffering from medical issues related to preeclampsia, I submitted my application for promotion to full professor. I definitely wasn't at my best—physically or cognitively—but somehow I managed to collect my materials and cobble together an accompanying personal statement. After my materials were submitted, I was on parental leave for the remainder of the fall semester, and my health continually improved. I was able to work on my research with my students and went to Southwestern at least once a week, always taking Lucy with me. Though I felt reasonably confident about the likelihood of being promoted, submitting my application gave me a level of professional unease for the entire academic year as I awaited the decision of my review. When I was notified the following spring that I had been selected for promotion, I was gratified and thrilled: I was the first woman to ever become a full professor in Southwestern's chemistry department.

Balancing Family and an Academic Career

There is no doubt that our family benefitted greatly from Southwestern's inclusive policies and campus culture. Following Lucy's birth, Maria and I were both granted a full semester of paid parental leave, which we staggered to allow us to stay home with our daughter for the entire first year of her life. Even as a newborn, Lucy was a regular and welcomed visitor to the Southwestern campus. We toted her around to meetings, set up play and nap areas in our offices, and took her to various campus events. Rather than being annoyed by the presence of a baby on campus, faculty, staff members, and students were all supportive and understanding, offering help and plenty of advice.

When Lucy entered day care in the fall of 2009, Maria and I began to more fully understand the many challenges of balancing our career commitments with a family. We are commuters, and our house and Lucy's day care were over 30 minutes from the Southwestern campus even on a good traffic day. We had to be creative and intentional. We also had to work collaboratively to juggle our teaching, research, and committee obligations in a way that was fair to both of us and also nurturing for Lucy. Thankfully, we were both full professors and had the privilege of determining our respective teaching schedules. In those first several years, however, we truly learned the importance of good communication and the art of compromise. Despite the most thorough planning, things sometimes went wrong: I would have an important meeting, Maria was teaching a class, and we would get a call from Lucy's day care that our kid was projectile vomiting. When those situations arose, we learned to work together more effectively, discussing and identifying options until we found a solution that satisfied both of us. Mutual support and respect, effective teamwork, and shared understanding have been key to allowing us to maintain successful academic careers while raising our daughter in a happy and healthy family.

Being a mother has transformed how I work, and the changes have been for the better. When I am in my office, work is my top priority, and as a result, I have become much more efficient. I have learned that carving out even small amounts of time—typically early in the morning or after Lucy goes to bed—really can make a difference, whether I'm responding to e-mails or ordering lab supplies for my research students. I also prioritize what is most important and feel empowered to say "no" to some work requests. Faculty are asked to do a lot in addition to their teaching and research: administrative work, serving on ad hoc committees, contributing to recruiting and admissions efforts, and assuming leadership positions on university councils. While some of these tasks are integral to our jobs, others are not. When asked to take on a new responsibility, I consider the following: is it central to my teaching, research, or university service, *and* is it worth the time that it will most likely take from my family? If the answer to these questions is no, then I pass—but with the full realization that my privilege as a full professor allows me that luxury.

This level of focus has helped me to not only become a more effective teacher since becoming a parent but also a more productive scholar. In fact, it has allowed me to really hit my stride professionally, increasing my publication rate and securing substantial external grant funding. In recognition of my achievements, I was appointed as an endowed chair at Southwestern in 2013, a prestigious position that also provides me with a discretionary annual research budget. As a first-generation college student who greatly benefitted from outstanding mentors, I am most excited that I have recently had the opportunity to create and direct new programs at Southwestern that aim to increase the persistence and success of underrepresented students in the sciences. These initiatives—which involve widespread curriculum revision, increased access to undergraduate research opportunities, and scholarship support—have shifted my career focus in a new direction while also providing me with immense personal satisfaction.

Final Reflections

There is no doubt that it can sometimes be challenging to balance my roles as a faculty member and a parent. Yet I am continually reminded that these two identities—chemistry professor and mom—are seamless, reinforcing rather than detracting from one another. When our daughter Lucy was 5 years old, she was fascinated by the depictions of molecules in my general chemistry text. We would spend hours searching for ball-and-stick and space-filling models in my book and then discussing simple molecules and matter on an atomic level. One hot summer day, as Lucy was sitting in the kitchen drinking a cold glass of water (H_2O), she looked at me and boldly proclaimed, "Mommy, I can really taste the oxygen!" Though I doubt that was actually the case, I realized that my 5-year-old daughter was thinking about the composition of water on an atomic level. Our discussions about molecules had fundamentally changed the way that Lucy viewed the world. Likewise, Lucy helped me understand what my students see when they look at pictures of molecules in their own textbooks—and the misconceptions such figures can cause—which ultimately led me to become a better teacher. For me, the synergy that exists from being both a chemistry professor and a mom was completely unexpected, yet it has also been tremendously enriching and extraordinarily gratifying (Fig. 1).

> **Reflections from Emily's Daughter**
> **Lucy (age 8)**
> Hi, my name is Lucy, and as you people know, my mom is a chemist. Let me just say that she is probably the best teacher of chemistry ever! She studies for exams, but she always finds time for me and my other mom. I love you, mommy!

Fig. 1 Lucy the scientist

Acknowledgments I would like to thank Dr. Maria Lowe and Dr. Amber Charlebois for their support and encouragement while writing this chapter. Their insights, edits, and thoughtful comments were extremely valuable.

About the Author

Education and Professional Career

1993	BS, Chemistry, Ohio Northern University, Ada, OH
1998	PhD, Analytical Chemistry, University of Buffalo (UB), State University of New York, Buffalo, NY
1998–2004	Assistant Professor, Southwestern University, Georgetown, TX
2004–2009	Associate Professor, Southwestern University, Georgetown, TX
2009–present	Professor, Southwestern University, Georgetown, TX
2013–present	Herbert and Kate Dishman Chair in Science, Southwestern University, Georgetown, TX

Honors and Awards (Selected)

2014	Distinguished Alumni Award, Elida High School
2013	Outstanding Texas Women in STEM Award from Girlstart
2009–2010	Exemplary Teaching Award from the Board of Higher Education and Ministry of The United Methodist Church
1997	American Chemical Society Division of Analytical Chemistry Graduate Fellowship

Emily was instrumental in securing a grant to Southwestern University from the Howard Hughes Medical Institute (HHMI) in 2012. As the Program Director for the HHMI-Southwestern Inquiry Initiative, she successfully instituted a transition to an inquiry-based curriculum across Southwestern's science and math departments while significantly expanding undergraduate research opportunities for underrepresented students.

Elements to Successful Motherhood and the Professoriate

Sherine O. Obare

Photo Credit: Western Michigan University

The Early Years

It was sometime in 1988, while sitting in my science class in middle school, when I remember falling in love with the concept of elements. I had always enjoyed science and math, but there was something special learning that chemistry was a central science and that everything surrounding us could be about the elements. As I

S. O. Obare (✉)
Department of Chemistry, Western Michigan University, Kalamazoo, MI, USA
e-mail: sherine.obare@wmich.edu

progressed into high school, my passion for chemistry increased. At the time, I lived in Germany as that is where my father's profession had taken him. I attended the British High School where during my last 2 years, I only had to take three subjects, and I selected to study chemistry, mathematics, and physics. I remember having several of my friends and family members being quite critical of my selection, often wondering what future would a girl have taking "difficult" science courses. Nonetheless, I found much joy in every chemistry class I attended and desired to have a profession in it.

As I began to consider college, I was encouraged by family to consider majoring in engineering. While I considered majoring in either chemistry or physics, it was unclear to those advising me what a career I would be able to pursue. In my freshman year, I took an engineering course, and I was in a course full of boys. I remember being made to feel like I did not belong in that classroom. My college roommate at the time had a friend who stopped by to chat with me one evening. She was a chemistry major. I was so excited to see another woman pursuing a degree in chemistry. She was a great inspiration. Later that semester, I decided that I was going to switch my major to chemistry. The rest of my undergraduate years were very enjoyable. I was engaged in research and I worked as a tutor in chemistry.

When I was in middle school, my chemistry and physics teachers were both women. While I did not realize it then, I now know that they truly instilled in me a passion for science in a way that no other male teacher would have. Not only was it a matter of me being passionate about the subject, but there is something to be said about the learning environment that women, as teachers, create for female students that is quite different than their male counterparts.

The Decision to Pursue Graduate Studies

It was the summer before my senior year when I sat with my family to discuss my plans after graduation. My father encouraged me to attend medical school, while I think my mother just wanted to see me get a job and someday be married and start a family.

During that Fall semester, I decided to follow my mother's advice and start looking for jobs. It was not long before I had a job offer with the US Environmental Protection Agency. While I was excited to have secured a job, I realized that there was so much more I wanted to do. I enjoyed tutoring and teaching and I felt more education was not a bad idea.

One afternoon, very stressed out about my future, I ran into one of my biology professors. She was the only female science professor I had had in my entire college career. She asked me what I was stressed about, and I explained to her that I wished I could do more with my life than simply take a job that did not guarantee career growth. She told me to look into graduate school. Just that simple short conversation with her and the amount of help she offered me made me realize there was so much more I could do with a career in academia. I had additional discussions with her about her career path, and soon I figured out what I needed to do. I quickly took the GRE exams and applied to graduate programs, and soon I was moving to Columbia, South Carolina, to pursue a doctorate degree at the University of South Carolina.

In sharing with my friends and family my plans, I was reminded of the time I was in high school making decisions about taking courses and how negative many were. I was again reliving those days where many told me I was being too ambitious to think I would get a doctorate degree in chemistry. I did not let that chatter bother me and decided to follow my dreams.

With not many role models, I arrived at graduate school with little understanding of the expectations. I knew that I had to work hard to get through the program. I had also met a great guy who I knew would be the person I would spend the rest of my life with. It was a little scary to me to figure out how I would balance a relationship while getting through graduate school. I was extremely lucky that Troy was highly supportive. Late nights when my chemical reactions just had not ended, Troy would pick me up to make sure I was not walking to my car late alone.

There were major decisions to make, such as selecting a research advisor. Many of my fellow graduate student colleagues would list several reasons as to why they would choose an advisor. What was important to me at the time was to be in an environment where I would be able to learn, work hard, and get along with my advisor. I interviewed a number of faculty in the organic chemistry division, and it was difficult to figure out who would be best for me. Soon someone encouraged me to speak with a faculty member from the inorganic division who may have some organic chemistry projects. I soon started speaking to inorganic faculty, and, after one short meeting with Professor Catherine J. Murphy, I prayed she would accept me to her group. She had exciting projects in her lab, a wonderful environment that I knew I would be able to thrive in, and was an excellent mentor.

As I began to accelerate through graduate school and make progress, my fiancé at the time and I were ready to get married. The idea of being married and having a family was very important to me. Though I was never sure when would be the best time to plan a wedding, I knew that the man I had chosen to spend the rest of my life with was someone who would stand by me and support me through the big and small choices I made in my career. We both decided that it was important that we were married, and that further made my mother a very happy lady. We had a very small wedding, but it was beautiful.

And Then the Kids Came Along

While we did not think about kids, I woke up pregnant while in my last year of graduate school (Fig. 1). I was terrified because I did not think being a parent and having a doctorate degree in chemistry went together. I quickly realized that there is never a good time to have kids and sometimes you just count your blessings and

Fig. 1 Sherine holding her son Amir Ross-Obare at the University of South Carolina's Commencement on May 11, 2002

keep going. I remember telling my husband that I would not be able to be both a mother and pursue a career. His answer to me was "nonsense"—we will do it.

I went on to pursue a postdoc at Johns Hopkins University in Baltimore, Maryland. I must admit I was terrified to move to a different city. My husband still had one semester to complete, which meant I would be moving alone with the baby until he was done and ready to join us. I had my mother stay with me for a month. Luckily she was a school teacher, and with summers off her schedule permitted her to do that. When I went to lab, she stayed home with the baby. Soon it was late August and she had to resume to her normal schedule. We found a home daycare, which we thought would be alright. However, I was fortunate that my aunt who had just retired offered to come stay with me and help me with the baby. She helped me for 1 month, so by the time I had to place my son in daycare, he was already 5 months. It was still very difficult to leave him in daycare and to plan my lab work around daycare hours.

It was important for me to be organized and to make sure that when I got to lab, I stayed focused and got as much done as possible. There were other things I could do at home, and I made sure I managed my time well.

The biggest challenge is that my child could never have a full night's sleep. I had to learn how to function effectively with little sleep. Good thing my body was able to quickly adjust to such changes.

Transitioning to Faculty Life

As a mother in a postdoc position, I was daily reminded of the tremendous demands that came with job. It was enlightening, however, to have undergone the postdoc experience while being a parent, because it opened and helped me learn how to better manage my time at work as well as at home.

At the end of my first year of my postdoctoral fellowship, I knew it was time to start thinking as well as searching for a more permanent position. I knew I wanted to pursue a career in academia, and at the time, I was confident that I wanted to do so in a small liberal arts institution, mainly because I felt that it would provide me with a better balance of work and life. I worked diligently to write down my teaching philosophy and to develop a detailed research plan that would be required as part of my job applications. Each week I would review job announcements posted in *Chemical and Engineering News* (*C&EN*), and I would make a list of institutions that were looking for inorganic chemists that I thought I would be a good candidate for.

Toward the end of the summer, I sat down with my postdoctoral advisor, Jerry, to share my plans about my job searchers. I remember sharing with him a list of small liberal arts institutions that I planned to apply to. Interestingly, Jerry was not very enthused with my plan. He began to discuss with me the need to add more institutions to my list and the importance of seeking institutions that would allow me to execute the research program I had developed as part of the application. He also discussed with me the benefits of being able to establish a strong independent research program and the need to be at an institution that would provide resources for me to succeed. I was not sure I fully understood or agreed with his point of view at the time, but in an effort, not to disappoint him, I decided to apply both to liberal arts institutions and to chemistry departments that offered doctorate programs.

Interestingly, I was invited to interview at only the doctoral programs that I applied to. I learned a great deal on my interviews. I received four faculty position job offers, and it became clear to me after the interviews that I really was pleased to have applied to those institutions. With four offers at hand, I had a difficult decision to make, and I knew as I made that decision, I was not simply deciding what was best for me and my career, but I also had to carefully think of my husband and his career as well as where I would be raising our son.

When I applied to Western Michigan University, I honestly had never heard of the university before, nor did I realize the town Kalamazoo even existed. My visit to the university was pleasant, but it was in the middle of winter, and the cold weather was very uninviting. But what was quite amazing during my interview is how my prospective male colleagues seemed to value family. Each one would tell me how well they were able to balance their careers and be engaged in the lives of their children. While I realized that it would be different for me as a mother, it was refreshing to see colleagues who valued family. It also helped to see a female faculty member, Dr. Susan Stapleton, who was very successful in the department and had managed to maintain a strong marriage, amazing kids, and a research program. So while I never imagined moving to Michigan, I found myself taking my family to start a new life in Kalamazoo.

Keys to Success as One Transitions to Faculty Life

Johns Hopkins University School of Medicine has an excellent program for postdoctoral fellows. While I was not in the medical school, a friend invited me to one of their seminars. The seminar was all about starting a faculty career. The first thing they went over is get your lab set up and start writing your grants. I immediately contacted the department chair and asked if there is any way for me to start ordering my lab equipment before I arrive. Furthermore, in my start-up package, I had budgeted for a postdoctoral fellow. I was very fortunate that I was allowed to work with vendors early and to post for the postdoc. By the time I arrived in Kalamazoo, my postdoc was there and so was my equipment. We worked long hours to get our lab running. I knew I had to get the lab ready because once the semester started, I would have added teaching responsibilities, and I wanted to be careful about managing my time.

Before I knew it, the Fall semester had started. I was assigned to co-teach two courses—instrumental methods and nanotechnology (a new course). While I knew the semester would be busy, I could not imagine how challenging it was. Writing lectures from scratch was painful. It took me 6–8 hours to organize the content in a way that I thought would be effective for my students.

To make matters even more challenging, my husband had found a job, but it was a night-shift job at a company about an hour away from where we lived. In addition, my son, at the time, had major difficulties sleeping through the night. In fact, he would wake up every 2 hours and that meant significant interrupted sleep for me. It was quite a struggle, and I honestly have no idea how I got through that semester. I thought I had made a bad decision and that pursuing a faculty career and being a mother were impossible.

But as the semester came to an end, I was quite pleased with the work I had done. With regard to my course, I had put together a significant package of notes, coupled with literature articles and material that would be useful for future instruction. I was even looking forward to teaching those courses again because I now knew how to do it.

As time progressed I learned how to better manage my time, be resourceful about helpful tips, and how to make things work at home. I again had to manage my hours at work around daycare hours, which meant all writing was going to happen at home. My son soon learned that there was something called "proposal writing time" where I would sit and write, while he would color and draw. I made sure to include him in everything I did.

I was very fortunate to secure some grants early in my career. I also was very lucky to have some amazing students join my lab. It was important for me to make sure that they understood my availability and expectations so that we could get as

Fig. 2 Sherine's sons Amren (left) and Amir (right) in December 2010

much accomplished during the time we were together. Slowly the papers got published, the grants were awarded, and things seemed to be moving in the right direction. The important thing I realized is that I had to be flexible and I had to make sure that my students were flexible in their schedules.

Sometimes, with no plans, things in life happen, and, while I was not prepared for our second son, he showed up in my fourth year of being an assistant professor. It was tough, but I was determined not to have anyone use that against me. I did not take maternity leave—instead I plowed through. It is amazing how strong our bodies are and what we can do. I do not encourage any woman to do it my way, but I think it is important to note that it is doable if that is what one has to do.

I applied for tenure after my fourth year and it was successful. The moment I found out that I would be tenured, I automatically started to outline my goals for promotion from associate to full. I now had two kids, age 6 and a baby. I also knew that together we would work hard to get what we needed to do. What I loved most about my position as a faculty member was the flexibility. I had sitters I could call on when needed, but most of the time, if it was necessary, I could always make things work.

I applied for full professor 4 years after being tenured. I was quite pleased to be promoted. Having been on a fast pace for promotion, I felt a little lost after it was final. I was not sure what big goal I had to accomplish next (Fig. 2).

Transitioning from Faculty Member to Administrator

In the Spring of 2015, I received a call from the Dean of the College of Arts and Sciences asking me if I would be willing to serve as Associate Dean of Research. While the call was completely unexpected, it did come at a time when I was ready for a new challenge. I was not sure if it would be the right career move for me either or how it would impact the work I enjoyed doing such as maintaining my research program and mentoring students in my lab. I discussed some of these concerns with my dean, and he agreed that it was important that I be given time to maintain my research program.

I became Interim Associate Dean for the College of Arts and Sciences in February 2015. During that time my responsibilities included enhancing the college's research enterprise and overseeing the mathematics and science departments, diversity, graduate education, and global engagement. I viewed the opportunity as a way to help move things forward within my college. I was quite honored to serve in that role for about 2 years and was able to implement several programs that will support faculty and students within the college. One of the great advantages to having served in that role was the opportunity to step outside my comfort zone and learn about the management and operations of other departments, not only in the sciences but also in the humanities and social sciences. It was extremely rewarding for me to be able to use the problem-solving skills I have learned as a chemist but broaden that skill and use it to solve other problems within the university.

I was also very fortunate to have much support as I transitioned into this role. My biggest concern was that with an administrative role, one's schedule is 8 am to 5 pm, and I was worried that I would lose the flexibility that had served me well as a faculty member. My husband, however, was very supportive, and my sons had started being a lot more independent, making that transition much more manageable. Therefore, I went on to start a career in academic administration while still running a lab full of students and being a mom.

After a little less than 2 years in my role as Associate Dean, a new opportunity presented itself—Associate Vice President for Research. I viewed this as an opportunity to take much of what I had done at the college level and apply it to the university level. I have been honored to serve my institution in this capacity. The role has also helped me work closely with faculty at my institution and to support them in building their research programs. I have managed to find a good balance between marinating research in my lab, mentor my students, and serve my institution.

My kids are now 15 and 9 years old. They have grown to be highly independent, and I think it has to do with me making sure that they were able to do their part just like I am able to do mine. They are accustomed to having a mom who will always be there for them and that everyday work will come home. It is part of life. At their current age and given how independent they are, I worry less about them. My older son is an athlete, and I make sure I attend all his events. My younger son is very passionate about science. For every talent show that his school has, he often asks me to help him come up with some fun science demonstrations that he can do for his peers.

Obstacles Encountered

The obstacles one encounters as a woman in science are numerous. The major obstacles are trying to balance one's career with family. Oftentimes in the workplace, one is given more responsibilities in areas of service simply because students find you more approachable relative to one's male counterparts. This results in several interruptions during the day and less time to focus on research activities.

From a personal perspective, my significant obstacles were encountered early in my career as I tried to find a balance between my career and my husband's career. It took a few years before we both felt that we were in the right place. Once he was able to obtain a good job and I got tenure, we felt very settled in our lives.

Reflections on My Career

I have been extremely lucky to have a highly supportive husband who has always said to me "don't say no to the opportunity—we will figure out how to manage our family obligations." Without his support, I think it would have been very difficult to envision myself as a faculty member at a research institution and being able to raise two boys. I have made sure to include my sons in many aspects of the work that I do. I share with them often the research that we do. I stay involved in their school activities and often volunteer to talk about science in their classrooms. Because of the support I have, I don't think I would have done anything differently whether I had a family or not. I do think my level of productivity would have probably been much higher had I not had a family. However, I am appreciative of being able to accomplish what I have so far.

As I reflect on my career now after having been in academia for over 13 years, I don't believe there is anything I would have done differently. I would not change a single thing about the pathway to my career. I have been extremely fortunate to have amazing strong women, who, through their support, small talks, or guidance, have made it clear that one can have a happy marriage, be a good parent, and be successful in academia.

When I started graduate school, and learned about the various challenges women in science face, particularly how hard it is to have a family, I thought I would never be able to have my own family. So, I remain grateful that I have managed to maintain a happy marriage and that together we have been able to raise two wonderful boys. The journey has not been easy, but I have been fortunate to have good support from friends and family and that I also landed at an institution that provided me an environment in which I have been able to thrive.

Advice to Women in Chemistry

A career in academia can be extremely rewarding. It is extremely important to always have colleagues within your institution and outside your institution who can support you and guide you along the way. It is also very important to be able to understand the value of the environment you are in and what is valued. Oftentimes we have our ideas of what it means to be successful, but as you go through the tenure and promotion process, understanding what is valued by your colleagues, the chair, the dean, and the provost, matters. Know that people in those leadership positions can change quickly, and so you have to be fully prepared to adjust to those changes. Here are a few tips that have helped me along the way:

- As a woman in science, you are considered a role model and the person many others will need to come to for help and advice. You are also that person that will be asked to sit on numerous committees at your institution. Guard your time. Time is that one thing that will never ever be returned, and you must pick wisely how you will invest your time.
- Always take advantage of opportunities that are presented to you. I have always tried to find opportunities that will lead to meaningful collaborations, both in research as well as in service and teaching.
- Be realistic about what you can and cannot do. Know that it is alright to say "no" when you know you truly cannot provide what you have been asked to do.
- Be protective of your time. There are only 24 hours in a day and 365 days a year. During the pre-tenure years, every single day matters, and you cannot afford to have anyone waste your time. Carve out time during the week to dedicate to "thinking" about your work and disseminating it in a timely manner. The work done in your pre-tenure years will shape your career and pave the path to your success.
- It is important to be truly passionate about your academic career because there will be difficult days on which you will need to remind yourself of the "why" that keeps you going.

Reflections from Sherine's Sons
Amir (age 15) and Amren (age 9)

We love having a mom as a professor. She is smart, and whenever she can she will volunteer to come to our class to put together an exciting science demonstration for us. All our friends think we have the coolest mom ever. Every year we have a talent show at school, and we have always done a science demonstration for the show with the help of our mom.

The only hard thing about having a mom as a professor is that she is always busy and she does a lot of work at home. Often, she takes us to work with her. Our favorite part is when we attend group meetings with all the graduate students because they enjoy telling us fun stories about life in college. She also travels a lot, but she always brings us gifts from the amazing places she visits. Even though she is very busy when she travels, she always calls to check on us. When she is traveling, she calls several times a day to make sure we are doing well. She also helps us with our homework when needed through FaceTime. Our weekends are always fun because we always try to do something enjoyable together. When we grow up, we want to be just like our mom—scientists. It would be great to be able to have a lab where we can do experiments and teach other students important science. We do think we will work a little less than her though.

Elements to Successful Motherhood and the Professoriate

Acknowledgments I remain extremely grateful to Dr. Catherine J. Murphy (University of Illinois), Dr. Sandra McGuire (Louisiana State University), and Dr. Vicki Grassian (University of California, San Diego), who have simply been phenomenal role models. I am also extremely grateful to male colleagues who have been highly supportive: Dr. Robert Ofoli (Michigan State University), Dr. James (Ned) Jackson (Michigan State University), Dr. Gerald J. Meyer (University of North Carolina Chapel Hill), Dr. Howard Fairbrother (Johns Hopkins University), Dr. Dionysios Dionysiou (University of Cincinnati), Dr. George Cobb (Baylor University), Dr. Clemens Burda (Case Western reserve University), and Dr. Donald Schreiber (Western Michigan University). Through their support, they have opened doors for me that helped paved the road to success.

About the Author

Education and Professional Career

1998	BS Chemistry (Minor in Biology) (*cum laude*), West Virginia State University, Institute, WV
2002	PhD Inorganic/Analytical Chemistry, University of South Carolina, Columbia, SC
2002–2004	Camille and Henry Dreyfus Postdoctoral Fellow (Environmental Chemistry), Johns Hopkins University, Baltimore, MD
2004–2009	Assistant Professor of Chemistry, Department of Chemistry, Western Michigan University, Kalamazoo, MI [Professional leave 2007–2008, Department of Chemistry, University of North Carolina Charlotte, Charlotte, NC]
2007	Adjunct Professor, Department of Chemical Engineering and Materials Science, Michigan State University, East Lansing, MI
2009–2014	Associate Professor of Chemistry, Department of Chemistry, Western Michigan University, Kalamazoo, MI
2009–2016	Graduate Advisor, Department of Chemistry, Western Michigan University, Kalamazoo, MI
2011–2016	Associate Chair, Department of Chemistry, Western Michigan University, Kalamazoo, MI
2014–present	Professor of Chemistry, Department of Chemistry, Western Michigan University, Kalamazoo, MI
2015–2016	Interim Associate Dean for Research, College of Arts and Sciences, Western Michigan University, Kalamazoo, MI
2016–Present	Associate Vice President for Research, Western Michigan University, Kalamazoo, MI

Honors and Awards (Selected)

2013	Named one of the top 25 women professors in the State of Michigan by *Online Schools Michigan*
2012	President's Award, National Organization for the Professional Development of Black Chemists and Chemical Engineers
2012–2013	Emerging Scholar Award, Western Michigan University
2012	Faculty Achievement Award in Professional and Community Service, College of Arts and Sciences, Western Michigan University
2012, 2011, 2010	Science Spectrum Magazine Trailblazer Award
2010	American Competitiveness and Innovation Fellowship, Division of Materials Research, National Science Foundation
2010	Lloyd N. Ferguson Young Scientist Award
2009	George Washington Carver Teaching Excellence Award
2009	International Union of Pure and Applied Chemistry Young Observer Award
2007	Carl Storm Minority Fellowship, Gordon Research Conference
2007	American Chemical Society PROGRESS/Dreyfus Lectureship Award
2007	American Chemical Society Leadership Development Award
2006–2012	National Science Foundation CAREER Award

In addition to her research and editorial service in the field of nanomaterials, Sherine is a member and active volunteer for the American Chemical Society and the National Organization for the Professional Development of Black Chemists and Chemical Engineers.

A Family Grows

Patricia Ann Redden

Photo Credit: Ernest Napolitano

When I first became a mother, I wrote a "baby book" for my daughter, titled "A Family Grows." When we think about growth, we often feel as if it follows a predictable path. My Labrador puppy's future appearance will be similar to her parents and most other labs. A plant seed will (unless my black thumbs thwart it!) become a mature plant with predictable characteristics. However, humans and their families take their own wandering paths. Let me tell you a bit about mine.

P. A. Redden (✉)
Department of Chemistry, Saint Peter's University, Jersey City, NJ, USA
e-mail: predden@saintpeters.edu

How Did This Story Begin?

I grew up in a fairly typical Irish-American family in upper Manhattan, Washington Heights to be precise, with two younger sisters. Dad's parents had both died when he was young, but mom's parents and her married brothers and sisters all lived in the neighborhood for most of my childhood, so we had a very vibrant extended family. My grandparents on both sides were Irish immigrants who met and married in New York City, and Irish music and history were an integral part of our lives. The families had the usual dreams for their children—a move from "blue-collar" (grandparents) to "white-collar" high school graduates (my parents and their siblings) to college graduates (my generation).

Growing up in Manhattan was unlike all the books we read in grammar school. We lived in a walk-up apartment building, with the three of us sharing a bedroom until I was in graduate school. We didn't have yards, gardens, or lots of pets, but we had the resources of the city at our beck and call. Washington Heights is at the far northern end of Manhattan, most easily located by the nearby George Washington Bridge and Fort Tryon Park, and made famous by Lin-Manuel Miranda's 2008 musical play, "In the Heights." None of our friends rode bikes in that neighborhood on the Palisades, but after school we played stickball, stoopball, and marbles on the street, roller-skated, or played basketball in the park. I was always an animal lover, and my biggest regrets were that I didn't have a dog or a horse (which I suggested, at the age of five, could be kept on the fire escape of our building). Weekends would include trips to the nearby Cloisters, the Bronx Zoo, the American Museum of Natural History, or Central Park Zoo—probably where my love of science was awakened—or we would walk across the George Washington Bridge and picnic on the New Jersey side of the palisades. Summers would be a rented bungalow for a month, alternating between Rockaway Beach and the Catskill Mountains.

When I was old enough for school, my parents made the decision to enroll me in a private all-girls Catholic school, Mother Cabrini, for grammar and high school. Looking back, I'm sure it was a stretch for them to pay tuition for me, and later for my two sisters, but they believed it would be a better environment and education than the local public or parochial schools. Since it was a private school, I actually began first grade on my fifth birthday, having skipped kindergarten. I loved the school and quickly found that math was one of my favorite subjects. Our nuns, the Missionary Sisters of the Sacred Heart, made it clear that we were not limited in our aspirations because we were girls. I do sometimes wonder how I wound up in chemistry, though, since our high school science courses didn't involve lab at all, and most of the spring semester was devoted to teaching to the mandated Regents exams (does that sound familiar these days?). The only time we actually were in the lab all year for either chemistry or biology was when it was time for yearbook pictures. It just goes to show that high school courses don't always predict your success in a major.

Why Become a Chemist and Professor?

When it came time to choose a college, I opted for Cabrini College (now University), a brand new women's college in Pennsylvania, founded 1 year earlier by the same order of nuns; the first president was actually my principal from New York. Even though it was *very* small (my graduating class totaled thirty-five, two of whom were chemistry majors) and *very* new, I felt comfortable in the single-gender environment and knew that I would thrive there. The big question was whether to major in chemistry or biology. At that time, the first year course in biology was usually a semester of botany followed by a semester of zoology. As a city kid, I was perfectly comfortable with classifying vegetation as tree, bush, flower, grass, or weed, and I really didn't look forward to a year of memorization and taxonomy. Chemistry was much more interesting and challenging.

I was the second in my extended family and the first girl to go to college, so I never really thought of pursuing a higher degree. However, the college president and my department chairman had a different idea, and they literally pushed me to apply to graduate school for the doctorate. My thought through senior year was to concentrate in organic chemistry, but that summer I had the opportunity to do research in physical chemistry at St. Joseph's College (now University) in Philadelphia, and I found it totally fascinating, so when I went to Fordham University in the Bronx, I studied physical chemistry, specifically thermodynamics.

As usual, I was a teaching assistant for 3 years, until I got an NSF fellowship to continue my research. When we were kids, my middle sister was always the teacher when we played school. I had absolutely no desire to teach—bench chemistry for me—but I did enjoy the teaching assistantship. Then in my last year of grad school, one of the Jesuit professors, who had become a very good friend, told me of a teaching job at St. Peter's College (now University) in Jersey City, NJ, another Jesuit college. I confess that I had the typical Manhattanite's knowledge of New Jersey, exemplified by the classic 1976 "View of the World from Ninth Avenue" cover cartoon of "The New Yorker" magazine—west of the Hudson River was absolutely "terra incognita." It turned out that Jersey City was literally across the Hudson River from Manhattan, so I applied and was hired for a year. That was in 1968, and I'm still there!

That first year was rough. Many of the students were returning from Vietnam and were about my own age, the college had just gone co-ed the year before my arrival and had very few women students or faculty, and I was teaching analytical chemistry as a required course to senior biology majors who didn't want to take it. However, I decided that I really enjoyed teaching, and they seemed to like me, so I stayed.

The next years were typical of faculty at a small liberal arts undergraduate college. I took a year's sabbatical at University College, Cork, Ireland, doing research in photolysis and additionally pursuing several somewhat unusual activities: becoming reasonably fluent in the Irish language, skiing in Austria, singing in an international choral festival, learning dressage and jumping while working weekends at a stable, and bringing home an Irish Wolfhound puppy. Back at the college, in addition to the usual committee work and developing new courses, I served as an officer of our Faculty Senate, won grants for new programs in the sciences and for entering students, and developed college-wide programs for faculty development and curriculum. Tired of fighting traffic, I moved to Jersey City, which has often been called New York City's sixth borough. I also served as department chair for a total of 30 years, cruel and unusual punishment to be sure!

When my nephew was in kindergarten (he's now 37 years old), I began to visit his classes to do science experiments with the students, and that expanded into workshops and demonstration programs for in-service teachers, students, and scout groups, ultimately leading to teaching the course in Methods of Teaching Science to the elementary education majors. Going back to my roots at the American Museum of Natural History, for 10 years I organized for the New York Section a full day of hands-on experiments and performed demonstration shows to celebrate National Chemistry Day, reaching about 1800 attendees each year. In 1986, I became a Scholar-in-Residence and then Associate of New York University's Institute of Fine Arts Conservation Center.

Several members of my department were active in the local section of the ACS, and they encouraged me to get involved as well. By 1977, I was a member of the Board of Directors of the Hudson-Bergen Subsection of the New York Section, and that led to further involvement in subsection and section, in particular with the safety committee and later as councilor and section chair. One thing led to another, and the section's safety committee brought me to the Division of Chemical Health and Safety, serving as program chair and divisional chair, and that in turn led to membership on the Committee on Chemical Safety on the national level. All in all, I've been active on the national level of ACS since 1983, representing the New York Section on Council for 32 years and serving on six national committees, and I've made some wonderful friendships and professional ties as a result.

Finally, Motherhood

So far, this has been about my professional development, but of far more importance to me have been the personal events that led to my inclusion in this book—motherhood and all that has meant. For one reason or another, I had never married, but I had always wanted children. One day I was watching my nieces and nephew playing in my sister's backyard, and I knew that it was time to do something.

My younger sister had adopted her younger daughter from Korea, and we talked that day about my dream of having a daughter. Fortuitously, she was going to an adoption program 2 weeks later, so I decided to go as well. There was a session on single-parent adoption, and I spent a lot of time with the moderator afterward. I had thought of adopting from Korea, but I found that they did not allow single parents to adopt. However, India welcomed them. I was then living in Jersey City, which has a large Indian population, so that sounded like an ideal choice. There were also some really weird connections between me, Ireland, and India—my Irish Wolfhound was

A Family Grows

named after an Irish goddess who also turned out to be an Indian goddess, the Irish and Indian flags both involved the same colors (one with vertical and the other with horizontal segments), my college roommate had served in the Peace Corps in India, and I had even been involved in a television discussion group about India in high school. It was meant to be!

I went to meetings of a single-parent support group in New Jersey and realized that becoming a single parent of a little girl was something I really wanted to do and could do. So I started the process with World Association for Children and Parents (WACAP), an adoption group in Seattle that several members of the group recommended.

After going through the application and home study, I received a four-page questionnaire asking what physical issues I would consider in a child and an additional two-pager on developmental and emotional issues. The potential issues literally ranged from a birthmark to a need for immediate heart surgery and from anxiety to pyromania. I don't think I ever had such a daunting form to complete. How do you make decisions like that, knowing that on the other side of your response was a little girl who could be your daughter or who would be left in an orphanage? After a lot of soul-searching, I completed the forms, listing a lot of the options as possible, some as definite yes or no answers.

Then it was the wait for a referral. I was asking for a toddler girl, since of course I would be working and didn't want to leave an infant in day care during the school day. WACAP, I had been told, would send information on a particular child and allow you to say yes or no, and I couldn't imagine how you made that decision. A friend who had adopted through them told me that, when offered her choice of four little girls, she immediately knew which one was hers, but that seemed impossible to me. However, I was sent information on two adorable little girls, and they didn't "feel" like the right match, so I thought maybe she was right. Then in September 1987, I got a phone call about another similar little girl, two and a half years old, paraplegic as a result of polio, and not walking. Without hesitation, I said, "Send me the acceptance papers for my daughter." They urged me to bring the health information to my pediatrician (now really, what nonparent has a pediatrician?) so I could make an informed decision. I agreed but knew that this was not a question—Sheela was mine!

Then it was the waiting game, but everything fell into place in a few months, and just before Christmas, I was on my way to Kolkata to meet my daughter, now named Margaret Sheela after my mom, and take her home. What a Christmas present! (Fig. 1)

Margaret (later to become Maggie) couldn't speak English, couldn't straighten her legs or walk, and was on prophylactics for a year for tuberculosis, but she was a joy and eager to learn. That spring and the following year, she was in an early-childhood program for children with disabilities so she could get physical therapy and learn English, but in order to be able to straighten her legs and wear braces, she had to have the Achilles tendons on both legs surgically released. That meant a cast on pelvis and both legs through June and July, but after that she got her first set of braces (hip to foot on both legs) and a walker, and she was off and running; she also

Fig. 1 Pat's daughter Maggie (3 years)

got her first wheelchair for longer distances. Maggie took her first steps at the age of three at a baby shower for a friend of mine, with all of us cheering and crying.

When she was 4 years old, she entered prekindergarten, again in a school for special needs kids so she would get physical therapy every day. By that time she was fluent in English and ready for school. Our school schedules were roughly the same, a major advantage for a single parent, and if she had no school on a particular day, I would bring her to my office and classroom. My middle sister's family shares a two-family house with me, and they were also invaluable in helping take care of Maggie.

At the end of Maggie's first day in prekindergarten, I got a call asking if I would let her participate in wheelchair track and field. Since one of my friends had a daughter who was a wheelchair athlete, I immediately agreed. Maggie loved it, and an athlete was born. She actually participated in her first national track competition when she was barely 5 years old, competing against 9 year olds in the 60 meter track event and wheelchair slalom. At that meet, I bought her first racing wheelchair, so small that I was able to bring it home in the front seat of our Mustang. That was the first of many competitions and the first of many racing chairs (Fig. 2).

When Maggie was ready for kindergarten, I faced a dilemma. We lived in Jersey City, and the public schools were generally academically poor, and both public and parochial schools were in old, non-accessible buildings. Maggie was using a walker and just beginning to transition to crutches, but stairs were going to be a real issue. The question was whether to find a private school that would work for her and be affordable or move to the suburbs where she could attend public school but not be

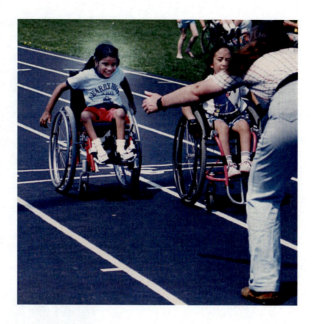

Fig. 2 Maggie at a track meet (4 years)

nearby in case of need, not to mention the higher cost of houses in the suburbs. Luckily I found Mustard Seed School, a private, nondenominational Christian school in Hoboken, the neighboring city. At that time, their kindergarten and first grade classes were in the basement of a church (still stairs!), but they hired a classroom aide to work with all the students but with the particular task of helping Maggie on the stairs. By the end of the year, she was using her crutches and confidently climbing stairs. The school moved the next year into a building that had an elevator, so her day became a little easier.

One Was Not Enough

My two sisters had always been very close to me, and I wanted that experience for Maggie. When she was about 8 years old, she and I discussed adopting a sister for her. She had already had two more surgeries on her hips, and I knew she was going to have another in the not-too-distant future for scoliosis, but after discussion we decided that we would look for a little Indian girl a few years younger than her, also a polio survivor. In March of 1994, after a long wait, Mariya Bridget joined our family, again thanks to WACAP, and our family again "grew." (Fig. 3)

Mariya was almost 5 years old and from Chennai (Madras). She had had her Achilles tendons released in India, so she was already walking with braces and crutches, and she had less motor impairment in her legs than Maggie. Like Maggie, Mariya spoke no English when she came home that March. The orphanage she had been in was very poor and had very little in the way of toys or even crayons, so she was really not ready to enter kindergarten in the fall. We also found out, in the initial

Fig. 3 Pat's daughter Mariya in prekindergarten (5 years)

Fig. 4 Maggie and Mariya just after Mariya arrived

medical screening when she arrived home, that she had a very high lead level, which could lead to learning disabilities. I made the decision to enroll Mariya in the same prekindergarten program that Maggie had attended, this time for a year, to allow her to acclimate, get some physical therapy, and gain academic skills. After that, she went with Maggie to Mustard Seed School in Hoboken for kindergarten (Fig. 4).

Add a Few Activities to Fill in Our Days

Balancing home and work became quite a bit more complicated as the girls grew older. I was heavily involved with ACS, attending two national meetings a year in addition to many local meetings, and the college and outreach activities took a very large chunk of time. I was reluctant to leave them home when I traveled, so the girls came with me at least until they were in high school.

I think each of them attended more professional meetings than most of my colleagues! I quickly learned to bring lots of books and toys and a sleeping bag, and one or both would stay in a corner of the room or at my feet during the meeting. When they were older, I would get a recommendation for a bonded sitter from the hotel, and the girls would stay in our room with the sitter while I was in the actual committee meeting. That always made me nervous, of course, but I would pop back in during breaks in the meeting to see how things were going (no cell phones in those days!). A number of parents like me were urging ACS to set up some sort of day care during the meeting, but that wasn't successful at the time (luckily, this is now more available). To balance the meetings, we would devote at least a full day to enjoying whatever city we were in, of course visiting every science museum and zoo. When Maggie was in high school and we were at a meeting in August or during school vacation, she would be in charge of Mariya, and as they both got older, one or both would stay with a classmate for the spring meeting so she wouldn't miss school.

Although both girls walked with long leg braces and crutches through high school, they used wheelchairs for longer distances and for various activities. Don't let the words "wheelchair-bound" even enter your mind. For them, there were very few limits, and the wheelchairs effectively gave them more freedom. In grammar school, their chairs were highly desired by their classmates in the park, all of whom learned to pop wheelies and do a slalom course. They used both crutches and wheelchairs, depending on the activity. When they each graduated from high school, they literally took off their braces for the last time, hung up the crutches, and transitioned totally to their chairs.

Of course, kids have afterschool and weekend activities, and mine were certainly no exception. When Mariya came home, she followed in Maggie's footsteps in athletics as well as in school. As the years passed, they expanded sports participation to include horseback riding, basketball, tennis, archery, skiing, and swimming; all but the horseback riding and tennis were on a competitive level. I never got to the Caribbean, but they each traveled twice to Grand Cayman Island to become certified scuba divers. This meant specialized chairs for each sport (other than scuba diving, riding, and swimming). At one point, our dining room table was circled by two extra everyday wheelchairs, two racing wheelchairs, a field chair, a tennis chair, two basketball chairs, and two monoskis! We moved up from a small car to a station wagon and then to a minivan so that we could travel with all the equipment (Fig. 5).

Their interests did diverge as they grew older. Maggie in college concentrated on track, Mariya on basketball. Mariya was on the ski racing team, Maggie became a ski instructor. Their sports also opened many opportunities for them. Both girls

Fig. 5 Taking a rest between runs on the ski slope

competed across the USA and in Canada, first with their local teams, the Lightning Wheels track-and-field team sponsored by Childrens' Specialized Hospital, and the NJ Nets wheelchair basketball team, and then later with national teams organized by Wheelchair Sports USA and the National Wheelchair Basketball Association.

International travel involved Canada and Australia for both girls while in high school; Mariya competed in South Africa as well. Maggie competed in track in the ParaPan American Games in Rio de Janeiro and in the 2008 Paralympics in Beijing. When Maggie was named to the US 2008 Paralympic team, I immediately called my dean and informed him I was taking time off from classes to go and watch her compete in the Bird's Nest—no discussion needed (Figs. 6 and 7).

If that isn't enough, both girls enjoyed music and were in the chorus and instrumental groups in grammar school as well as in high school. Mariya played percussion, violin, and saxophone, and Maggie bells, piano, and harp, including performing with Frank Paterson in Carnegie Hall and the New Jersey Performing Arts Center. Meanwhile, not to be left out, I was also singing with a New York classical chorus, performing twice a year in Carnegie Hall.

How did I spend my time outside of driving and professional responsibilities? When Maggie was in kindergarten, I became a Girl Scout leader, continuing with that until Mariya was in fifth grade. While they were skiing, I volunteered with the ski program and ultimately found myself in charge of the equipment room for the Adaptive Sports Program at Windham Mountain, adjusting ski bindings and setting up monoskis. At track-and-field meets, I moved up from helping as a volunteer to becoming a USA Track & Field nationally certified official, certified on the Paralympic level as well as for ambulatory sports. When the girls were both out of the

A Family Grows

Fig. 6 Maggie racing in the Bird's Nest at the 2008 Paralympics

Fig. 7 Mariya playing basketball for Edinboro University

house (Mariya at college and Maggie on her own after college), I started raising puppies for Canine Companions for Independence, the oldest and largest service dog organization in the USA; I'm now on my seventh pup.

Fig. 8 Mariya and Maggie on Maggie's wedding day, December 2016

Fig. 9 Mr. and Mrs. Abram Waugh on their wedding day

Where Are the Girls Now?

Maggie attended Penn State, earning a bachelor's degree in communications, and then received a Master of Public Administration degree from St. Peter's University in May 2015. After volunteering for a year at a home for orphan girls in Kolkata, she worked as a transition counselor, then for Social Security, and now for a non-profit in Washington, D.C. In December 2016, Maggie married Abe Waugh, a young man she has known for years through wheelchair sports. Mariya completed her degree in communications at Edinboro University in 2016 and has begun a new job. Both are on their own—Maggie in New Jersey and Mariya in Pennsylvania—and enjoying life. I am incredibly proud of both of them and can't imagine what my life would be if I had not attended that first adoption meeting (Figs. 8, 9, 10).

Fig. 10 Mariya and Chris Peterson

As I said in the beginning of this article, life takes its own path, but I had the support of family and friends along the way. Each step had its own inevitability, and I love where it has led me.

Lessons Learned

Take a chance and follow your heart. When I first decided to adopt, I was in my forties, and logic would probably have said (1) you're too old to raise a little girl; (2) as a single parent, you're going to have to change your life completely; and (3) what happens to your child if you're unable to work for some reason. Thank God I didn't listen to the logic, and I went to that adoption meeting.

To be honest, a college professor, particularly one like me who was pretty well established professionally at the time I first became a mother, is in a lot better position to raise a child, even as a single parent, than someone working a job with more regular hours. School schedules are usually not the same for colleges and other schools, but they do conform somewhat when it comes to major vacation periods, so you can have time to spend with the family at those important times. I also had the flexibility to bring the girls to my office if our schedules didn't match, and I was able to arrange afterschool programs at their schools when I had a late afternoon lab or meeting.

I did make sure I had a will drawn when I adopted, and I do, even now that the girls are on their own, think a bit about what would happen if my financial situation changes significantly because of illness or retirement. I am truly lucky to have a strong family support system, however, and that was crucial to the adoption decision.

Some would say that I also took a major chance in adopting two daughters who were paraplegic and not infants. The reality is that you can't always predict problems that can arise with birth children either, and I had time to think about how to deal with at least some of the issues before I applied for and accepted the girls.

The road has not always been smooth. There were hurdles with health issues, dealing with accessibility or lack thereof for the girls, even the adjustment of the girls to a new family, culture, and language.

The first year after Mariya came home was difficult financially, with adoption costs and tuition for both girls in private school, but that year my college realized that there were real inconsistencies between salary, rank, and years of service (particularly for senior women!), and we had a very significant salary adjustment. Keeping up with the cost of sports equipment as they grew, when a new chair for a sport (needed every 2 or 3 years as they outgrew the old chair) would typically cost at least $3000, made big dents in the budget, but the benefits to the girls far outweighed the need to cut back on other things, and the girls were never the type to want the most expensive or fashionable brands of clothes. Both girls opted to pass up free tuition at my college or other Jesuit colleges when they selected their universities, but their choices were right for them, even if far more expensive for me.

You can do most of it, if not all, but it requires a lot of juggling. I look back now and wonder what I did with my free time before I became a parent. When I first broached adopting a second child to my dad, he suggested that I would have to give up something, probably my NYC chorus, but that was the thing that kept me semi-sane, so nothing was dropped.

When the girls were growing up, sports had us traveling all over the mid-Atlantic region and the USA for practice and competition. Winter weekends were at the ski area in the Catskills, but the entire year found us traveling in all directions from Jersey City for practices and competition in one sport or another. One year we actually drove to Philadelphia early Saturday mornings for basketball practice, then up to the Catskills overnight for skiing on Sunday, and finally home. In 10 years I put over 250,000 miles on my van, and none of that was for commuting to work! In addition, there were travel and meetings for ACS and work at the college.

Talk about juggling! I guess I could be labeled as a type A person. Sitting around while the girls had music or sports activities drove me a bit nuts, so I got involved as a volunteer in each of their activities.

Now that they're out of the house, I still don't have any free time, with my chorus, the service puppies, and continuing volunteer work fills the "free" time after teaching, college committees, and ACS commitments. Travel for the girls' activities has been replaced by travel of my own—the Amazon rainforest, canoeing and camping on the Canadian border, building walls and studying elephants in Namibia. I wouldn't have it any other way, but I do wonder how others find time to relax! (Fig. 11)

A Family Grows

Fig. 11 Pat in the Amazon rainforest in Peru with ET (actually a baby sloth)

Fig. 12 Maggie, Mariya and Pat

The juggling is worth it. The best two decisions I ever made in my life were when I said "yes" to my daughters. No matter how difficult the day, I have never for a moment regretted those decisions. A poem that hangs on our wall says, "Not flesh of my flesh nor bone of my bone, but still miraculously my own. Never forget, not for a minute, you weren't born under my heart, but in it." (Fig. 12)

Acknowledgments None of this would have been possible without:

- My two wonderful daughters, Maggie and Mariya, the joys of my life, who have grown into such remarkable women
- My parents and sisters, who showed the way and supported me through thick and thin
- The birth parents of the girls, who gave them life and love in their early years but who could not provide the help they needed after contracting polio
- WACAP in the USA and numerous individuals and organizations in India, all of whom cared for the girls until allowing me to adopt them

- All those who didn't see only a disability but who gave Maggie and Mariya opportunities to grow and expand their horizons and their potential—doctors, therapists, teachers at every level, coaches, and friends, all far too numerous to name
- My Saint Peter's colleagues and those in ACS who didn't flinch when the girls came to a meeting and who have continued to ask about them

About the Author

Education and Professional Career

1962	BS Chemistry, Cabrini College, Radnor, PA
1968	PhD Physical Chemistry, Fordham University, Bronx, NY
1968–1973	Assistant Professor, Chemistry Department, Saint Peter's College, Jersey City, NJ
1972–1973	Research Professor, University College, Cork, Ireland
1973–1980	Associate Professor, Chemistry Department, Saint Peter's College, Jersey City, NJ
1977–1995	Chair, Chemistry Department, Saint Peter's College, Jersey City, NJ
1986–1995	Scholar-in-Residence and Faculty Associate, New York University Institute of Fine Arts, Conservation Center, New York, NY
1980–present	Professor, Chemistry Department, Saint Peter's University, Jersey City, NJ
2001–2013	Chair, Chemistry Department, Saint Peter's University, Jersey City, NJ
2010–2013	Director of Environmental Studies, Saint Peter's University, Jersey City, NJ

Honors and Awards (Selected)

2015	Recognition for serving on the American Chemical Society (ACS) Council for 30 years
2011	Fellow, ACS
2001	Fellow, ACS Division of Chemical Health and Safety
1994	Outstanding Service Award, ACS New York Section
1993	Distinguished Service Award, Rockland Chemical Society
1991	Tillmanns-Skolnick Award, ACS Division of Chemical Health and Safety

Pat became an ACS member while in graduate school. Since 1977, she has chaired the Hudson-Bergen Subsection, the New York Section, and the Division of Chemical Health and Safety and served on six national committees. She is currently on the Committee for Chemists with Disabilities.

Teacher to Scientist and Back Again

Rachel E. Rigsby

Photo Credit: Sam Simpkins, Belmont University

Early Life

I always wanted to be a teacher. I remember as a young girl arming myself with chalk, an old-fashioned easel-style chalkboard, various books and supplies, and even a stick to use as a pointer and heading to my playhouse to teach imaginary students the basics of reading and math. I adored my elementary schoolteachers and couldn't imagine doing anything else. Summer was always too long; I was inevitably ready to go back to school by July. Long days were filled with reading, mostly mysteries,

---

R. E. Rigsby (✉)
Department of Chemistry and Physics, Belmont University, Nashville, TN, USA
e-mail: rachel.rigsby@belmont.edu

© Springer International Publishing AG, part of Springer Nature 2018
K. Woznack et al. (eds.), *Mom the Chemistry Professor*,
https://doi.org/10.1007/978-3-319-78972-9_31

although any series had to be read to completion—classics like *The Boxcar Children*, Nancy Drew mysteries, and even books about the nurse Cherry Ames kept me entertained. I read short stories aloud to invisible ears, asking follow-up questions about characters and events. It was an enjoyable way to grow up in my rural Kentucky home.

I had many role models growing up. Definitely my parents were supportive and impactful. They worked hard and raised my brother and me without the help of grandparents or extended family. They taught me to pursue my dreams no matter what and that no amount of money could substitute for a satisfying career. Many of my teachers encouraged me as well. Early in elementary school, my teachers encouraged my love of reading. My first grade teacher chose more advanced books like *Charlotte's Web* for me from the school library because she recognized that I needed to be challenged. When I was a bit older, another teacher asked me to read to her class some days after lunch. While I'm sure it was a much-needed break for her, it allowed me to gain confidence in a leadership role. I of course idolized my teachers and wanted to be an elementary schoolteacher for several years.

Everything changed my sophomore year of high school. I walked into my chemistry class and sat in the back of the room (being a shy introvert who didn't like being called on in class), and I can clearly recall my teacher's first words. He claimed that he could assign course grades right then—those in the front would get As, and those in the back, well, wouldn't fare so well. I thought to myself, "Mister, I am going to prove you wrong!" And I did. I got an A and never looked back. I took AP chemistry and then physics with the same teacher, and I decided to pursue a chemistry major in college due in part to his inspiration. I had no concrete plans on what I would do with the degree. Originally from central Kentucky, I chose to attend Kentucky Wesleyan College in Owensboro, Kentucky. It is a small, Christian college not far from where I grew up, and, for a first-generation college student, it felt very safe. My life was profoundly affected by the caring faculty there, both in the sciences and in other areas. It was definitely the right place for me.

During my freshman year, I really questioned my decision to be a chemistry major—chemistry was hard, and it really wasn't as much fun as I thought it would be. I remember thinking, "Why exactly am I a chemistry major?" I was taking biology courses as well, and for a while, I considered a career in medicine.

My sophomore year of college was again a transition point for me. I knew I couldn't stand the sight of blood but started volunteering at the local hospital anyway. Passing out after watching an IV insertion procedure put thoughts of a medical career out of my mind. Additionally, I found myself enjoying my organic chemistry course. It was challenging, but I had always enjoyed solving puzzles, and it was exciting to see how chemical reactions worked. I found it very engaging to try to understand something I couldn't see. My questions felt answered. This was why I was a chemistry major. I finished my last biology course and felt good about continuing my chemistry major.

In my junior year of college, I discovered biochemistry. Studying the complexity of biological processes, especially enzymes and the reactions they catalyzed, was awe-inspiring. I thought, *"This* is why I'm a chemistry major!" I was hooked. It was chemistry or bust. However, I didn't know what to do with my major.

When my professors started talking about graduate school, I didn't really know what that meant. I didn't know anyone who had continued their education after college, but my professors explained that I wouldn't have to pay for it due to the availability of graduate teaching and research fellowships. I began to consider applying to some graduate programs.

One of my professors, Dr. W. L. Magnuson, had a significant impact on my decision to pursue an advanced chemistry degree. He was a "tough love" kind of guy. One day he asked me if I had applied to any summer research programs; when I responded that I hadn't thought about it, he bluntly told me he would "kick my butt" if I didn't. Not wanting to disappoint, I applied for and was accepted to a National Science Foundation Research Experience for Undergraduates program at the University of Louisville. That experience taught me that I could do graduate research, but I didn't really want to live in a city like Louisville.

After returning to Kentucky Wesleyan College in the fall for my senior year, I took the GRE and looked at graduate programs in chemistry at the University of Kentucky and Vanderbilt University. Ultimately I was accepted to the chemistry program at Vanderbilt and moved to Nashville after receiving my bachelor's degree. During my final year of college, I served as a lab assistant for organic chemistry. Through these courses, I learned about managing students in the lab as well as how to grade with a rubric. Many of these experiences shaped the foundation of my own teaching career.

From Graduate School to a Career

At Vanderbilt, I joined the lab of Richard Armstrong, a renowned enzymologist, and worked diligently my first year to balance coursework and time in the lab. It was very challenging, but I felt academically prepared to succeed in my courses.

I met my future husband in my second year of graduate school. We decided to get married quickly rather than waiting until I finished school, partly because unlike undergraduate degrees, a graduate degree in chemistry didn't have a set time endpoint. We got married in August as I was starting my third year. At that point, I was finished with my coursework and was working full-time in the lab. This was helpful because it felt less intense than the grind of homework and studying, and I was able to build flexibility into my work schedule to be able to spend time with my husband. It was challenging, though, and took an understanding spouse to deal with some really long days and weekends in the lab when experiments were just not going the way I wanted them to. Overall, it was a very good decision to be married during graduate school.

I became focused on a desire to return to teaching and decided to pursue a career as a college professor. I was fortunate that my graduate advisor valued teaching and was himself one of the best teachers I had ever experienced. He allowed me to participate in a Future Faculty Preparation Program offered through Vanderbilt's Center for Teaching. This program was invaluable and gave me opportunities to learn about teaching and faculty life. Experiences ranging from attending a Vanderbilt Faculty Senate meeting, sitting in on classes, and interviewing faculty at other local universities to being introduced to ideas of teaching gurus such as Ken Bain were instrumental in my decision to pursue a career in higher education.

As I finished graduate school and began looking at career options, I was fortunate to apply for several positions in the Nashville area. I was offered a job teaching general and biochemistry at Belmont University, which I happily accepted. The first few years were grueling but very fulfilling. I enjoyed being in the classroom, supervising undergraduate research, and most of the things that went along with being a college professor.

Becoming a Mom

My husband and I decided to wait to start our family until after graduate school and getting a full-time position. Around the end of my second year at Belmont, my husband and I started thinking about starting a family. Here we hit a roadblock. I didn't get pregnant and discovered some undiagnosed health issues.

We moved into a different area fairly quickly—adoption. After attending initial adoption seminars, we felt that domestic adoption was the right choice for us. We were open to the race and gender of our child, and after filling out a mountain of paperwork, getting background checks, and having our home inspected, the wait began. This was a very challenging time—often being sad or frustrated in my personal life but moving on professionally.

Thankfully, our wait was short, just over a year, and we drove to Mississippi to adopt the sweetest little African American boy you have ever seen! He was the perfect addition to our family, and as a preemie who at two and a half months old at adoption didn't weigh quite 8 pounds, he was a handful (Fig. 1).

I'm glad we didn't have our son until I had been teaching for a few years—the first few years were brutal! I would not have been able to survive new course preps and service expectations during those first few years had I had a little one.

Adoption presented obvious challenges in my personal life as I transitioned to being a working mother, but it also brought a unique challenge to my professional life. Colleagues I knew who were having babies typically planned a semester of leave and stayed home. As I learned, there was no planning with adoption. I had hoped for an ideal situation—getting placement in May so I could have the entire summer off with my new baby. What a dream!

It was not to be. We received the call that we had been chosen to adopt our son James on a Friday, drove to Mississippi the next Wednesday, and came home on Saturday. It was the third week of January, just after the start of the term, in my mind the worst possible time. So, I didn't take any time off, and my husband decided to initially stay at home. If I could change anything, it would be to take time off the first

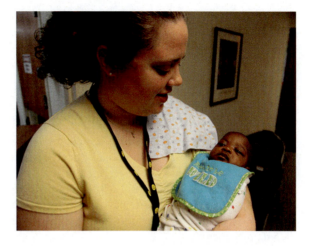

Fig. 1 Rachel with James, the new addition to her family

few months with James. As difficult as it would have been to arrange to have my classes covered, I should have found a way to do it.

Thankfully, I had great colleagues who picked up a lot of departmental duties that first semester so I could focus on teaching my classes (12 hours' worth, a full load) and then go home. In truth, the semester was a blur as I adjusted to life as a working mom. As with many professional moms, James often came to Belmont as my husband and I juggled work and childcare. We handed him off many days in the hallway as I finished class or lab and headed home. I appreciated that my position did allow a lot of flexibility in my schedule so I could come and go as needed.

In future semesters, we had a variety of childcare situations. In the first year, we arranged a nanny share with a friend and had a nanny 2 days a week. For the next 2 years, James was in childcare not far from Belmont. Then, he moved to a preschool in the area. We periodically evaluated our childcare situation and made changes as James grew.

Making It Work

My work/life balance really shifted after we had our son. I found myself taking advantage of time management strategies and being as productive as possible at work so I could focus on being a mom at home. I kept an ongoing "to do" list on my desk so that even 10–15-minute chunks of time were used effectively. I rolled off of some nonessential committees and limited my time on email. One thing that helped with email was getting a smartphone—this allowed me to scan emails quickly on the go and delete unimportant ones so that when I did sit down at my desk, I could focus on what needed to be done.

Achieving a sense of balance was a struggle. I have found it helpful to set boundaries, as otherwise I feel constantly torn between home and work responsibilities and not doing well at either. Currently, I make every effort to not check email or do any work between 5 and 8 pm when I am home preparing dinner and going through the evening routine. After James goes to bed, I try to catch up on anything urgent, mostly email and grading. My time spent working on weekends is also limited. I am often up early on Saturday morning preparing lectures or writing exams, but I have learned that relaxing and spending time with my family help me be rested and ready for work again on Monday.

I would say I have been lucky to not have experienced many of the obstacles I've seen others experience. Working at a liberal arts college and having the focus be on teaching rather than research are huge. I have a reasonably flexible schedule and don't have the pressure to pursue research funding. I have used resources from our teaching center and the support of a writing group to successfully complete publications. Also, we have many female faculty in the sciences at Belmont, and our chemistry department faculty ratio is 4:2 (female/male)! The environment is definitely family friendly.

One of the most surprising things I've experienced as a working mother in academia is the challenge of transitions, especially the transition from working mom to stay-at-home mom. This happens every May as I am off contract for 2 months. Every summer we pulled our son from day care/preschool, and I stayed home. It was hard! I needed the routine of work and a schedule to (honestly) not feel crazy, so I learned to add routine to our life at home during the summers. For me, it has always been easier to shift back into "working-mom mode" in August, but the transition at the end of May remains difficult every year. I think this is partly because every summer is so different—in the first summer, our son was still tiny and took two naps each day. In the next summer, it was only one nap, and by the third summer, naps were nearly gone—no more personal time during the day! On the bright side, though, no naps meant it was easier to plan day-trips without worrying about missing nap time. Now that James is in school, the summer is a much-needed break, and we enjoy going to the zoo or heading downtown to the library and eating ice cream at our favorite shop (Mike's Ice Cream) (Fig. 2).

Reflections and Conclusions

Being a mother is the most rewarding experience I have ever had. It is also by far the most challenging. A close-knit group of colleagues and the strong support of my husband make it possible for me to continue working. As a mother working in academia, I enjoy the flexibility of being able to stay at home during the summer. My working schedule is flexible so I can volunteer in my son's classroom and participate in a variety of activities with him. Having achieved tenure and promotion to professor, I feel successful professionally. If I could do one thing differently, I would have figured out how to take maternity leave. The first few months as a mom were much too stressful!!

Fig. 2 Rachel with her husband and son. Photo Credit: Tiana Lee

I feel very fortunate to have landed where I did. For others, especially those very early in their career path, I would recommend that you start early with planning—consider what type of institution you want to end up at (research vs. teaching focused) and where geographically you want to end up. If it's important to you to be near your family when you have children, that may take considerable planning. Think about research vs. teaching postdoctoral avenues and how to make yourself a great candidate for your dream job. Talk with your spouse about both of your career plans and how those will fit in with having a family. Finally, consider options for taking a break. I know women who have taken untraditional routes—one PhD biologist taught high school for a few years and then took a few years off while her children were very small. She began teaching night classes at a community college before moving back into a full-time position. Anything is possible with enough planning.

A career in academia is certainly full of challenges. Being a working mother adds another layer of complexity but also many, many rewards. I am very happy to enjoy the life of a "Mom the Chemistry Professor."

Acknowledgments I would not be where I am today without the help of so many wonderful teachers and chemists. My high school chemistry teacher, Arthur C. Hale; my college advisor and professor of biology, Dr. Rob Kingsolver; and my chemistry professors, Dr. W. L. Magnuson, Dr. Henry Connor, and Dr. Bob Flachskam. I am forever indebted to my graduate advisor, Dr. Richard Armstrong, who encouraged me when I wanted to quit and who allowed me to pursue teaching endeavors while in graduate school. You will notice that all of my teachers were men. I did not have a female teacher for any college- or graduate-level mathematics, chemistry, biology, or physics course. Not one. But all of these men encouraged me in many ways, and not one of them ever doubted that I would be successful. I hope I am as encouraging to future generations of men and women as they were to me.

About the Author

Education and Professional Career

2000   BA Chemistry, Kentucky Wesleyan College, Owensboro, KY
2005   PhD Chemistry, Vanderbilt University, Nashville, TN
2005–present   Professor of Chemistry, Department of Chemistry and Physics, Belmont University, Nashville, TN

Honors and Awards (Selected)

2012   Belmont Student-Athlete Advisory Committee 2011–2012 Inspiration Award

Rachel is a member of the American Chemical Society (ACS) and an active member of the Tennessee Academy of Science (TAS). The Belmont University ACS Student Chapter, which she serves as a faculty co-advisor, has received chapter awards since 2008. Rachel also serves on the TAS Executive Committee as the managing editor of the *Journal of the Tennessee Academy of Science*.

A Divinely Ordered Path

Renã A. S. Robinson

Photo Credit: Anne Rayner/Vanderbilt University

Current Situation
9/1/15 1:33 am
    I was invited to write a chapter about being a mom as a faculty member recently. The start of writing this chapter was not supposed to happen until

(continued)

R. A. S. Robinson (✉)
Department of Chemistry, Vanderbilt University, Nashville, TN, USA
e-mail: rena.as.robinson@vanderbilt.edu

Christmas break. However, I thought what better time to start writing than in this moment right now. It's 1:33 am on a Tuesday morning. Today is the first day of class for the fall semester. I teach my first class tonight at 6 pm. After working for the last 3 hours on my syllabus and lecture notes (albeit with interruptions), I was finally ready to call it quits. In fact, I just closed my laptop, and as I was about to get up from this worn-out corner on the futon in my bedroom (my 2nd home office), my 2-year-old daughter starts screaming. First it was a low whine that followed with a "Papi" that could be heard one floor up that then turned into a loud scream of "Mommy" "Papi" that echoed through the audio monitor in our room. As I got up to see if it would pass and she'd fall back asleep, the sound of her jumping up and down frantically on the bed let me know that I would have to take this one. I looked over to my left, and my husband was asleep in no man's land even with the room half lit from the ceiling light over the futon. No budge. I waited 2 more seconds. No movement. Within all of 4 minutes, I tiptoed downstairs, swooped her up in the dark, took her to the potty, changed the pull-up, washed hands, and consoled and held her for 1 minute before laying her down to rest assuring her that her bedroom door was still open. I tiptoed back upstairs to keep from waking up my 6-month-old son. Tonight I let go of my anxieties about not nursing my son for one feeding and instead gave him a bottle of breast milk for his 12:30 am feeding. The goal is to get everyone more rest through the night. Well, we'll see if I get to sleep until 6 am to know if indeed it worked. Nonetheless, amidst these interruptions, I have an updated course syllabus and template for my lecture slides. I even added some new slides after a 4-year hiatus from this course and can peacefully go to sleep knowing that I have at least reviewed the lecture notes once through. It's now 1:44 am. I'm wrestling with continuing to write or stopping to lay my mind to rest for what I hope will be at least 4 hours as this is going to be a fun chapter to write.

Getting Here

Reflecting on the past has always been informative for helping me to see how I have arrived at my current destination but also to realize that life is not random. My being a chemistry professor, mother, wife, and all the other hats that I wear is not by chance. I strongly believe that God has ordered all of my steps to date and will in the future according to His plans and purposes for my life.

It is no accident that I am teaching chemistry and have a research laboratory. The opportunities that I have been afforded since elementary school have shaped me in this way. I was born and raised in Louisville, KY, and spent most of my K-12 career in my home city with the exception of some summers.

A Divinely Ordered Path

My elementary school teachers took notice of me beginning in kindergarten and recognized that I had advanced or exceptional understanding of the classroom material. This led to my first exposure to race, expectations, and opportunities (or the lack thereof). I attended a neighborhood elementary school, Martin Luther King Jr. Elementary, in a Black neighborhood where all of the students and many of the teachers looked like me. Because I was performing at a high level in my classes, the teachers thought that I should be moved to another school that had a curriculum that would be more challenging. I was bussed out of my neighborhood, more than 10 miles away to Chenoweth Elementary, a school that was predominantly full of White students and teachers. After flying through the curriculum in those classes, I was encouraged to take the "advanced" placement test and was bussed again to St. Matthews Elementary to take advanced classes. I don't necessarily remember having a passion for science and math during that time. My summers were spent in "Quick Recall" programs where I, along with now close friends, learned as much information as possible about Greek gods, vocabulary, trigonometry, and other random facts, so that we could beat other teams that we competed against.

Moving on to middle school, I was very fortunate that my mother worked with my counselors and principal and encouraged me to apply for after-school programs aimed at college readiness. I became a member of the 2nd class of the Whitney M. YOUNG (Youth Organized to Understand New Goals) Scholars program founded by Dr. Samuel Robinson of the Lincoln Foundation in Louisville, KY. This program has a mission of preparing and providing direct opportunities for minority students to graduate high school and attend college. The Scholars program was intense but extremely fun. Many weekdays after school or Saturday mornings, I, along with more than 50 other students, spent my time learning reading skills, math preparation from algebra to calculus, and science preparation from biology to chemistry. I was able to see how valuable it is to recognize that other students are gifted in these areas as well and was humbled to learn that it took more than smarts and intellect to be successful. We all were gifted and smart.

I also was fortunate to have an African-American science teacher, Mr. Smith, in middle school. Every morning he had a large cake size mixing bowl from which he used to eat his Frosted Flakes cereal with a serving size spoon. Hilarious. While it was his daily ritual, I loved the comedic relief and the fact that his humor carried over into his teaching us the importance of moles and atoms. Perhaps he helped to foster my initial interest in science. I am not sure.

I certainly remember that high school became one of the defining periods in life shaping me into the individual that I am. I attended Louisville Central High School Magnet Career Academy, which had magnet programs in medicine, law, computer technology, business, and veterinarian science. My most fun days at school were days when I was in band class, laughing nonstop at our drummers and trumpeters, while playing melodies of The Gap Band on the flute. I remember days when I cried from laughing so hard in chemistry class because of a class comedian who could do phenomenal impersonations of everyone from Muhammad Ali to Mr. Emory our chemistry teacher. There it is. Mr. Emory was able to teach us how to perform titrations while encouraging us to have fun at the same time.

Chemistry came easy in this class because it was really just math taught under a different name, at least that's what I thought, and math skills I had inherited from my father.

While continuing to participate in the Whitney M. YOUNG program every week, and every summer attending college campuses and classes at Kentucky state colleges, I was once again encouraged by my counselors to apply to another special program for gifted minority students, African-Americans, Hispanics, and Native Americans. The program, Math and Science for Minority Students (MS^2), took ~30–50 students from across the USA and placed us on the Phillips Academy Andover campus, near Boston. We had our own curriculum and classes that were taught by college faculty and recent doctoral school graduates across disciplines. I remember having a math class called Discrete Math that didn't come with a book but rather a binder full of notes from a young math professor who had just created this math class. I took college-level calculus as a sophomore high school student and was exposed to college-level chemistry and biology classes. While things still came easy sometimes, I had to put into action study habits that I was taught earlier on, because I was around some extremely smart peers and the classes were challenging. I was impressed to see so many other students of color that were gifted and felt humbled to be among them but admit that at times I felt doubtful about how I had managed to be there.

I share a few of these memories because it's been insightful for me to look back on what experiences have shaped and caused me to gravitate toward this particular path in chemistry. The common themes in my past have all been centered on achieving academic success in stimulating environments (mostly around math and science), among diverse groups of people. I have had to always take tough academic concepts and make them relatable and accessible to diverse groups of people. Likewise, I have always been placed around gifted and smart individuals who thrived on being intellectually stimulated and challenged, sometimes in lighthearted competitive ways. I grew up in an environment and was exposed to different environments that showcased the disparities that exist in education for minorities and the poor but in contrast others that highlighted the successfulness that minorities and persons from low-income backgrounds can achieve. And while this chapter is not about race, it is important for me to acknowledge that my being a mom, a chemistry professor, and an African-American female comes with a unique set of challenges.

Defining Moments in College

I entered my freshman year focused on becoming a cardiac surgeon rooted around the loss of my father to heart failure and the monetary attractiveness of being a doctor. I do not remember anyone ever telling me I could become a college professor or should be, except for my graduate school advisor. College continued to open my eyes to the possibilities that exist in the world and also how critical it is to surround one's self with like-minded individuals.

I continued to perform well in math and science classes and became a chemistry major. Chemistry was a great option because it fulfilled the premed requirements and also was not a bad choice for a plan B in life. I heard that chemists at companies could make decent money.

What is most interesting about my college experience is that despite how well I performed academically in most of my classes, I was not "in the know" or advised well about life after college. I decided that medical school was not a practical option because it would take another 12 years before I could become a practicing surgeon and actually make money.

Inspired by my close friend, Danelle Stevens-Watkins, I took on extra credits each semester and summer classes in order to graduate a semester early. The spring semester of 2001, after graduation, was just spent laying low. I worked jobs in banking, retail, and financial planning that were unrelated to my degree. When searching for chemistry jobs, I was unenthusiastic about the local jobs as a chemist working on paint formulations or doing quality control for McDonald's headquarters.

It was then that Danelle told me to get up off my butt and apply to graduate school. Huh? Graduate school? I didn't know much about it or why I would want to go, but because of her frankness, I decided to explore chemistry program options. When I realized that you were paid to learn instead of paying to learn like in medical school, I was totally on board.

Current Situation
9/17/15 12:43 am

I have to get some sleep. The little one or shall I say the youngest one has a doctor's appointment in the morning. But I had to share what today, oh yesterday, looked like as I just completed 30+ reviews in a virtual NIH panel throughout the course of the day.

- 7:30 am—I woke up to the sun shining in my face, the quietness of the kids asleep, husband gone to work, and no dogs barking and thought "I can work with this."
- 7:50 am—My getting ready is rushed by the youngest one stirring. Time is up. I grab my bags and head downstairs.
- 8:05 am—The youngest one is ready, and he and I go sing our "Buenos dias" song to wake up my daughter. She had a fit momentarily until I put the baby in the crib to kiss her. Lots of patience is required to get her ready, but by 8:30 she's dressed, hair combed, and ready for school. Because she's having a moment of really needing mommy, I take half the flight of stairs down with gym bags and baby in car seat. I come back to the top of the stairs, pick her up, and carry her down. I repeat the next flight and come

(continued)

down to both dogs excited to see us. In the midst of all of this, the young one screamed off and on. He's teething and wanted a bottle, but I recently learned from a friend that "backtracking" is not productive or efficient. So no trips downstairs until everyone is ready.

- 9:10 am—The kids are dropped off to school.
- 9:35 am—My time was short in the gym but I ran a mile and did an abs workout.
- 9:58 am—I just arrived at my office in time for a 10:00 appointment with one of the newest undergraduates to join our research team.
- 11:30 am—At this point I'm already late to my virtual panel but need to log in and get going before my 11:30 meet with a seminar speaker. I love when seminar visits are easy and there are mutual interests and great discussion. I learned something about immunology and encountered my first male visitor that wasn't ashamed to ask for a restroom break.
- 12 pm—I had lunch with a female colleague, in her office, multitasking service work and catching up on life—refreshing.
- 1 pm—I spend 1 hour participating in the virtual panel and 30 minutes meeting with a second new undergrad researcher.
- 2:30 pm—Sadly to say, I struggled for most of the weekly seminar to stay awake not for lack of a good seminar but because this was the first time I sat still all day. Sleep is coming over me, and my diluted almond milk coffee hasn't kicked in yet. All the encouragement I needed came from another female colleague in the department who remembers how rough these years can be with little ones, and a simple "just hang in there" after the seminar helped.
- 3:30 pm—I get back to the virtual panel and peruse emails.
- 4:15 pm—This is the point in the day that I begin course prep for tonight's class and updated spectroscopy slides, trying to remember the material prior to lecture.
- 8:00 pm—Lecture and tonight's recitation are over and, at least for me, rocked!
- 9:00 pm—I arrive at home, the kids are down, and my husband is cleaning the kitchen. I hate when I don't get to see my children before they go to bed. My dinner is warming (praise Him), I eat, and the upside to kids being down is we can talk about as much as we can. This has become our nightly routine most nights, chatting and cleaning the kitchen prep bottles together—it's working for now.
- 10:00 pm—Being mindful to prepare for the next day, I pack my gym bag for tomorrow's personal training session, shower, and complete my last comments on the virtual panel.
- 1:08 am—I really just had to share this while it's fresh. I am done here and shutting my eyes.

Fig. 1 Celebrating Sunnie Myung's PhD: (from left to right) David Clemmer, Renā, Sunnie, and Stormy Koeniger

Still on the Path

Students ask me now what is the hardest thing I have ever done in my life or what was the hardest thing about graduate school. At the time, I thought that graduate school was the hardest thing that I had ever done in my life. I mean really. I had never been stretched, pulled, pushed, and challenged in this way at all. Things that used to come easy required me to work harder than ever before, and there were times when hard work and smarts did not equate to success. Research is sometimes unpredictable and doesn't always go as planned. Of course, this is both the beauty and the curse of scientific research. The biggest lesson that I learned in graduate school was that my successes and failures could not be measured against other's successes and failures. I also learned that life's joys and hardships are best shared with close friends (Figs. 1 and 2).

Graduate school was initially what I thought was the "means to an end" of owning my own cosmetics company and being a highly successful and paid entrepreneur. However, graduate school became the place in which I learned what and who I was made of and recognized that the world of chemistry was fun, exciting, and diverse.

My mother has always been my biggest role model because she understands what is necessary in order to make things work, she isn't afraid to do the work to make it work, and she will do the work without complaining about the work. She does all this while keeping a quiet confidence and smiling. I most certainly followed and still follow my mother's example of work ethic in order to be in this profession.

Fig. 2 Sunnie Myung (left) and Renã (right) at the Indiana University commencement ceremony in 2007

I did not know while in graduate school that I wanted to become a professor. In fact, I hated the idea because I felt like it would be graduate school times a thousand all over again. Some of the professors seemed really stressed, and I felt I had such unrealistic expectations of students and science. I did not see at the time how I could or why I would want to fit into this world. On the other hand, I really enjoyed all of the opportunities for learning new science, concepts, and methods and being able to become an expert in a topic.

My dissertation work focused on using analytical chemistry to study proteins and aging in fruit flies. This was a new area for our research group and really hit close to home in terms of my personal interests in aging and age-related disease.

I appreciated the opportunities to collaborate and share my research progress at conferences and travel to other cities such as Orlando, York, and Montreal. In fact, one of my collaborative projects studying aging in worms gave me the opportunity to travel to the University College London.

I have a number of highs and lows that I can share about graduate school; however I am extremely grateful to my advisor, Dr. David Clemmer, for taking me into his research group. David is my academic father but more importantly has become a lifetime friend and supporter. He has taught me so much about science such as how to master fundamentals, make connections to the big picture problems, execute successfully in the laboratory, and turn my unrealistic ambitions into manageable projects. David and his wife Wendy and three daughters also modeled for me what it looks like to manage successful careers in chemistry while keeping your family a priority.

As I drew to the close of my dissertation writing and applied for industrial positions, I realized that a postdoc would give me more time to figure out what I really should do next in life. I was very strategic in choosing a postdoctoral position that would be in a fairly new area that could expose me to new skills but also be able to contribute scientifically to the group and keep me geographically close to my family.

I was offered a postdoctoral position at the University of Kentucky with Dr. D. Allan Butterfield. I was fortunate to receive the Lyman T. Johnson postdoctoral fellowship from the University of Kentucky that provided me with financial support. The scientific expectations were high in this laboratory; however the atmosphere was very family-oriented and relaxed. I found that I could really be productive while still heading home by 6 pm to attend Bible study.

My research project was a logical continuum of my graduate work as I focused on Alzheimer's disease and other age-related disorders, using a new area of proteomics called redox proteomics. I was most excited by bridging chemistry research to the people affected by disease, a concept termed "translational research."

I continued to mentor graduate students and undergraduate students and realized that it wasn't too far-fetched for me to generate scientific ideas. I was able to reflect on my time in Dr. Clemmer's laboratory in a different way and realized that I never would have known what I was made of (some really tough skin and potential to learn really hard things if I kept at it) had Dr.'s Clemmer and Butterfield not kept the expectation bar so high for me. I realized that being a professor while a demanding position was also potentially very rewarding. Primarily, I saw the fun that my advisors had with their science and students but also the flexibility they had to enjoy their lives and families. This helped with planting the seeds for my deciding to become a chemistry professor.

Desires for Family or Lack Thereof

For a majority of my life, I never spent time envisioning what I would look like as a bride walking down the aisle, planning my wedding, rubbing my belly with a growing thing inside, or going to school events with little children. The truth is that I did not want children or a husband growing up, especially as a younger adult. I was always very driven, and embedded in my subconscious was this push to move up the ladder. As far as I knew, a husband and children would mean I would either have to take a detour from climbing the ladder, climb the ladder at a slower pace, or continue to climb the ladder with someone or something always trying to pull me down the ladder. Therefore, I made it up in my mind that I would not go this route. It was very narrow thinking, and I never took the time to consider other options because I enjoyed the path that I was traveling. It was a path that allowed me to advance with my education in hopes to secure a career that would set me up in life. And then I met Antonio one Sunday morning.

The Beginning of My Family

I was 6 months into my postdoctoral fellowship, and Antonio was 6 months into his new job at Lexmark, Inc. Both of us were not from Lexington, so we were newbies trying to get connected socially, professionally, and spiritually. From our first conversation after church, we realized we had a lot in common and wanted to learn more about each other. We pretty much were inseparable, and as we learned more about each other, we became connected in our church and community.

I started a spiritual quest to figure out what it is that the Lord wanted me to do here on this earth. And of all the ideas, plans, and dreams that came to my mind, being a professor kept coming to the forefront. I rationalized that there were a lot of upsides to having a flexible schedule, scientific freedom, traveling, and teaching the next generation of scientists. Most importantly, I realized that if I wanted to see more female and underrepresented minority students pursue chemistry as an option, I had to become a part of the solution. Becoming a chemistry professor would give me an opportunity to be a role model and help influence and create opportunities for female and underrepresented minority students to pursue chemistry and STEM. Antonio was totally on board with my pursuing an academic position, and thus I began the process of applying for jobs.

I'll never forget my 30th birthday. 30 was a crazy milestone to reach in life because it took me from a young adult to a "real" adult. At this time, Antonio and I were growing closer, and, while we talked about marriage as a topic of conversation, it wasn't clear how my accepting an offer for an academic position would affect us. I certainly didn't fully expect him to move wherever I landed, but I was hopeful, so he went along with me for second visits to see what opportunities might exist for him. I realized that until he "put a ring on it," there were no guarantees.

Antonio influenced me greatly with new and refreshing ideas about what it could look like to have a husband, especially a godly husband. He definitely desired children, and, because I wanted to make him happy, I started to consider the thoughts of little ones. To my wildest surprise, Antonio proposed on my 30th birthday, only a few days before I was planning to drive to Pittsburgh to start my position at the University of Pittsburgh in the Department of Chemistry as a professor. I said "yes." And to get more tears rolling, he had already secured a position at a company in Pittsburgh, being flexible and open-minded. This was not a part of my original plans, but I'm thankful that I opened my mind and heart to God's plans for me (Fig. 3).

Navigating the Path

Fast forward from getting married to renting a house, to buying a house together, and to taking on two sibling puppies, it became more real that I had to get serious about thinking about children and planning. Antonio and I had some discussions about timing but realized that really it wasn't about my tenure clock as much as it was

A Divinely Ordered Path

Fig. 3 Renā (left) and her spouse Antonio Robinson (center) at their wedding reception, celebrating with her graduate advisor David Clemmer (right)

about God's timing. We made a joint decision after 3 years of just the two of us (and two dogs) to try to add on to our family and prayed really hard for twins. One shot, two babies, craziness but then we'd be done.

We did look at the calendar to plan when to start and were very blessed that we got pregnant quick with our first child. Reina was born in June, which was absolute perfect timing on an academic calendar. This gave me the summer and the fall terms to adapt to being a new mom, to nurse for a long period, and to navigate work, being a wife, and having a new baby girl. There might have been a way that we would have been able to handle twins but we're thankful that it worked out how it did.

I was torn between wanting to protect and know how my baby was being cared for all the time and getting back to the office to keep my research program going. We had part-time help from a church member that became our Nana, initially starting out with in-house care while I tried to work from home. That quickly became difficult for me to get a lot of things done, so I started leaving the house and going to my office to work a few hours. After months of working part-time during my official maternity leave, I realized that I really enjoyed my work and wanted to be back full-time. If this were going to happen, then I had to be at peace at work in order to be productive, which meant that we had to make sure Reina was in good hands. We did a lot of research and touring and found a church daycare that worked well for Reina. She started part-time in the daycare and part-time with the Nana. This was a good system that allowed me to get back to chemistry full swing and the business of research and mentoring students (Fig. 4).

Fig. 4 The Robinson laboratory group with friends and family after a cruise on the Gateway Clipper along the three rivers of Pittsburgh in 2016

Because we did not have twins the first time around, we were hoping that the second time our prayers would be answered. Antonio and I didn't really have a set number of kids that we wanted so we were open to three, four, five, or maybe more?!!!! Twins would definitely get us there.

And once again after trying to get pregnant and being blessed to get pregnant quick, we had our son a year and a half later. There's a lot that happened toward the end of this pregnancy that showed me how as a mother and a woman you really have to fight for what matters most for your family. I had some pregnancy complications during the last trimester, which were not a part of my plans, and they caused me to be restricted in what I could do physically as well as shut down my travel and lots of activity.

A colleague took the first month of my course to help out and ended up teaching it the whole semester since it was too risky for me to be in the classroom. I am so grateful to him for doing that; it definitely freed up one major responsibility from my plate.

But I also was trying to finish an R01 proposal before the baby came. And with being in and out of the hospital, it was not certain that I was going to meet the submission deadline. And while there were other deadlines later in the year, based on my tenure clock (even with another year delay), I really needed to get this grant in to have good feedback at the minimum for the second resubmission. Not that I was stressed (the baby didn't deserve that), I did have to pray a lot and put things in the proper perspective.

Grants, courses, labs, and everything else aside, the most important thing at the time was that I do everything I could to protect the life inside of me until my son arrived. And we didn't want him to arrive too early so I had to settle down. I have a huge sense of humor as well as calm, so when I had unexpected trips to the triage unit of the hospital, I was always sure to grab my laptop. I worked on my grant while in the hospital. I mean what else was I going to do while getting poked all day. I submitted my grant during that cycle as crazy as it may sound. It was only possible because of the large support that I had from my department chair, senior colleagues, and administrative staff. My students were very responsive and helped to get in information for the grant through email, and all in all it was in and submitted. My son, Antonio II, arrived in mid-February, weeks before his early March due date, however right on time.

> **Chemical Safety**
> While I was pregnant, I was intentional about reading MSDS of all chemicals that I might be exposed to when visiting my laboratory and wore the proper personal protective equipment at all times. I also identified an appropriate space (my office) for breastfeeding, pumping, and storing milk when on campus and requested a space when traveling to other institutions and conferences.

Learning to Ask for and Take Help

I want to say loud and clear that I have learned that **I can have a family and be a chemistry professor**. I realize that women can start their families at any stage of their career and things will work out how they are supposed to. I understand the importance of being flexible, open, and honest with myself and others about what and who matters most to me and what I am willing to compromise, sacrifice, or do without, and be okay with knowing that I cannot humanely or possibly do it all, especially all very well. Support systems and networks are extremely valuable for succeeding in this and any other profession, including being a stay-at-home mom. I have a tremendous network system full of people that are supportive in helping me to succeed.

My support system has been critical for my own sanity and helping my husband and me to work through the logistics of family, careers, and maintaining a godly marriage. This support system includes our family, especially my mom, in-laws, and godparents, who, while not living in the same city as us, are willing to travel miles to visit and help with the little ones. Our friends and church family give us the occasional date night and also show us different systems for managing careers and families.

Fig. 5 The Emerging Leaders in Bioanalytical Mass Spectrometry Symposium at Pittcon 2016: (from left to right) Ying Ge, Amanda Hummon, Renã, Heather Desaire, and Yu Xia

Being in a supportive environment academically has been tremendous, and I could not imagine how much extra stress would be added to my plate if I were not. I also have a growing and secret support group of female colleagues in chemistry and academia that empowers and inspires me to keep pressing forward (Fig. 5). This includes those female colleagues without children who have sacrificed to move up in their professions, those female colleagues who are my peers going through the thick of it in terms of having children at similar stages in their careers, and those who have made it to high levels in their careers and can attest to the challenges and triumphs that come from having a family and being a chemistry professor. While at conferences and seminar visits, I am regularly reminded that, while the struggle is real, it is possible and doable. In the course of 2 weeks, I was in the presence of at least ten other female chemists who are very accomplished, and 80% of them have families and children at home, little ones to say the least.

Current Situation
Today I had to respond to an email from a collaborator asking for an update on a manuscript that we owe her (it's beyond any deadline past due; in fact it's more than a year and a half since we've gotten our act together on this). I tried to craft my response many times and then resorted to the short but truthful statement about "in addition to unforeseen issues and complications, this was low on my priority list. And in fact there's so few hours left in a day that this will not make it to the top in the near distant future." But to my relief, her response was one of "I totally understand"; she gave some updates on her own

(continued)

> situation and missing deadlines with kindergarteners (twins in separate schools) and still offered to help with the manuscript once we got there. What a relief! But still this is our reality.

Shifting Priorities

Since having children, my priorities have shifted regarding my career, and it falls lower on the list of things that matter most to me. God being first, my husband second, my children and family next, and then my work. This is not to say that my career doesn't matter but more to say that my priorities have shifted (Fig. 6).

My career has influenced my family in several ways. First, my career, even before children, required that my family try to understand the requirements of my profession and that I have a lot of demands to balance. This has been helpful for fostering family discussions around the topic. Second, it's important for my family to welcome and be a part of my academic family consisting of my research students. One morning on our way to school, Reina at 2 years old told me that she wanted to go to Boston with me. Boston? Yes, Boston. I was trying to figure out where that came from, but I had just come from Boston recently for a "work conference." I realized in that moment that my children want to be involved with my career and work. She always wants to go to "work" instead of school and especially if she can go with mommy. I have to give her and my son these opportunities, and so sometimes they

Fig. 6 Thanksgiving 2016 family photo shoot: Renã (left) holding Antonio II and Antonio (right) holding Reina

travel with me to conferences or seminar visits. I bring help or have them meet us at the airport. They love it. Sometimes they come to the department for after-hours student organization meetings or to hang out in the office, and my research students love the opportunity to interact with them.

And on Being the First

It would honestly take another chapter or even book in its entirety to fully share with you more about what this path has been like specifically as an African-American female. There have been obstacles that have come up at each stage of my career, some big and small, that I know have only come up because I am African-American and because I am a female. However, what is most important to me is that I have been able to overcome those obstacles or have the barriers moved out of my way because of my faith. I may have been the first female African-American PhD in chemistry from Indiana University (or the first one in a long time) but am not fully certain. I am now the first and only tenure-track assistant professor that is an African-American female in my department. This is progress on many levels, considering that I had only identified four African-American female chemistry professors across about 50 major research institutions when I searched as a postdoc. Ironically, I came to know all four of them and several others throughout my career. I was not the first female in my current department with children, but I am now the only one (out of six female faculty members) with children.

Being a pioneer has required having alligator tough skin, faith, and compassion to look beyond ignorance. I am still on this journey but am encouraged to keep going, especially when I see the positive impact that my presence in this field has on the next generation of scientists. My impact is not just limited to women and minorities, as there are men who are also benefitting from the diversity and different thinking that I bring into the field. One of my friends likes to call me her "unicorn" because of all the firsts that I represent. I am flattered by this every time but am more so humbled to be alongside other minorities and women who are breaking barriers in order to make STEM fields more diverse and inclusive. Regardless of being a "first," it is more critical that I am not the last.

Wisdom Gained

There's no right time. There's only the right-now time. I have grown so much wiser since having children and know that my future is not guaranteed or promised and it will not always go according to my plans. Because of my willingness to be flexible and work smart as well as hard, I have figured out how to navigate my profession with the addition of a family. To whom much is given, much is required, and I have had to expand my bandwidth in order to accommodate both a family and career. I am so much better as a person and as a manager, since I've had to learn to manage both family and career and keep them in the proper context of my life.

There are a growing number of women and men alike who are having families and making it work. I have been on two graduate student committees of females who have been pregnant, and they continued to move forward in their careers. I have had male students in my class who have returned to graduate school after starting their careers in industry and having families, I have male colleagues who have children and some who are the sole caregivers of their children, I know of wives of chemistry professors who make it work with their professions and others who are proud to make it work at home, and yes there are those who decided that being a professor was not the right career choice for them, and they moved into other careers. Regardless of the path we have taken, this book serves as a testament that it is possible to be both mom and the chemistry professor.

Acknowledgments I first give thanks to God in whom I believe and know that all my blessings and strength flow. I would like to especially thank my husband Antonio and children for their unconditional love, support, and patience with me. I am grateful for my mother, my in-laws, siblings, godparents, and all family who shower me with love, laughs, and grounding. I thank my especially close friends, distant friends, and folks who make my world better and let me be a part of theirs. I also would like to acknowledge my former advisors who have become lifelong mentors and friends, David E. Clemmer and D. Allan Butterfield. I finally thank the many supportive colleagues, mentors, and teachers I have obtained throughout my educational and professional career, both nationally and internationally, including those at the University of Pittsburgh.

Editor's Note

Between this chapter's submission and the publication of this book, Dr. Renã Robinson has changed institutions. She is now an Associate Professor of Chemistry and the Dorothy J. Wingfield Phillips Chancellor's Faculty Fellow at Vanderbilt University.

About the Author

Education and Professional Career

2000	BS Chemistry with Business Concentration, University of Louisville, Louisville, KY
2007	PhD Analytical Chemistry, Indiana University, Bloomington, IN
2007–2008	Lyman T. Johnson Postdoctoral Fellow Postdoctoral Fellow, Department of Chemistry, University of Kentucky, Lexington, KY
2008–2009	UNCF/Merck Postdoctoral Fellow, Department of Chemistry, University of Kentucky, Lexington, KY

2009–2017	Assistant Professor of Chemistry, Department of Chemistry, University of Pittsburgh, PA
2011–2017	Faculty Member of Biomedical Mass Spectrometry Center, University of Pittsburgh, School of Medicine, PA
2015–2017	Core Faculty Member of MD-PhD Scientist Training Program, University of Pittsburgh, School of Medicine, PA
2017	Associate Director of Outreach, Recruitment, and Education, Alzheimer's Disease Research Center, University of Pittsburgh, PA
2017–present	Associate Professor of Chemistry, Vanderbilt University, Nashville, TN

Honors and Awards (Selected)

2017	Pittsburgh Conference Achievement Award
2016	Women of Excellence Award, New Pittsburgh Courier
2016	C&EN's Talented 12, *Chemical & Engineering News*, American Chemical Society
2014	Lloyd N. Ferguson Young Scientist Award, National Organization for the Professional Advancement of Black Chemists and Chemical Engineers
2012	Keystone Conference Symposium Early Career Travel Award
2011	American Society of Cell Biology Travel Award
2010	Summer Research Internship Best Mentor Award, University of Pittsburgh EXCEL Program
2010	Society of Analytical Chemists of Pittsburgh Starter Grant Award

Renã is an active member of several organizations including the American Society for Mass Spectrometry, American Chemical Society, Society of Analytical Chemists of Pittsburgh, Spectroscopy Society of Pittsburgh, Association of Merck Fellows, US Human Proteome Organization, and Women's Chemist Committee. She also is a co-founder of the National Organization for the Professional Advancement of Black Chemists and Chemical Engineers (NOBCChE) local Pittsburgh student chapter and is a NOBCChE at-large board member.

Raising Three Children Across Three Continents

Omowunmi A. Sadik

Photo Credit: Jonathan Cohen

The Early Years in Lagos, Nigeria[1]

I was born in the bustling city of Lagos, Nigeria. Nigeria is a country with an estimated population of nearly 200 million, which makes it the most populous country on the continent of Africa. Eighteen million people inhabit the city of

---

[1] I have discussed the early years in a short autobiography published in *Legacies: A Guide for Young Black Women in Planning their Future*, by Constance Gipson and Hazel Mahone, pp. 6–9, ISBN:978-0-9897114-0-1, 2013, Vision 2000 Educational Foundation Publishers.

O. A. Sadik (✉)
Department of Chemistry, Binghamton University, State University of New York, Binghamton, NY, USA
e-mail: osadik@binghamton.edu

© Springer International Publishing AG, part of Springer Nature 2018
K. Woznack et al. (eds.), *Mom the Chemistry Professor*,
https://doi.org/10.1007/978-3-319-78972-9_33

Lagos, and growth estimates suggest that the population of Lagos will be nearly 22 million by the year 2020.[2] That will make Lagos the seventh largest city in the world.

I met my husband, Abiodun (Abi) Sadik, when I was a freshman at the University of Lagos (Unilag). Abi was in his senior year as an electrical engineering major. After graduation, I served on the mandatory, 1 year National Youth Service Corps (NYSC)—a paramilitary service for all Nigerians who graduated college before their 30th birthday. During my NYSC year, which I completed at the Nigerian National Petroleum Corporation, Abi and I got married; and with it was the arrival of our daughter, Adeola (Ade). I returned to Unilag for my master's degree in Chemistry, and thereafter we had our first son, Ademola (Dem), a little over 2 years after his sister Ade.

After completing my master's degree, I worked briefly as a high school chemistry teacher at Gbagada Grammar School in Lagos Mainland and then later joined the Nigerian Institute for Oceanography and Marine Research (NIOMR) located in the Victoria Island section of Lagos, as a research scientist.

I never planned to go into academia, even though I enjoyed my job as a research scientist. After graduating at the top of my class, my professors suggested that I should enter a PhD program. I decided to earn my master's degree first, because I was actually unsure of whether the career paths that came out of a doctoral study appealed to me, despite my love of chemistry itself.

I decided to enroll in a master's course, which I completed in 1988 while I was also teaching high school science classes (Fig. 1). I eventually decided to pursue a PhD after joining NIOMR after realizing that one could not be a successful researcher without one. During our weekly NIOMR-wide seminars, I noticed that the most knowledgeable invited speakers and guest lecturers were PhD holders or current doctoral students. I loved how their keen intellect and vast knowledge did not prevent them from breaking the science and math of their research down into its simplest parts, and I was very impressed by their superior command and interpretations of their research data. I also loved the fact that they were not merely rattling off memorized textbook knowledge, as many of my classmates had resorted to doing during my undergraduate years. These people *actually understood* the science! They were applying their knowledge to solve problems in our country and the broader world around us.

Promotion at NIOMR was contingent upon successful publication of either a monograph or at least two peer-reviewed publications. As a result, I decided that perhaps I needed to do a PhD so that I could continue to gain promotions in the workplace, as well as for the love of solving real-life problems using my chemistry research skills as I saw my PhD colleagues doing.

In that sense, I could not say that I was the traditional PhD degree holder: I did not care about pedigree, impact factors, being at the top in the field, or having some titles

---

[2]The world's largest cities and urban areas in 2020. http://www.citymayors.com/statistics/urban_2020_1.html

Fig. 1 Wunmi's master's degree graduation picture from the University of Lagos in 1988 taken with daughter Ade

behind my name. It was much later in my career that I learned about these things and discovered how your publications or where you earned your PhD would impact how your colleagues perceived the significance of your work; factors like these would eventually get you a job or even help in landing you major research grants or continuing research funding. In my case, I just wanted to tackle real-life problems and to find ways to apply my research experience to address these problems. This was my main motivation for enrolling into a PhD in chemistry, and this has continued to form the basis of the research that I do until this day.

I published my first monograph titled *Heavy Metal Contaminants in Some Nigerian Marine Fishes of Commercial Importance*, NIOMR technical paper No. 63, ISBN 978-2345-063, in November 1990.

To Wollongong, Australia

After receiving an Australian Merit Scholarship, I left Nigeria on February 10, 1991, and headed to the University of Wollongong, in New South Wales, Australia. The scholarship was awarded on the basis of high academic achievements and provided

Fig. 2 The Sadik family at the Opera House in Sydney, Australia, in July 1993 (Back row from left to right: Wunmi, and Abi; Front row from left to right: Dem and Ade)

funds for up to 4 years of PhD program. This competitive scholarship acknowledges and rewards students who have the drive and determination to succeed academically despite financial restraints. The scholarship covered my full tuition of AU$15,000/ year, and it also paid annual stipend of AU$12,000.

In Australia, PhDs typically take 3–4 years after a master's degree. The condition of the scholarship required that I depart for Australia ahead of my family so that I could get settled before my family could join me within a few weeks. The promise of "a few weeks" turned to months. After about 5 months of waiting for my family to join me, I decided it was better for me to return to Nigeria than to live separated from them. As I was getting ready to return to Nigeria, the Australian Embassy in Lagos, Nigeria, decided that it was time to grant the visa applications for the rest of my family. Altogether, it was more than 6 months before I could set my eyes on my children (Fig. 2).

Wollongong, informally referred to as "The Gong," is a seaside city located in the Illawarra region of New South Wales, Australia. Wollongong lies on the narrow coastal strip between the Illawarra Escarpment and the Pacific Ocean, 51 miles south of Sydney. Wollongong's Statistical District has a population of 292,190 (2010 est.),[3] making Wollongong the third largest city in New South Wales after Sydney and Newcastle and the tenth largest city in Australia.

Our early days in Wollongong were difficult. My husband, who had previously served as personal technical assistant to the managing director/CEO of Nigeria's National Electric Power Authority (NEPA), was unable to find employment. Given these less-than-ideal circumstances, it was not easy for the four of us to settle into a new city, country, and continent.

---

[3]City of Wollongong, https://en.wikipedia.org/wiki/City_of_Wollongong

Combining Marriage, Motherhood, and Research

My daughter, Adeola (Ade), was 5 years old when she arrived in Australia, while my son, Ademola (Dem), had just turned 3. We enrolled Ade into the local public school, where she started from kindergarten. We struggled with major decisions about our future and were torn between returning to Nigeria and simply forgetting the idea of a PhD and staying the course. It was not an easy decision, but somehow we decided to stay.

I started my research with Professor Gordon G. Wallace in conducting polymers. The Wallace Research Group consisted of 25–30 members, with graduate students from different parts of the world, including Nigeria, Iran, Australia, China, Indonesia, and Papua New Guinea. There were also postdoctoral researchers, as well as paid research assistants. It was a microcosm of the United Nations, and the group was jokingly referred to as the "Evil Empire."

Since my PhD was fully funded through the Australian Merit Scholarship, I was solely devoted to my research and didn't have to worry about the additional responsibilities of working as a teaching assistant. The Wallace Research Group, officially known as the Intelligent Polymer Research Laboratory (IPRL), actually prepared me for the competitive academic life that I would eventually enter when I moved to the United States. At IPRL, each student had to compete for everything ranging from instrument time, reagents, and electrodes. The IPRL was very competitive and rigorous. We were required to write monthly technical reports, a practice I still maintain with my group to this day. Half of the group worked on polypyrroles, while the other half worked on polyanilines. There were sub-teams working on different applications of conducting polymers such as sensors, separations, and membranes. My project was focused on the new sensing techniques based on conducting polypyrroles.

We met with Gordon once a month in a series of meetings that could take between 15 and 30 minutes. In order to make the meeting meaningful, each student was required to submit a monthly report ahead of time. It was not uncommon for some of these reports to eventually form the basis of a full manuscript.

Although there was competition among group members, we also collaborated often, as Gordon had the knack for connecting work from each student and creating technical papers out of our reports. The group was very productive and we had access to international journals and visitors. I met Alan MacDiarmid, the co-winner of the 2000 Nobel Prize Winner in Chemistry for his work in conducting polymers, during one of his visits to our labs at the IPRL. The IPRL also hosted workshops and conferences attended by many influential researchers from around the globe.

Gordon and the team leaders wrote proposals and competed for the Australian Research Council grants, with a majority of the staff being paid from soft money. Most students at IPRL brought their own scholarships or were on Australian government support grants. We also had a sizeable number of Iranians who were being supported by their home government.

This was an incredible period of global exposure for me. I credit the makeup of the IPRL for the strong belief in the benefits of diversity of thought, background, and origin that I hold to this day.

My daughter, Ade, attended Keiraville Public School, and my son, Dem, attended the campus daycare, which was known as "KidsUni." A typical day began with Abi and I getting the children ready before taking them to school. Ade took the public school bus, while Dem went with me to campus.

My husband, Abi, initially could not find a job in Wollongong and had to commute to Sydney, where he found a part-time job at the University of Technology Sydney (UTS) as a data analyst. Sydney is about 1 hour of train ride from Wollongong.

Ade would join Dem at KidsUni after her elementary school day finished, and I usually pick them up around 6:00 pm. Sometimes, I would have experiments running and would pick them up before we took the last bus ride home to prepare dinner while waiting for Abi to return home from his Sydney job and long train ride. Many times we could not have dinner together because Abi would return home around 8:30 pm and leave before 6:00 am the next morning.

Weekends were our best times as we were able to spend time together, taking walks to campus. Oftentimes, I had experiments running while the kids played on the beautiful Wollongong campus with my husband.

I remember an incident shortly before my family joined me. I was running an experiment late into the night and was forced to walk about 2.5 miles to my home. I noticed a man following me, but it didn't bother me because I had no money on me. As with any large city, walking late at night in certain areas of Lagos created the risk of being robbed, but it never crossed my mind that rape was an issue. As I turned the corner and up the hill, the man continued to follow me, but when I got to the street just before our flat, there were dogs barking, and the man disappeared. The next morning, I read in the local newspaper that a woman was raped around the area where I walked the night before. I mentioned this to one of the graduate students, and he told me that I should let someone know if I needed a ride home instead of walking home when working late. That incident shook me to my core. I could have been the unlucky one.

I experienced some memorable friendships in Wollongong and at the IPRL. While some of the friendships were quite genuine and fulfilling, others befriended me in order to assuage their curiosities about living in Africa. Most people expressed their ignorance of the continent by asking me if we lived on trees. Luckily, I had plenty of pictures that I took with me from Lagos to diffuse their ignorance. One individual in particular expressed his surprise that in one of my pictures, we actually had Coca-Cola! We also participated actively in our church, known as The Potter's House Christian Fellowship, and we made many lasting friendships that way.

My husband's part-time contract job ended after several months and he was again unemployed. As an intelligent and ambitious individual who had found considerable professional success during our time in Nigeria, he was understandably unhappy about this turn of events. After a few months, my husband found a job as a janitor on the same campus where I was a PhD student. It was a major change in our lives, but

Raising Three Children Across Three Continents 449

we quickly embraced it. I became much more productive on my research after my family joined me, and by 3 years, I had co-authored eight peer-reviewed publications and even had two US patent applications.

Life as a Graduate Student

Life as a graduate student was very much predictable: wake up and get the kids ready for school, set off to campus, and work in the lab until noon. Then I would return home. On Fridays, we had group seminars, and one or two individuals would give group seminars on their research.

The Chemistry Department at Wollongong was run very much like a community. For most of the staff, their day began at 9:00 am, and by 10:30 am, it was the first teatime. Most members of staff and students were also members of the Tea Club, and teatime was held in the departmental lunchroom. Lunch took place around noon and we had afternoon teatime between 3:00 and 3:30 pm. Most staff and a few students would leave by 5:00 pm. The only people working around the clock were the international graduate students due to the fact that most of them didn't have their families with them. The only option for them was to keep working, even on weekends.

My routine was sometimes interspersed by midweek church services on Wednesday evenings. On Sunday mornings and evenings, we went to church, and our children enjoyed interacting with others of a similar age.

The Department of Chemistry had a community newsletter called "Methyl Mercury" with Dr. Gary Mockler as the editor. He wrote satirical articles about some members of the department regarding their weekend hiking or body surfing (a common sport in Australia) or other fun activities. He had a way of writing that made every member of the community look forward to a new edition of his weekly write-ups. I was nearly at the end of my first year before I realized that Dr. Mockler was actually a member of the faculty. He was incredibly humble and mixed freely with students and faculty alike.

On Mondays, Gary would go to each lab to collect AU$2.00 from Tea Club members. The Tea Club provided unlimited coffee and milk, as well as snacks. Occasionally, cake was served to celebrate a birthday. These interactions made the working environment more pleasurable. In addition, there were plenty of outdoor barbeques, an Australian tradition, and other celebrations.

On Fridays after 5:00 pm, the entire department would meet for a weekly bar crawl. Gordon would take a few of the Australian students and staff to enjoy Australian beer at the local pub. Even though the international students were invited, most of them didn't accept such invitations or simply didn't feel conformable enough to join in.

Most international students found the Western culture to be quite different from what they were used to. Our Iranian colleagues, in particular, absolutely refused to address Gordon by his first name. They claimed that in Iran, not only would they not call him by his first name, but they would use the prefix "Professor" in addressing him, as well as "sir" when he gave directives on what they should do, with attendant bowing of their heads. I could relate to these cultural differences as we have somewhat similar situations in Nigeria.

As a graduate student in Australia, I had the opportunity to attend conferences in New Zealand and the United States. In 1993, I attended the Pittsburgh Conference ("PittCon") on Analytical Chemistry and Applied Spectroscopy in Atlanta, GA, where I gave a presentation and made numerous contacts through vigorous networking. I met some of my advisor's friends at Dionex Corporation—an American equipment company that develops and services chromatography instruments used in medical research, environmental monitoring, and food testing, and I also learned about the US National Research Council's (NRC) associateship programs, which eventually paved the way to moving my family to the United States after the completion of my PhD degree.

Coming to America

My PhD chemistry degree was awarded in December of 1994 even though I completed my degree by August of the same year. I was also awarded a postdoctoral fellowship through the NRC. The award was tenable at the US Environmental Protection Agency (US-EPA) on the campus of the University of Nevada at Las Vegas.

My family and I moved to Las Vegas in November 1994, where I continued as a postdoctoral fellow. Ade and Dem were enrolled in grades 4 and 2, respectively. At US-EPA, I initiated a new project and was very productive, publishing several more papers.

I began applying for full-time faculty positions during my second NRC fellowship year at EPA. I joined the faculty at Binghamton University, State University of New York, in September of 1996. My father, whom I last saw before I left Nigeria in February 1991, suddenly took ill and passed away in March of 1997, less than a year after we came to Binghamton.

Preparing for an Academic Career

Ade started middle school when we arrived in Binghamton, while Dem started fourth grade at Vestal Hills Elementary School. The life of an assistant professor coming from my background was challenging but very interesting.

In 1996, the Department of Chemistry at Binghamton University had three analytical chemistry professors before I joined. The senior professor retired, the mid-career-level associate professor resigned to focus on his pharmaceutical analysis company, while the junior assistant professor was denied tenure. It was not the most reassuring environment in which to start my career, but I was determined to succeed.

I learned very quickly that to obtain tenure, one was required to receive federal grants, excel in teaching, and be seen as a good citizen at the departmental and university levels. Thoughts of success dominated my thinking, and I was determined not to fail.

I actually began looking for research funding opportunities in October of 1995 after I accepted the position of assistant professor at Binghamton University. Around January of 1995, the EPA intramural call for proposals was released, and my research advisor, Dr. Jeanette van Emon, asked me to respond to the call. I spent a week reading through the announcement. The jargon and the environmental terminology were all new to me, but I learned very quickly. My proposal submission was made much easier because I did not have to create a budget or justify any expenses. Unbeknownst to me, this round of intramural grants was what secured my NRC's second year renewal to continue my research at EPA. Having secured my second year as a postdoctoral researcher at EPA, I started looking for more grant opportunities. The Internet was not in widespread use at the time. Because I did not receive my PhD in the United States, I had never heard of the National Science Foundation (NSF) or the National Institutes of Health (NIH) or about the avenues of funding for most career researchers. Most career announcements came through the newspapers or through the *Chemical & Engineering News*, and paper copies of journal articles had to be obtained at institution libraries.

By early February of 1996, another EPA call for proposals was released. This time, it was for the extramural Science to Achieve Results (STAR) grant program. I read the call from cover to cover and highlighted the areas of interest. I also called the program manager to ask for specific questions that would enhance my chances when I responded to the advertisement. I called the Chair of Chemistry at Binghamton University to see if I could submit a proposal, as I had not yet started my appointment. I was given a preliminary research assistant professor position that enabled me to submit the proposal with the support of Binghamton University.

I spent 2 months in early 1996 working on the proposal and fine-tuning my ideas. Since my proposed research submitted to Binghamton University was going to be on sensors, I focused on sensor arrays using conducting polymers. I submitted my first EPA STAR grant shortly before leaving for Pittsburgh Conference of 1996.

Meanwhile, life continued as usual at home. Although I was legally allowed to work since I was on a US government work visa, my husband needed to wait until our family received permanent residency in order to work, so he was again unemployed. We found a lawyer, who helped us put together a strong application, and within 9 months, we were approved, and he was legally able to seek employment. He first worked at a discount retailer called McFrugal's before transitioning to a computer sales job at a Circuit City.

In early spring of 1996, I was contacted by the Chemistry Department at Binghamton University about putting together a course syllabus. I began by counting the number of weeks in the semester and using the table of contents of one of my favorite texts, *Principles of Chemical Sensor*, by Jiri Janata, to put together my first syllabus. When I eventually met Professor Janata in the fall of 1996, I asked him to sign my personal copy of his book since this was what really got me started in academia.

By summer of 1996, I learned that I would be teaching Quantitative Chemical Analysis in spring of 1997, and I needed to know what textbooks were being used. The US-EPA buildings were located on the campus of the University of Nevada, Las Vegas (UNLV). I sometimes attended seminars on that campus and had met some of its chemistry faculty. I went to speak with one of the analytical faculty at UNLV and asked for a copy of his course syllabus. I used his sample course syllabus to write my own, and I sent this to the department secretary at Binghamton University.

Binghamton and Professorate

My family and I moved to Binghamton in July 1996, after the last day of school for our kids. We first spent a few days in New York City, where two of my sisters lived. We then rented a U-Haul and drove to Binghamton for the first time together.

Our first night in Binghamton was a chilly night for all us. It was cold and dark, and it did not look anywhere close to the modern buildings I had seen during my initial interview visit. It certainly didn't look anywhere close to Las Vegas, where everything was bright and full of energy. I feared I had made a terrible mistake.

Upon arriving in Binghamton, I started working on my lectures. I was assigned to teach a special topics course in my area of expertise. I chose chemical sensors, and I immediately received two e-mails from students wanting to join my group. I had ten graduate students enrolled in my graduate class in quick succession. I swiftly learned how to prepare lecture notes, plan office hours, and set up my lab simultaneously.

My usual days began with dropping my daughter off in middle school before 7:00 am and arriving at the Chemistry building soon after, with my husband taking my son to his elementary school which started much later at 8:45 am. My days often continued until midnight. Oftentimes, I returned home for dinner with my children already in bed.

My office is located in the seventh floor of the Chemistry Department, and I was initially alone on that floor, with the exception of two of my graduate students, who would stroll in at around 9:00 or 10:00 am and leave before 4:00 pm. It was years later that I learned that other departments actually provided some form of mentoring. I was left alone to either sink or swim, with the Chemistry Department chair occasionally dropping in to chat about my progress. I continued to learn and teach, performed research, organized the lab, and ordered chemicals and equipment. I had no mentor and I was entirely unskilled in the American system.

After continuing with this routine for about a month, I contacted the program manager of the EPA STAR grant and was told that my proposal had been approved for a major funding for 3 years! I started my tenure-track position at Binghamton officially on September 1, 1996, and I received my first federal grant as a principal

Raising Three Children Across Three Continents

investigator on October 1, 1996! This set a record at Binghamton University that no other junior faculty has ever managed to beat since.

Abi got a job at a local Fox TV Station as a broadcast engineer. After 1 year, he transitioned to another ABC TV station, where he is currently the assistant chief engineer/IT specialist.

As I continued the daily routine of teaching, writing proposals, setting up the lab, buying equipment, and getting quotes, my two children were growing up very quickly. Ade completed middle school and graduated from high school. Upstate NY was a different ball game for a growing Nigerian-American youngster.

I vividly recall an incident in April of 1999 when a high school student, who unfortunately had many of the same classes as my daughter, decided to do a school project about Ku Klux Klan and came to school dressed in a white sheet and hood and carrying a noose. He wore that outfit throughout the entire school day until he physically accosted my daughter in the school hallway and threatened to "lynch" my daughter. Until a physical altercation between the student and my daughter occurred, no Vestal Middle School teacher or administrator saw fit to instruct this student to remove his white sheet, white hood, and noose. In fact, this student's nominal reason for wearing his KKK attire was that he had chosen the KKK as his mandatory eighth grade English research project, with the approval of his English teacher. As a point of comparison: my son, Ade, chose to do his eighth grade English research project on the sport of soccer. At this point in time, we were not aware of the fact that Binghamton was formerly the official headquarters of the New York State KKK.[4] The Vestal School District made an attempt to address this incident by giving students talks on multicultural education.

I learned a lot from my children through their interest in American history and social studies. Most of our discussions took place in the kitchen and during dinner time. My children often competed for my attention to tell me about the books they had read, and my husband took them to the public library on a weekly basis. After school, Ade would take the school bus to my office on campus. After snacks, it was time to pick up Dem from school, and they would both go home to complete their homework, with their dad's supervision.

We continued to encourage them to rise above their social and racial issues at school and to focus on their studies. After we found a church in Binghamton, we began to settle in to make new friends. My husband helped the children in mathematics, while my role was to take care of the sciences and social studies. Once their homework was done, my kids loved to read in the evenings. The TV was only turned on for the evening news, and the kids also occupied themselves with their piano lessons and school league sports.

---

[4]Ku Klux Klan Reference Deleted From Pamphlet. http://www.nytimes.com/1993/09/12/nyregion/ku-klux-klan-reference-deleted-from-pamphlet.html

Off to College!

Very soon, it was high school graduation time for Ade. She received an undergraduate full tuition scholarship to the University of Pittsburgh and left in 2003, just as I left for my sabbatical as a Radcliffe Fellow at Harvard University. Dem also performed well in school and turned down the option of joining his sister at Pittsburgh in favor of another full (tuition and board) scholarship to the University of Maryland, College Park.

Ade received her bachelor's degree in Chemistry in 2007 and was accepted into over 20 medical schools. She chose to stay at the University of Pittsburgh School of Medicine, where she obtained her MD in 2011. She then went on to Harvard's Massachusetts General Hospital for her medical residency in anesthesiology. She completed her residency in May 2015 and 1 year of medical fellowship, specializing in Critical Care Medicine at Johns Hopkins Hospital in Baltimore, Maryland. Ade is now a double-board certified, practicing physician in anesthesiology and critical care.

I was promoted to Associate Professor of Chemistry in September of 2002, following an extensive process of external reviews with a unanimous decision. In 2003, I did my first sabbatical leave at Harvard University in George Whitesides Laboratory, and I received the Radcliffe Fellowship during the same year. I was promoted to full professor in 2005.

Dem received his bachelor's degree in Finance from the University of Maryland, College Park, in 2009. He moved off to New York to work for the Boston Consulting Group for 1 year before attending Yale Law School. He earned his JD in 2013 and worked as an attorney for *Wachtell, Lipton, Rosen & Katz in New York City*. Dem received his MBA at Harvard Business School in May 2017 and now works in commercial real estate development and acquisitions. Both Ade and Dem have decided that academic careers are not for them.

The New Addition

On March 25, 2005, Abi and I had another son, Olaolu (Ola for short), who is the last addition to our family, coming nearly 17 ¾ years after Dem, his next-oldest sibling. Ola's arrival marked another chapter in our family. I had gotten used to the independence of being a mother with adult children.

I continued my active lifestyle and even taught a large General Chemistry class in December of 2004 while I was expecting him. The class enrollment was around 650 students, with 24 teaching assistants. I learned that Binghamton University had no maternity/parental leave policy beyond "sick leave" or "disability." With no possibility of any time off, I continued teaching until March 23, 2005, 2 days before Ola arrived. I used my accumulated sick leave to take 4 weeks off after Ola's birth. Meanwhile, I arranged for my postdoctoral fellow, Dr. Silvana Andreescu, to continue teaching my graduate biosensor class during these 4 weeks. I returned to continue my class until the end of the semester around the middle of May of 2005.

Ola and Dem overlapped at home for only about a few months before he graduated from high school. Both Ade and Dem continued developing warm and close relationships with their younger sibling.

Ola is 13 years old now. He had a fast track in elementary school and is now attending middle school. He skipped one grade in elementary school and continued to be in the top 3% of the NYS standardized exams. He recently participated in the 2015 Mathalon, a competition organized by British Aerospace Engineering Systems (BAE Systems). Ola is in competitive swimming, and he is a master pianist.

As scientists, we are the embodiment of enlightenment, and we have enthroned reason. Sentiments and emotionalism are kept at bay, and we must be true to the process of science and scientific reasons. I have defied all stereotypes, and I can say that through the years, a Bible passage in Philippians 4:13 rings true in my life. My pathway has defied all conventional labels, and I have been able to accomplish more than I've ever dreamed. With God's help, and the support of my family, I was able to achieve success in academics and research (Fig. 3).

Fig. 3 The Sadik family in May 2013 (from left to right: Abi, Ade, Dem, Ola, and Wunmi)

Reflections and Advice to Young Women Planning Careers in Academia

"Despite all odds, I can testify to the goodness of God, a lot of hard work and tenacity." Regardless of your environment, background, and the odds against you, persevere and lift your head high no matter the circumstances.

My advice to young women planning careers in academia and also seeking to have a family is not to be deterred by the challenges and seemingly insurmountable odds. As with any other career path, it is a balancing act between your role as a scientist, professor, mentor, wife, and mother. Your ability to cope with several situations requires an effective time management, persistence, and a true love for what you do. As the mother, wife, colleague, teacher, scientist, and everything else in between, you will need to develop the ability to see beyond any limitations. I believe that perseverance, risk-taking, and luck play an important role in discovery and that, as scientists, you should not be afraid of challenging the conventional wisdom.

Acknowledgments I would like to acknowledge Abi, my husband of over 30 years, for being there with me throughout the long and windy roads to success. I also like to express my gratitude to my children (Ade, Dem, and Ola) for fulfilling our lives and for making the journey worthwhile. Without their constant support, love, and friendship, the journey would have been impossible. I would also like to acknowledge all members of the Sadik group, past and present, including undergraduate students and visiting scholars from Australia, Romania, Jamaica, Nigeria, Turkey, and South Africa.

About the Author

Education and Professional Career

1985	BSc Honors Chemistry, University of Lagos, Nigeria
1987	MSc Chemistry, University of Lagos, Nigeria
1987–1989	Science Teacher, Gbagada Grammar School, Lagos, Nigeria
1994	PhD Chemistry, University of Wollongong, New South Wales, Australia
1994–1996	National Research Council Postdoctoral Fellow, US Environmental Protection Agency, National Exposure Research Laboratory, Las Vegas, NV
1996–2002	Assistant Professor of Chemistry, Binghamton University, State University of New York (SUNY), Binghamton, NY
2002–2005	Associate Professor of Chemistry, Binghamton University, SUNY, Binghamton, NY
2003	Radcliffe Fellow, Harvard University, Boston, MA
2004–2016	Director, Center for Advanced Sensors and Environmental Systems

2005–present	Professor of Chemistry, Binghamton University, SUNY, Binghamton, NY
2012–present	President and Co-founder, Sustainable Nanotechnology Organization
2016–present	Director, Center for Research in Advanced Sensing Technologies and Environmental Sustainability (CREATES)

Honors and Awards (Selected)

2018	Brian O'Connell Fellow, University of Western Cape, South Africa
2017	Jefferson Science Fellow, State Department, Bureau of East Asia and Pacific Affairs
2016	Nigerian National Order of Merit Award for Science
2012	Fellow, American Institute for Medical and Biological Engineering
2011	Chancellor's Award for Scholarship and Creative Activities, SUNY
2010	Fellow, Royal Society of Chemistry
2005	NSF Discovery Corps Senior Fellow
2004	SUNY Research Recognition Award
2003	Harpur College of Arts & Sciences' Dean Distinguished Lecturer
2002	Chancellor's Outstanding Inventor Award, SUNY
2000–2001	National Research Council Collaboration in Basic Science and Engineering Fellowship Award
2000–2001	Harpur College of Arts and Sciences' Dean's Research Semester Award, Binghamton University, SUNY

Wunmi's research areas are in surface chemistry, chemical sensors and biosensors, and in their application to solving real-life problems in biological systems, energy, and the environment. She is promoting the responsible growth of nanotechnology around the world through research, education, and outreach.

Ebony and Women and Science... Oh My!

Darlene K. Taylor

I'm a woman, a black woman, with a doctoral degree in physical chemistry. I challenge the norm as a black, female, physical chemist. I'm not sure when I figured that out, but somewhere along the way, I realized that my career path was unique. And just like Dorothy in the Wizard of Oz, who encountered amazing adventures as she followed the yellow brick road, I have traveled a long way and accomplished a lot on my journey.

D. K. Taylor (✉)
Department of Chemistry, North Carolina Central University, Durham, NC, USA
e-mail: dtaylor@nccu.edu

Before Kids

I attended a Catholic school from kindergarten through eighth grade at a time when nuns still served as grade school teachers and were permitted to use paddles to keep kids in line. My experiences at this Catholic school gave me a real sense of service, hard work, and discipline as well as isolation. I didn't really fit socially in this predominately white environment where the kids were quick to label you as "the black girl." Despite this reality, I thrived in my academics and extracurricular activities and developed some really good friendships. These formative years taught me how to interact with diverse people, and to this day, I choose to see conflicts in contexts other than black and white. I don't recall my science classes during these early years as much as I remember my art and religion classes.

However, I do remember that I loved to make things. Perhaps it was my father's influence. My childhood memories are filled with images of my father leaving for work before sunrise to start the pastries at the local bakery where he was employed. He learned how to produce pastries long before he acquired the formal education of the science behind his cooking and baking techniques. His desserts were beautiful. I always had the most amazing cakes at my birthday parties. In hindsight, growing up with all this baking may have spurred my interest in chemistry. Whatever the reason, I enjoyed creating things even to this day as a chemist in the lab.

After grade school, I attended a midsize suburban high school known for its women's basketball team and not particularly recognized for STEM education. A special recognition for excellence in physical science, a freshman science competition on conservation, a sophomore fellow in the Intensive Summer Science Program sponsored by the Ford Foundation, a junior biology trip to the island of Andros in the Bahamas, and a variety of senior chemistry club activities were my high school science highlights. I gave a lot of time (pinned for 100 hours) to volunteering as a candy striper for the local hospital. This seemed like the perfect way to nurture what I thought was my dream job as a medical doctor. No one in my family was a medical doctor, but this was the career I thought I would pursue. This all changed when I entered college.

I choose to attend a small college in Towson, Maryland, when I wasn't accepted at my in-state pick of the University of North Carolina at Chapel Hill. I attended Goucher College during a unique time in the history of this college. This women's college transitioned to a coeducation college after my graduating class of 1989. I'm still unable to quantify my personal growth and development as a result of attending this liberal arts women's college. My dreams of pursuing a medical degree slowly changed as I took biology courses at Goucher. I realized my mastery of biology was a real struggle yet I really enjoyed chemistry. I had an excellent physical chemistry teacher who I could frequently engage due to our very small student-to-teacher ratio. I didn't realize until many years later his impact on my career.

As a graduation gift to myself, I took a trip to Canada with my college roommate. This turned out to be a significant event in my life because I met my future husband, Erwin. After I returned from the weeklong trip, Erwin and I kept in touch and frequently met midway between our homes in Baltimore over the course of 5 years. He finally proposed to me after flying me up to Canada for one of the Christmas holidays. We had a yearlong engagement and finally were married in 1993. I was 2 years into my doctoral studies when we married, but I'll pick up on that later. Erwin relocated to North Carolina after we were married, and we began our journey together. He is an amazing partner and so supportive of all my endeavors.

Once "the trip" was over, I returned home to North Carolina to start my master's degree at North Carolina A&T State University. This was the first time in my educational experience that my peers were predominately black. I again felt a little like an outsider: my introvert personality—honed to perfection growing up without siblings—coupled with my years of education in predominately white schools where I survived socially by not standing out, didn't naturally fit into the culture of an historically black college or university.

I didn't have to figure this problem out right away because my career took another turn. During the first few months of my master's studies, I was selected to be the first Scholar in Residence at Rohm and Haas Company in Norristown, PA. I worked at Rohm and Haas for 18 months to complete my master's research. It was at Rohm and Haas that I encountered physical chemistry for a second time. I developed a deep appreciation for this field working closely with physical chemist, Dr. Rebecca Smith. Dr. Smith was a smart and creative manager at Rohm and Haas who served as one of my research mentors. I liked her mathematical approach to scientific problems.

I successfully defended my master's thesis and applied to the University of North Carolina at Chapel Hill, where this time I was admitted as a doctoral candidate. I chose physical chemistry as my major. I was the first graduate student in the laboratory of a newly hired assistant professor. However, after joining this research group, my research interests shifted from the photophysical characterization of large cyclic molecules to the design and characterization of polymers. I completed a second master's degree and then joined the research group of the father of polymeric liquid crystals, Dr. Edward Samulski, to complete my doctoral studies. In 1999, I completed my doctoral studies in physical chemistry. I was one of few females, let alone African-American females, to do so.

Then Comes Mary

My husband and I had been married 5 years before we had our first child, Mary. There really is no ideal time to start a family as a female scientist, but for us, it was important for me to complete my doctoral degree before we thought about becoming parents. I was 5 months pregnant when I accepted a postdoctoral appointment with the highly accomplished scientist, educator, and entrepreneur, Dr. Joseph DeSimone.

I like to think that I remained productive yet balanced, even while Mary was an infant. If she wasn't with me, Mary was with my mom. During the weekdays, I drove from Chapel Hill to Burlington to drop Mary off with mom. On weekends, when I needed to go in for a few hours, I put up a playpen in the office for Mary. I felt that I was giving her a good start. Mary was breastfed for her first 8 months. A co-worker in the lab, who was also a new mom, set up a lactation station in the women's restroom. It amounted to a black metal chair placed in the handicap stall. This was our lactation station. Our female colleagues were very understanding during those first 8 months.

Then Comes Justin

Two years later, I was pregnant with my second child. The kids stayed with my parents or my husband when I traveled. The lactation station was upgraded to the convenience of a winged armchair in my office.

My research project changed as well as the circle of mentors. I clearly remember one mentor in particular who was rude, confrontational, and verbally abusive yet clearly the expert and leader in their field. I needed their knowledge to advance my research. I was forced to collaborate. Maybe it is a result of growing up in the South or perhaps my Christian upbringing; whatever the root, I firmly believe that respectful interactions go a long way in establishing a common ground. When respectful interactions are not possible, as was the case with this mentor, then you must find ways to move forward. I did move the research project forward. I attended international (France and Germany) and national (Hawaii and California) scientific workshops and technical conferences. I published a few manuscripts. However, I kept several of my email correspondences to remind me of this challenging period in my academic career.

Then Comes Skyler

Another 2 years later, my husband and I were expecting our third child. My postdoctoral appointment morphed into a managerial position, wherein I directed the outreach activities of a NSF Science and Technology Center. As outreach manager, I arranged for more than 30 industrial companies to tour the laboratory facilities, enhanced collaborations between center principal investigators and international scientists through management of external travel and research awards, engaged middle school teachers in hands-on activities and exhibits at the local museum, and organized the summer research experience of an average ~50 interns over multiple summers. I co-authored draft business plans for startup companies, and organized industrial symposiums including graduate student poster sessions. These activities were my way of staying involved in my field while the kids were still very young.

I commuted 45 minutes one way from home to work each day. This may seem like madness, but in reality, it was my peace of mind. My husband and I built a house right behind my parents in Burlington. My husband was the manager of a wireless cell phone chain and his hours were crazy. So I relied on my parents a lot. My daughter stayed with my mom for the first 3 years and then entered preschool at the community college where my father worked. Dad went back to school and traded his bakery job for a teaching job in the culinary department of the community college. And the best part of this deal is the home-cooked meals that we typically eat together (parents, daughter, two sons, and sometimes hubby) around 6 pm.

This family mealtime has continued to this day even though the kids' extracurricular activities often interfere with a set time for dinner. The meals prepared by my dad range from typical southern dishes (like collard greens, corn bread, fried chicken, and chocolate cake) to gourmet meals (like roasted red pepper and goat cheese tart; beet salad with toasted walnuts, feta cheese, and arugula; Wiener schnitzel (cutlet of veal); German potato salad with bacon dressing, asparagus, and rye bread (sourdough preferment, caraway seeds); and crème brûlée). You see, I have the great fortune that my father is a certified pastry chef and he cooks dinner for my family every weeknight. To be honest, I'm not sure I could do my job and be a mom without grandparents like my mom and dad who live so close. So the commute is really like an opportunity to clear my mind before I get to work or before I get home. It's my breath before I begin my respective jobs.

My Next Career Move

When I began looking for academic positions in 2005, my goal was to remain in North Carolina close to my family. I wanted to find a university culture that fully utilized my unique skill set acquired through (1) years of experience as outreach manager; (2) expertise in the design, synthesis, and characterization of specialty polymers; and (3) professional training at a liberal arts women's college, a teaching-intensive historically black university, and a research intensive institution. I found such a position at North Carolina Central University. NCCU's motto (Truth and Service) resonated deeply with me.

I joined the Chemistry Department at NCCU in 2005. I consider myself as someone who has successfully maintained significant research activities at an institution whose mission is undergraduate education and (increasingly) research. NCCU is a comprehensive university where faculty members traditionally maintain 12 teaching contact hours per semester. I chose this career path because I wanted a "balance" between my professional life and my personal life as a wife and mother of three young kids. I felt my ethnicity, training, and interdisciplinary approach to research ideally equipped me to thrive in the environment at NCCU.

My broad interdisciplinary experience in analytical chemistry, theoretical modeling, physical chemistry, and polymer chemistry allowed me to develop a niche in the materials field. In 2007, I was awarded my first independent external grant from the

Petroleum Research Fund and, as a result, published my first independent research investigation in the well-regarded *American Chemical Society Journal of Physical Chemistry B*. I have raised approximately \$1.5 million in external funding on projects for which I serve as principal or co-investigator.

NCCU offers me a good home for my scholarship and teaching agenda. This agenda emphasizes the application of interdisciplinary materials research to reduce health disparities and diversify our nation's renewable energy sources. The proximity of NCCU to Duke University, North Carolina State University, and University of North Carolina at Chapel Hill as well as the debut of NCCU's Biomanufacturing Research Institute and Technology Enterprise (BRITE) and its successful processor, Biomedical/Biotechnology Research Institute (BBRI), garners a wide support base for research collaboration.

For the most part, my career choices played out exactly as I anticipated. There are aspects of my career that I'm not particularly proud to own (like two master's degrees, the length of time to finish my doctoral degree, etc.), but I'm not sure that I would change anything. All of my experiences have shaped me, and I think each one was instrumental in my maturation. People develop at different rates, and my winding, often uncertain, career moves offer me a unique perspective to share with students that I advise at NCCU.

Reflections

Working as a chemistry professor at a small teaching university has afforded me a somewhat balanced life of raising a family while maintaining my career. However, balance is still a tricky aspiration.

First, it's a balancing act as a tenure-track professor at a small university. There are choices to make regarding the teaching, research, and service roles of my job. A few examples that I encounter on any given week include publication impact balance (i.e., high-impact manuscripts typically co-authored with research intensive collaborators or lower-impact communications co-authored with undergraduates), research/teaching balance (i.e., collection and analysis of data for an abstract submission or development of a new inquiry-based module for tomorrow's lecture), and student success rate balance (i.e., high pass rate afforded through critical thought preparation or "spoon-fed" anticipated answers).

Second, it's a balancing act to be a chemistry professor and a mother of three. I would advise aspiring women chemists to be true to themselves and follow their dreams. My daughter has expressed an interest in combining chemistry and law for her studies. I didn't really encourage this because it is tough being a female scientist. Yet I didn't discourage this career choice either. If Mary really wants this "black, female, scientist" lifestyle, she will find a way to reach her goal... and I'll be one of her biggest cheerleaders. Oh my!

Reflections from Darlene's Daughter
Mary Taylor (age 17)

One of my earliest memories of my mother as a chemist is when I was in fourth grade and she came to my school to do a science demonstration. For our class, she did the elephant toothpaste experiment and oobleck. This would be the first time I hear the words "polymer and exothermic reaction." I remember the amazement on everyone's face during the reaction and the amazement on mine thinking that my mother is the coolest person ever. This thought would soon spark my love and curiosity for chemistry.

Growing up with "Chemother" (as I like to call my mom) was a bittersweet relationship. Sweet in the sense that I was strong in science and had an upper hand of grasping science content more than most kids my age. On the other hand, bitter, because we did bump heads a lot due to our common personalities and ideas. Mom's high expectations for me can be very stressful, and she can be demanding especially in the math and science field. I know deep down that she only wants to make me stronger no matter how much I fight her.

All and all, my mom is one of my biggest role models and motivators. I strive to be similar to her one day (Figs. 1 and 2).

Fig. 1 Darlene's daughter Mary at the Uterine Fibroid Gala 2017 with Dr. Friederike Jayes, the winner of the cake Mary made for the silent auction

Fig. 2 Mary, Justin, Marvin Kimber (Darlene's dad), Skyler, and Evie Kimber (Darlene's mom) posing with dogs Brittany and Sassy at Darlene's dad's 80th birthday

Acknowledgments I could not have made this wonderful journey without guidance from my Lord and Savior Jesus Christ. In addition to my faith, I have enjoyed the companionship of my husband Erwin for the better part of this journey. Finally, the support of my parents has been priceless.

About the Author

Education and Professional Career

1989	BA Chemistry, Goucher College, Baltimore, MD
1992	MS Analytical Chemistry, North Carolina A&T State University, Greensboro, NC
1994	MS Physical Chemistry, University of North Carolina at Chapel Hill, Chapel Hill, NC
1998	PhD Physical Chemistry, University of North Carolina at Chapel Hill, Chapel Hill, NC
1998–2002	Postdoctoral Fellowship, University of North Carolina at Chapel Hill, Chapel Hill, NC
2002–2008	Outreach Manager, NSF STC Environmentally Responsible Solvents and Processes
2005–2012	Assistant Professor, Department of Chemistry, North Carolina Central University, Durham, NC
2006–2008	Consultant, Fuxin Hengtong Fluorine Chemicals Co., China
2009–2011	Clinical Research Training Program, Duke University Medical Center, Durham, NC

2011–present	Adjunct Member, Department of Obstetrics and Gynecology, Duke University, Durham, NC
2012–present	Adjunct Member, Biomanufacturing Research Institute and Technology Enterprise, North Carolina Central University, Durham, NC
2013–present	Associate Professor, Department of Chemistry, North Carolina Central University, Durham, NC

Honors and Awards (Selected)

2011	Technology Development Award, North Carolina Central University
2008	Building Interdisciplinary Research Careers in Women's Health Scholar Award, Duke University
2007	Supercomputing (SC07) Education Program Award
2007	Excellence in Research Award, College of Science and Technology, North Carolina Central University
2007	Duke University/North Carolina Central University STEM Partnership Award
2006	Faculty Fellow, Computational Science Education Reference Desk, Shodor
1995	Hoechst Celanese Fellow
1993	Department of Education Fellow
1990	Scholar in Residence at Rohm and Haas, Norristown, PA

Darlene engineers oligomers and polymers at the molecular level to study their structure-property relationships for advanced applications, such as materials for the active layer of solar cells and stimuli-responsive materials for drug delivery.

From the Periodic Table to the Dinner Table

Danielle Tullman-Ercek

Photo Credit: Peg Skorpinski Photography

How Did I Become a Professor?

Not until I was nearly finished with high school did people suspect I would someday become a chemical engineer or professor. I had a natural aptitude and love for math and chemistry, and since chemical engineering combines these two subjects, in hindsight it was an obvious choice. I also enjoyed my high school job of tutoring others in math and science and especially loved that indescribable feeling when I helped a student achieve an "A-ha!" moment: the dawning of recognition on their faces and the almost tangible clarity they suddenly seem to emit was a moment of excitement for me as a teacher as much as for my students. A teaching career was therefore a clear option by then as well. Without the aid of hindsight, my career plan follows my interests from earlier in my childhood.

Growing Up

When I was about 10 years old, my parents took me on a visit to a log home company from which they were considering buying a vacation/retirement property. I saw how the structure was made, including the many options for corners, insulation, seals, and roofing. Someday I would design and build houses; I decided I would be an architect! Even at 10, I was never one to do something halfway, so I poured over planbooks and tried to learn how to draft by hand.

My uncle, a builder, encouraged me. With his guidance I drew up countless floor plans and elevations over the next few years. Even after years of practice, my elevations resembled the drawings of a talented kindergartner. It was probably obvious to everyone that I was not cut out for architecture. Nonchalantly, my grandfather, who was a mechanical engineer, suggested I also consider an engineering career, since it would require less artistic ability. Just before I started high school, I was logical enough to agree that engineering was a better fit for me.

I transformed all the passion I had for architecture into exploring other careers, especially civil and architectural engineering, and attended all the local workshops and outreach events I could for aspiring engineers. The next year I had my first detailed exposure to chemistry, and it became my favorite class. I began to tutor others in the subject and quickly expanded to tutor students in any math or science course. I nonetheless remained undeterred from my goal of a civil engineering career until I took physics and found force balances far less intuitive than any aspect of chemistry. In this way, by the time I applied for college, I was set on a degree in chemical engineering without knowing what, exactly, a chemical engineer did. It was enough for me to know that it involved math and chemistry, and designing, well...something.

The College Years

My father was in the US Air Force, and as a result of his career, my family moved all around the country. Perhaps due to this upbringing, I was fiercely independent and eager to go out on my own. I also wanted to see a part of the country where I had not yet lived, so I applied primarily to universities in the Midwest. I chose to go to Illinois Institute of Technology, a small engineering college in Chicago. IIT offered

From the Periodic Table to the Dinner Table   471

me a full merit-based scholarship as part of their new Camras Scholars program, and that made my decision of which university to attend an easy one.

Once in college, I never wavered from my decision to major in chemical engineering. IIT offered an excellent introductory course to make students aware of the careers available in the profession, and I was delighted to learn that chemical engineers were needed to design processes (I knew that all engineers had to design something!) to make almost every consumer product imaginable, including pharmaceuticals. I still wanted to use my talents to do something positive for the world, so making lifesaving medicine seemed like the perfect fit. It wasn't quite as tangible as a completed building, but it was close.

My engineering courses were difficult and filled with both theory and practical examples. I remember coming out of my first exam in my materials and energy balances course thinking that I was going to change majors because I had done so poorly. As it turned out, I did do poorly on an absolute scale, but so did everyone else; we all failed to appreciate at first that engineering is not a topic to memorize if you wish to do well—a common mistake for entering engineering students. I stuck it out, of course, and as I began to understand the importance of treating each problem as a puzzle to be solved from basic principles, I performed much better. I even served as an undergraduate assistant for some calculus and chemical engineering courses over the next few years and served as a tutor in the campus resource center for 3 years. These experiences served to enhance my love of teaching, and I began to strongly consider a career in teaching. At that point, I thought a professor was only responsible for teaching courses, a vision that would soon be corrected.

During my junior year, I was encouraged by one of my professors to try an undergraduate research project in the area of mass transport and polymers. The project was theoretical in nature (no experiments or bench work), and I struggled with what was my first attempt at solving an open-ended problem that drew a little bit on everything I had learned so far. The project was unrelated to my favorite area of biotechnology, but I loved working on a problem that nobody had ever solved before. With this experience, I learned that research was about solving current real-world problems in creative ways and did not resemble my standard homework problems. I also spent a lot of time with both the professor that advised this research and his group of graduate and undergraduate students. I saw how much time my professor spent mentoring everyone and thinking about research and realized for the first time that *this* was what professors did in the many hours they had each week outside of the classroom. I also learned about myself: after spending so much time on theory but surrounded by experimentalists in the group, I knew I would prefer a project that required laboratory experiments. I also had yet to try working on a biotechnologically relevant problem and set that as a future goal.

As graduation loomed, I considered the idea of a PhD because I was so interested in someday teaching at the university level. The expense of graduate school seemed far too high—there was a reason I had gone to the university that offered me a scholarship, after all—and so I resigned myself to at least starting out in industry to earn money for a higher degree. I rationalized that this was for the best, because I would be gaining valuable experience to pass on to my future students. Then I received a flyer in the mail advertising a PhD program in chemical engineering. The flyer highlighted the fact that PhD students would have their tuition paid for and would receive a stipend on top of that. I wish I had saved that flyer because it certainly marked the turning point on my career trajectory. After speaking with my professors about it, I learned that most graduate programs in engineering were set up in this way. They expressed surprise that I didn't know this already—how many others were unaware? I promised to spread the word. Among my close friends in chemical engineering at IIT, about half ended up pursuing a graduate degree.

Graduate School and Beyond

I applied and was accepted to a program at the University of Texas at Austin with Dr. George Georgiou as my first choice for dissertation advisor. Dr. Georgiou was already well-known for his protein and antibody engineering achievements and fortunately agreed to advise me. At the time, I chose the lab because the projects were interesting and I knew Dr. Georgiou was highly respected for his work. To this day, I cannot believe I was so lucky given the criteria I used. It turns out that it was much more important to my career that Dr. Georgiou is an excellent mentor and genuinely cares about the lifelong success of each of his academic children.

My dissertation project required me to engineer a system for transporting proteins across the inner membrane of bacteria. This began my long obsession with the transport of materials across cellular membranes, a topic that eventually formed the basis of my own research group. (This is not so far from what I envisioned as a freshman undergraduate student: I design the microscopic factories—bacteria and other microbes—that turn sugar into medicines and other useful chemicals and make it easier to get the products out.)

As I neared graduation, Dr. Georgiou and I discussed potential postdoctoral research areas and advisors. It was the early 2000s, and he told me that I should think about getting involved in biofuels and/or synthetic biology. I stared at him blankly, having not heard either term yet. Both, of course, became quite popular within the next few years, and I marvel still at Dr. Georgiou's ability to stay a little ahead of the crowd in this way.

I followed his advice and ultimately joined the laboratory of an up-and-coming synthetic biologist: Dr. Chris Voigt—another rather lucky decision, in hindsight. At the time I worked with him, he was at the University of California, San Francisco, and pre-tenure. I watched as he molded the laboratory research portfolio from a rather disparate set of projects into a focused research group renowned for its ability to engineer genetic circuits and program cells for applications varying from nitrogen fixation to spider silk production. Dr. Voigt also taught me a suite of valuable skills for success in the academic world, including grantsmanship, the importance of understanding politics, and effective communication.

By this point, I knew academia was the right place for me. I would engineer living microbial factories, and I would have even more impact by teaching thousands of

Fig. 1 The Tullman-Ercek Group, November 2015, at the wedding of two group members. Photo Credit: Han Teng Wong

others how to do it. Less than a year into my 2-year postdoc in the Voigt lab, I began to apply for faculty positions. I was invited for several interviews and (admittedly, to my surprise) received multiple offers. A lot went into the decision, as is always the case when deciding on a job, but I found myself at University of California, Berkeley, in Fall 2009 and feel incredibly fortunate to have had my haphazard path lead me here (Fig. 1).

How Did I Become a Mother?

Alongside my career, my personal life proceeded, with just as many meaningful decisions: marriage, children, and what type of kitchen countertop I wanted in my life. A major theme was motherhood. I have always loved children, and the thought of not raising children of my own never crossed my mind. I did, however, worry about the logistics of raising a family—will I find the right partner? When is the right time to start a family? Should one of us stay home to care for the children while they are young? I also wanted to prove I was self-sufficient before settling down, and I wondered if this would be possible. Could I wait until my career was established before marrying and having children?

In college, I met the man I would marry within the first month of arriving on campus. If you had told me then that he was going to be the love of my life, I probably would have laughed—he wasn't my type, or so I thought. Jim appeared to be a typical Southern California surfer boy who lived in the moment and didn't take much seriously. Though we had many mutual friends, we rarely hung out together or even spoke. During junior year, we had our first real conversation—about life, and not just the weather or how classes were lately—via an online messaging service and had an instant connection. It turned out we had a lot of shared views of the world, and his ability to live in the moment was good for me; he taught me the benefits of relaxing once in a while.

We dated for the next couple years, but I was still clinging to the idea that I had to prove I could do everything on my own. So when I went to graduate school, I made the decision of where to go entirely on my own. I then told him he could follow me to Austin, but I would not follow him back to California. Not surprisingly, he returned to Southern California after hearing my position on the subject. Living apart, I figured out that while I was more than capable of going through life independently, it would be a much more enjoyable life with him in it. He came to a similar conclusion, and when he lost his civil engineering job in the downturn that followed 9/11, he took the opportunity to move to Texas. From then on, he would have an equal say in our decisions. We married 2 years later and adopted a rescued cocker spaniel. It was the start of my new family.

Jim found a job with a company based in Austin, and as my graduation neared, we had to make choices about both of our careers. His boss had children with advanced degrees—one of his daughters was even a scientist working in academia—and he understood the challenges we faced as a dual-career couple. He offered to let Jim work from home, wherever we ended up. This made the decision about where to postdoc much less stressful because we now had only a standard one-body problem.

Jim was the sixth of eight siblings and had dreams of an equally large family of his own. I negotiated him down to three or four children, but the "when" was still a problem. Jim wanted to have children early, and I wanted to wait until our careers were stable. I also felt guilty about going on maternity leave; my career and that of my advisor depended on staying ahead of the others in the field, and having a child would slow this process down, affecting everyone on the project.

Soon after moving to the Bay Area, I figured out that I would be in my mid-30s before my career was anywhere near stable, and my biological clock kicked in. I also observed that an assistant professor had more stress and demands on her time than a postdoc and that it only increased as a person moved up the ladder. It dawned on me, for the first time, that the adage "There is no best time to have a child" is absolutely correct. And if there isn't a best time with respect to career, then I reasoned that there was no point to waiting until I was older, which can certainly lead to more complications biologically. We made the decision to start our family while I was still a postdoc and were confident we would make it work. The saying "We plan while God laughs" describes what happened next. We tried for over a year, with no luck.

Meanwhile, I was choosing between multiple job offers—one at UC Berkeley and the others all from universities on the eastern side of the country. Each had advantages in terms of the program—Option A would be a supportive environment with terrific students, Option B would offer the opportunity to collaborate with incredible researchers, and so forth. Jim was not especially excited about staying in the Bay Area, nor about the high cost of living there, and wanted me to take the offer from the university nearest to his mother, who had relocated to the east coast. My family also was rooting for us to move to the east coast, as most of our extended family still lives on the eastern seaboard. However, I knew that the position at Berkeley offered me the highest chance of success because the university and its chemical engineering department are ranked among the best in the world. This meant that I would be able to work with the most talented researchers—professors and students—and I would have access to an extensive set of resources to help me get started. More intangibly, the community simply felt like the best fit for me and my research field. After a few weeks of debate, Jim agreed that Berkeley would be best for me and our future family, and I accepted my faculty position offer from UC Berkeley with his full support.

That same summer, I found out I was pregnant. We had given up on natural conception by this point, so we were surprised but thrilled. The timing was not what we would have planned, as the baby was due just a few months before my agreed upon start date at Berkeley, but we appreciated the miracle for what it was.

Our son was born in the spring of 2009. This life change triggered some involved conversations between Jim and myself about how I was going to balance a demanding career and an infant. Our strategy was (1) to have his brother, who lived with us at the time, serve as a part-time nanny and occasionally make dinner and (2) to have Jim always carry out duties such as errands, cooking dinner on weeknights, and taking care of most household chores (bathrooms were still my responsibility) while cutting back on his hectic business travel schedule. It was hardly the 1950s model of gender roles, but this division of labor worked remarkably well, and it was a good thing—after a relatively easy pregnancy and delivery, we faced one new challenge after another with the baby. None of the challenges were out of the ordinary but combined made for many sleepless nights and far too many day-interrupting doctor's appointments. Fortunately things calmed down as our son grew older, and we felt blessed to have such a wonderful little person in our lives.

As much as we loved being parents, we also quickly adjusted our plans for future children; Jim and I agreed that two children would make our family complete. We welcomed our second child, a girl, in the summer of 2012. This time around, things went more smoothly, though her personality was the opposite of our son's in nearly every way, which meant we had another steep learning curve to overcome.

Fig. 2 Danielle's family at her son's birthday party in February 2014

Despite the challenges, being a mother is the most rewarding job I can imagine. After a stressful day at work searching for funding and troubleshooting experimental protocols, I come home to children that make me forget everything with a smile or hug. Anytime I have a rough moment at work, all I do is think about my family, and I instantly feel better (Fig. 2).

How Do I Balance Career and Family?

I am often asked how I am able to handle my tenure-track position and my young children. To be honest, it would not have been easier if I were tenured, so the real question is how to balance any demanding career and still make time for family. In fact, many of my female colleagues, especially those that are pre-tenure like me, have also decided to start their families. It seems to me that even 10 years ago, women in academia were much less likely to have children pre-tenure, and a shift in the academic culture is finally not only permitting but enabling this change. Thus I think I have it pretty easy compared to other working moms. I have a supportive spouse, for example, and a job that permits some flexibility in hours. The University of California has mechanisms for stopping the tenure clock for all new parents (mothers and fathers) and even participates in a plan that provides emergency backup childcare both at home and while traveling, providing more options should a perfect storm of problems arise (Fig. 3). Nonetheless, I admit that my schedule is

Fig. 3 The Ercek family in Istanbul, September 2012 (I was there for a conference, but we were able to make that trip double as our family vacation.)

extremely hectic and far from that of a 9–5 worker. Thus I outline below the balancing act that I call "Monday:"

My days always start at 6:30 am, when my 1-year-old daughter wakes up with the precision of an atomic clock. We haven't needed an alarm clock for months! She also wakes up my 4-year-old son, so I (or my husband, if he has a free moment) feed them and get them ready for daycare/preschool. Once they are satisfied, I will turn on an episode of Mickey Mouse Clubhouse to entertain them while I get myself ready for work. We leave the house by 7:30, and they are in daycare/preschool by 8:00 am.

I take the train to work and use the commute to respond to emails that were neglected the previous day. Barring a transit delay, I am settled in behind my desk by 9:00 am, just in time to review the notes for my lecture one last time. The hour-long class starts at 10 am, and I hold my office hours immediately following lecture. During the lunch hour, I try to eat with other faculty as often as possible, but sometimes cannot avoid dealing with any crises that arose in the morning instead. At least once every 2 weeks, I also make time to have lunch with other female faculty members on the tenure track; we share successes, failures, and strategies for balancing work and family, and these lunches never fail to put me in a good mood even on a bad day.

At 1 pm, I have the first of my scheduled meetings with my graduate students. Each meeting revolves around a particular research topic and involves two to four graduate students and the same number of undergraduate researchers. We discuss the previous week's progress, or lack thereof, and troubleshoot when necessary. We also devise a plan for the next week. These meetings are the highlight of my work week;

it is the time when I get to interact with all of my academic "children," and I love talking about and brainstorming ideas for their projects. It is also gratifying to watch them grow into independent researchers.

By 4 pm, if I am on time (a rare event!), the last of these meetings ends. These meetings with graduate students are only on Mondays, but other days are equally filled with activities such as attending committee and faculty meetings, serving on graduate student qualifying exams, advising undergraduate students about careers and coursework, or serving as a peer reviewer on grant proposals or manuscripts.

Typically, the 4–6 pm block of time is used to take care of any issues that arose during the day—perhaps finding someone to fill a hole in a seminar speaker's daylong visit or working with my lab safety officer to make sure the standard operating procedure for a new chemical in our inventory is complete and accurate. Occasionally, I meet with a student for whom I will serve on the qualifying exam committee, or with my teaching assistant to discuss a homework problem, and if I am lucky I can use this time to get advice from my more senior colleagues. I also go to an on-campus yoga class from 5 to 6 pm at least once per week; it helps me deal with the stress in my life. I am noticeably grumpier and more tense if I skip yoga for an entire week! I leave by 6 pm so that I can be home for dinner by 7 pm, again using my commute to take care of email correspondence.

My husband typically picks up the children from their respective daycares and makes dinner for us all. My son says grace and does a round of "Cheers!" and we eat together. It is my favorite part of the day. My son might tell me about how he found five acorns on the playground that he saved for the squirrels or how he didn't get any time-outs that day. My daughter will say "Mmmm!" with each bite, even though only about 10% of it ends up in her mouth. She will then dance in her high chair as my son hums the theme of Jeopardy, which is on in the background, and by then it is time to clean up. I do the dishes as my husband coordinates bath time. By 8:30 we are finishing up the nightly routine of eating dessert, brushing teeth, reading a story, and tucking into bed. By 9 pm, the children (and often my husband, as well) are asleep, and I will use my last hours of the day to catch up on my writing. At any given time I have two or three grant proposals in progress, and four or five manuscripts to edit for submission to a journal for publication, and these hours are my most productive for working on these tasks.

I have had trouble falling asleep for as long as I remember, and reading works better than any sleeping aid. Each night I attempt to read a chapter of a "fun" book, such as a Jasper Fforde novel or James Rollins thriller, but invariably I fall asleep after just a page or two, always by midnight, and the routine begins again. . .

Side note: I write about my typical Monday, but I have found it is helpful to carve out an entire day to simply work on my writing tasks—writing grant proposals and manuscripts on my research—in order to accomplish these two vital tasks for my career. These writing days for me are usually on Tuesday and are held sacred on my calendar; if I did not treat them as such, the time would be gobbled up by additional meetings. Since the writing must still go on, that means it would have to happen on weekends, which in turn would mean much less time spent with my family—not a viable alternative. There are other ways to build writing into the workweek, such as daily writing hours, but I found that, for me, devoting an entire day to writing is the best way to ensure it actually happens.

I love my jobs—both motherhood and professorship. It may be obvious from the narrative of my typical day that I love these jobs so much that I don't leave time for anything else. Not counting Mickey Mouse and Jeopardy, I rarely watch non-recorded TV, and grocery shopping and social networking are squarely in the domain of my husband. But this is my (and our) preference; if I didn't want it this way, I would have switched careers or decided not to have a family. This is my life, and I love it. The funny part about it all is that until I was in each role, I didn't have any idea what I was signing up for. So in a way, I was extremely lucky—twice.

Update: Perspective as an Associate Professor

Just 2 months after the rest of this chapter was written, I was approached by Northwestern University about a tenured position in chemical and biological engineering, as part of their effort to make Northwestern a top destination for synthetic biology research. Everything about this offer was attractive for me and my family, and ultimately I moved to Northwestern with tenure about 18 months later—the process was a long one! I now have more time to spend with my family, which was extremely important to me, because my commute time is reduced to a total of 30 min per day. My new department also consists of a strong cohort of faculty with similarly balanced priorities, making it a supportive environment. In summary, the transition made me love my two jobs even more!

Reflections on...

...The Gender Gap

My experience with recruitment/tenure at Northwestern simultaneous with retention/tenure at UC Berkeley was a trying one and quite eye-opening for me. I realized that gender played a considerable but often invisible role in how I was treated and evaluated throughout my career. While I love that I did not see it earlier in my career, and think my optimism was an important component to success, it is nonetheless discouraging to see how pervasive and ubiquitous all such biases are. The implicit bias affected evaluations and comments made by both students and colleagues at my universities and around the world. For example, one colleague showed me a note by someone I held in great respect. The note was approximately a page long and described how the downfall of academia is due to the forced inclusion of women and minorities! I also saw how much the impact of implicit bias and gender or racial discrimination can vary among different departments and universities; no place is immune, but employing antibias practices can greatly decrease the effect it has on the culture. Seeing this, I now actively work to minimize bias wherever I have influence. As well, I feel fortunate to have found a department and school that actively works to minimize the effects of implicit (and not-so-implicit) biases.

... *Deciding When to Have Children*

My career initially strongly influenced our decision about when to start a family. Though we married relatively early in my time in graduate school, I felt strongly that having children during grad school would send a message that my career was unimportant. Few women in my program had children during grad school, and other students joked that it was the only way to be guaranteed you could graduate on time (because the faculty advisor would not want to pay a student that was out on maternity leave, presumably). Thus while in graduate school, I did not think I should have children until I had tenure. I revised this decision when I realized that a tenured professor has even less time and many commitments that cannot be pushed back due to maternity leave (obtaining funding for graduate students or publishing papers in a highly competitive field, for instance). Ultimately, I had my first child just prior to starting my faculty position. Surely those around us felt this was not ideal timing, but we made it work, and I would not do anything differently if given the chance.

My family continues to have an enormous impact on my career, as well. I am learning to be a mother at the same time I am learning to manage a group of young researchers, and despite the age difference and skills to be learned, there are many similarities in the two functions. More than that, though, my family requires time, which requires an efficient schedule and that I turn work "off" completely for a minimum number of hours per day. I cannot stay late 1 day to finish writing a proposal, or I simply do not see my children that day because I get home after they are asleep. I also rarely work at home in the evenings until after the children are asleep, both lessening my guilt at working and making the time much more productive and free of interruptions. Overall, having a family enforced set of times during which I could not be deeply thinking about my research, and (as taking a step back often does) this did wonders for putting certain problems in perspective (Fig. 4).

Fig. 4 The Ercek family on a camping trip in fall of 2013 at nearby Pinnacles National Park (Hiking with the family is a wonderful way to truly turn "off" and reduce the stress of a hectic life!)

... *Important Female Role Models*

I write of the influence my grandfather had on me in the narrative, but my grandmother was also an inspiring role model for me. She often recounted the stories of her time in nursing school and of how she met my grandfather and planned their wedding while simultaneously completing her nursing degree. She loved especially to tell me about one of her professors, who chided her: "Are you working on your M-R-S or your R-N?" She maintained that she could do both and did. She and my grandfather were together nearly 60 years and raised three children, and she worked as a nurse (often on the night shift) even when the children were young. They did not need the money, given that my grandfather was a mechanical engineer with a productive career designing sewing machine parts for Singer. She worked all those years out of a love for her job, and I plan to do the same.

... *The Critical Need for Institutional Family Benefits*

There are now many ways that universities can protect young faculty that need time off to care for loved ones, whether that means infants or ailing parents. I was given 1 year of teaching relief and 1 year of tenure clock stoppage for each child. For my second child, I also had the good fortune of giving birth in June, affording me a lighter load not only during the subsequent fall and spring semesters but also in the summer months such that I did not officially return to full duties until my daughter was 14 months old. You cannot always plan these things, but that was definitely fortuitous!

... *Letting Go of Perfectionism*

The best advice I ever received was from my mother. As I child, I hated to get anything other than the top score on every test and constantly compared myself to my classmates. My mother probably correctly worried that my self-worth would plummet when (not if) I was no longer the top student. In direct contrast to the trending parenting strategies of the 1980s, which seemed to be devoted to making every child aware that he/she is special, my mother went out of her way to remind me each time I brought a perfect test or straight-A report card home that "there are always going to be many other people in the world who are smarter than you, and that is okay." She was proud of me as long as I did my best and did not expect me to be *the* best. This didn't change my instinctive overachieving nature and perhaps even drove me to work harder to prove her wrong. But it also had its intended effect: it allowed me to more easily accept those times when I fell short of my goal, and it helped me to live with the fact that while I am a perfectionist, I am not perfect. This, I believe, was crucial to survival as a graduate student because failure in research is a necessary step on the path to success. Unlike many brilliant scholars, I knew how to accept failure and keep trying before ever setting foot in the research lab.

The mantra is easily reapplied to other aspects of life, such as motherhood. For instance, I know there will always be better mothers than me, so I cannot worry about it if I don't create a theme for each birthday party or if I cheat and use premade cookie dough to make Christmas cookies. As long as my children know I love them, that is all that really matters.

> **Reflections from Danielle's Son**
> My son is only 4 but told me this: "My mommy is a very good teacher. I like to go to her office because it is fun to see her students.[1] I also like to draw on her chalkboard" (Fig. 5).

---

[1] He comes to the campus a few times a year and whenever we do outreach events for families.

Fig. 5 The Tullman-Ercek lab group at the Monterey Bay Aquarium (We took time from our camping and hiking retreat to look at the jellyfish and sea otters, and I brought my son along as an honorary group member. He is the one holding a newly purchased small stuffed penguin, which became his favorite stuffed animal and must-have toy.)

About the Author

Education and Professional Career

2000	BS Chemical Engineering, Illinois Institute of Technology, Chicago, IL
2006	PhD Chemical Engineering, University of Texas at Austin, Austin, TX
2007–2008	Postdoctoral Research Associate, Department of Pharmaceutical Chemistry, University of California, San Francisco, CA
2008–2009	Postdoctoral Research Associate, Joint Bioenergy Institute, Lawrence Berkeley National Laboratory, Emeryville, CA
2009–2016	Assistant Professor, Department of Chemical and Biomolecular Engineering, University of California Berkeley, CA
2016–present	Associate Professor, Department of Chemical and Biological Engineering, Northwestern University, Evanston, IL

Honors and Awards (Selected)

2015   Searle Leadership Award, Northwestern University, Evanston, IL
2015–2016   Merck Chair of Biochemical Engineering, University of California Berkeley, CA
2012   National Science Foundation CAREER Award
2009–2013   Charles Wilke Endowed Chair of Chemical Engineering, University of California Berkeley, CA

Danielle is an active member of several professional organizations, including the American Institute of Chemical Engineers and the American Chemical Society. She is also an editor at mSystems and serves on the editorial board of *ACS Synthetic Biology*.

The Bigger Picture: My Journey to a Purposeful Life and Career in Academia

Idalis Villanueva

Photo Credit: Donna Barry, Utah State University

I am a chemical and biological engineer teaching introductory engineering courses; I am an engineering education researcher with an interest in broadening opportunities for diverse students; I am the only Latina woman assistant professor in my college and a minority in my community; I am a mom of two small children (7 and 3.5), and I am married to my husband of 12 years and many more to come; I am powerful and impactful, my life is purposeful and part of a bigger picture.

It Is All in the Subtleties

Nearly every day of my professional life, phrases about my fit crosses my head. I can't tell you the numerous times I have been reminded of "my labels" within my profession. Subtleties that, although many times come with no harmful intention, lead me to doubt my abilities and future as a tenure-track faculty member.

I still remember when I came back from maternity leave after having my second child. It was between the first and second year of my tenure-track position. I was opening my office door for the first time after 8 weeks of maternity leave, when another professor down the hallway told me, "Oh, you had a child. You must be devastated." My heart sank. I thought to myself, "What is so bad about coming back to work? Why do I need to be 'devastated' about having my children in day care while I work?" Luckily, that individual apologized for his words at a later time, but those words stuck with me and reminded me of the central theme that has surrounded my entire professional and personal life: I am a mom in engineering within academia.

Humble Beginnings

When I was a small girl, I was raised in a small fishing town on the island of Puerto Rico. While people think that being raised near the ocean is wonderful, it can be a very hard and depressing place to live. Crime and poverty is predominant in these regions as many people struggle to find food and sustenance for their families. I grew up seeing many of my childhood friends go into jail, prostitution, or drug dealing. I knew early on the consequences of this lifestyle. So, I focused on my education instead.

My dad was forced to drop out of high school to help his father maintain the family. As such, he matured at a very early age and had a strong work ethic. He worked in a bread bakery called a "panadería." Every day, he would bring us freshly baked bread and candy. At night, he would go to the basement to read. He loved learning despite his lack of a formal education. I always admired that about him. My dad was a lifelong learner and a brilliant man who completed whatever project he set his mind to. One of my father's goals in life was that my siblings and I would get a bilingual education. He would always say that knowing a second language would open doors for us in the future. His love of learning was the best gift he ever gave us.

My mother, a stay-at-home mother and college dropout, would help us with our assignments whenever she could. She always regretted not continuing school, so she would always tell us to strive higher and shoot for the stars. She instilled in me a love for reading and writing that has followed me throughout my career. She also was a brilliant woman and a person with a keen eye to detail, something I learned at an early age and still use to this day.

The Bigger Picture: My Journey to a Purposeful Life and Career in Academia   487

When I was 15 years old, I became interested in engineering. I am not sure when or how I found out about engineering, but I knew I wanted to be an engineer. Maybe the idea of a career in engineering came from my involvement in my dad's household construction projects or from my mother's ability to "engineer" her way out of broken items around the house. I remember mixing cement, measuring dimensions for electrical outlets, stitching broken clothes, and participating in multiple household and car repair projects. In a way, my "handy-woman" experiences ignited my interest in engineering.

Ironically, when my family first heard about my desires to become an engineer, the support for this career was limited. My parents did not like the idea of engineering, but also did not deter me from pursuing it, which I am grateful for. Other family members more vocally discouraged this career prospect. I still remember my grandmother saying to me, "You don't have to study engineering. This is a man's field. You should focus on getting married, learning how to cook and clean, and have kids. That is what you are meant to be." I would often disregard these comments, knowing that I had more to offer. I wasn't opposed to the idea of having kids and getting married but not until I got an education first.

Engineering or Math as a Career?

I remember being torn between a career in engineering or math. I didn't know who to talk to about which path to take. I am a first-generation college student, and no one in my family had ever received completed a degree. I felt lost. So I went to my high school counselor who provided me with a career placement exam. The results were that I was not good at math or engineering, but rather that my skills were in the social sciences. I thought that the results did not make sense as I deeply enjoyed math and I liked tinkering too.

When I was in eleventh grade, I was offered a scholarship to attend a weeklong engineering workshop hosted by a nearby university. It was the first time I had ever been to a university and was amazed by how big the research labs were. All the students selected for the workshop worked on small experiments representative of the different engineering departments. After the workshop, I became interested in mechanical and chemical engineering.

In my senior year of high school, I discovered that my family did not have the financial resources to send me to a university. I wasn't sure what to do, but I knew that I had to find a way to attend college. Around that time, my chemistry high school teacher pulled me aside and said that I had a lot of potential in chemistry. She asked if I was interested in representing our high school in some chemistry competitions throughout the island. I agreed, not realizing the positive implications this would have on my college admission. Having this experience under my belt not only opened the doors to attend a university, but it opened my eyes to scholarships and ways to attend a university without the need to rely on student loans. Thus, I opted to

488   I. Villanueva

pursue a degree in chemical engineering in the same university where I attended my engineering workshop the year prior.

My Engineering Undergraduate and Graduate Years

Early in my college years, I met Neil, who later became (and still is) my husband. In my culture, getting married early and having children (before the age of 25) is common. However, we both decided that, since we were students, it would be sensible to wait until we both got our bachelor's degrees first.

After I earned my B.S. in chemical engineering, I was offered a paid M.S./Ph.D. fellowship from the University of Colorado. Neil, my fiancé at the time, needed one more year to graduate. Thus, I was hesitant to take the fellowship and move to the U.S. mainland. Reluctantly, I moved to Colorado, and, a year later, I got married at the age of 25.

Our move to Colorado was not an easy one. We both struggled to get used to the culture and the environment. We were questioned often about our professional paths and our decision to be married at "an early age." To us, being married was not a deterrent, but rather a support that helped us rely on each other more, especially *in lieu* of no family nearby. We relied a lot on phone conversations with family and friends to cope with the loneliness. After 2 years of living in Colorado, we met a group of friends that became our support for the remainder of our time there.

My graduate school years contained some of the hardest and most eye-opening experiences I have encountered. As the only Latina female graduate student, I felt very isolated. There were not a lot of people I could connect with at a personal level who would understand my culture and my desire to become a mom. Despite understanding the importance of an education, to me, family came before profession.

As an aspiring academician, I felt hopeless. I recall looking around the hallways of the professor's offices and seeing only three women faculty, none of whom had children at the time. The only female with a child was a lecturer and her child was in her early teens when I met her.

In the afternoons, many of my female graduate school lab mates would talk about family. All of them wanted to have a balanced family life, but none had expectations of having one until their mid-30s. I, however, wanted to have children before my mid-30s.

One of the most pivotal moments in my decision to become a mother came from one of my graduate mentors. One day, during one of our meetings, the mentor blurted out, "I don't want you to drop your career and become a housewife." I must admit that these words offended me greatly but, at the same time, opened my eyes. I realized that I didn't want to marry my profession nor did I want my epitaph to say "Her job came and went and with it, no one to love or reciprocate that love." At that moment, I made a decision: I was going to have a family and a profession. What I was still uncertain about was if this meant that a tenure-track position would be out of the question.

The Bigger Picture: My Journey to a Purposeful Life and Career in Academia 489

I completed my dissertation on biochemical and biomechanical cues for cartilage cells encapsulated in synthetic polymeric hydrogels that were compressed in a custom-created bioreactor and earned my Ph.D. soon after. I was offered a postdoctoral position in a government research lab in Maryland, so my husband and I moved.

My Postdoctoral Years

In engineering, postdoctoral experiences are not common but growing in popularity. I figured it would give me an edge for research, if I decided to work in industry or government. My postdoctoral work focused on biological research primarily studying mitochondrial membrane potentials for neurons under ischemic stroke conditions. Again, I was the only female Latina in the lab. There was a male postdoc working in the lab, and he had young children and a stay-at-home wife. There was a female scientist working in the lab with no children and another female scientist with a single adult son.

I was beginning to get more and more discouraged about the prospect of combining research and academia into my personal life. One day, I met a researcher who was a woman scientist married to another researcher. She had two small children. She told me that in order to make her schedule work, she would work hard all day long, have dinner with her children and put them to sleep, and then resume work until midnight. Seeing the stress in her face was disheartening. I didn't want to be stressed and feeling guilty for having a family. I wanted to enjoy my profession but also know when to dedicate time to my family.

Soon after this encounter, I became pregnant and all the words and fears instilled in me by my former experiences surfaced. Would I be able to maintain a balanced career and have a family? What will people at work think of me? Surely, they are going to think that I don't take my research seriously enough. I heard these thoughts repeatedly in my mind. I thought of abortion many times and cried myself to sleep for the first 3 months.

The stress was affecting my work too. I couldn't concentrate and I didn't want to read any more journal articles. All I wanted to do was leave my career and not worry about anything else. Nonetheless, I decided to stick with my research and made arrangements with my advisor at the time.

One adjustment I had to make was that I would not be exposed to certain chemicals and equipment that may be deemed harmful. We made an arrangement that I would prepare laboratory experiments and handle confocal microscopy samples, but experiments where chemicals would be deemed harmful through Material Safety Data Sheets would be run by other laboratory members.

One day, while working with some samples in the confocal microscope, I felt a kick! Oh, what a wonderful kick! I was reminded that life was growing in me: a life that I can nurture and inspire. After that, I found renewed strength.

I began reading books of encouragement for mothers. I asked other women in forums and looked at online chat boards about how to best talk to my postdoc advisor about my pregnancy and navigating this path. Here is what I learned:

1. Get acquainted with all the maternity and paternity leave policies before you talk to your advisor and be prepared to share with them a copy of these policies.
2. Find out if your medical insurance will cover the birth of a child and the deductible you will have to pay, so you can plan ahead.
3. Seek orientation sessions and classes about birthing, breastfeeding, and partner support.
4. Have a calendared work plan in place.

Another agreement made with my advisor was that I would finish my experiments during the pregnancy and that my maternity leave would be used to begin analyzing data and writing papers. Having a plan and policies at hand seemed to put my advisor at ease, as he had never dealt with a pregnant postdoc!

I also knew that my postdoc was short-lived and that I had to think about the future. I still was hesitant about a tenure-track position in academia, so I began to explore other career options.

During my pregnancy, I found an opportunity to do an internship in a career service planning and development office. I figured that if I participated in orienting others about careers in and outside of academia, then I would be better prepared for a future position afterward. During my internship, I conducted phone interviews, wrote career service protocols, and created presentations. I learned a lot about myself, my career interests, and options.

It was through this internship that I discovered science education, which later led to engineering education, my current career. I thought to myself that this is a perfect path to balance my family and work. I didn't have to worry about working with radioactive or hazardous chemicals anymore. I also would not have to come to lab at odd hours in the day or night to take care of my cell lines or experiments. So, I began to conduct informational interviews to find out more about the needed skills for a career transition.

After the birth of my daughter, Kayla, I used my maternity leave to meet the agreed stipulations with my postdoc advisor but also took advantage of the times when my newborn was sleeping to look for career opportunities and refine my resume and application materials. As I explored science and engineering education careers, I realized that I needed quite a bit of knowledge in educational methodologies and methods, both of which I did not have. Thus, I decided that for now I would focus my job search for primarily teaching positions while building up my experiences in engineering education research.

A few weeks after my return from maternity leave, I had a meeting with my advisor. We talked for a little bit and soon he said, "You know, you should begin to look for a job now that you had a baby." I was confused by his comment and a little distraught by it. I didn't know what to make of the comment. Was he trying to say "Get out" or were his comments well-intentioned? I should have asked for clarification, but my mind went blank. Either way, it didn't matter.

A few days after this conversation, my husband was laid off from his job because of the recession. He was fortunate to have found a temporary job quickly, but we knew it was not going to last long with the economy being so unstable. I knew I had to do something; a postdoc salary was simply not enough to cover health care, childcare, and other household expenses. I aggressively continued my job search.

As suspected, my husband's temporary position was short-lived, and soon, money was getting tight. I didn't realize how many vaccinations and doctor visits were needed for my daughter during that first year.

A few weeks later, I received an email for an interview as a lecturer for a bioengineering department at a tier 1 research institution nearby. I knew that this may be my only chance for a stable position, not be in the tenure track, but certainly a job that would help me earn more income for my family. So, I did the campus interview, and a week letter, I received a job offer. They wanted me to start before the fall semester. My husband and I decided that this was the best option for our family.

When I spoke to my postdoc advisor about the new position, I could see the disappointment in his face. It was almost as if I had failed him and failed science. I was torn between my career and my family. In the end, my family came first regardless of what others thought of me. Within 2 months, I had left the postdoc and started my work as a lecturer. Kayla was 6 months old when I started my lecturer position.

My Experience as an Engineering Lecturer

Classes began 2 weeks after I got hired, and I was given the responsibility of teaching 2 sections of 150 freshmen engineering students. I was a wreck. I didn't realize how much energy teaching large undergraduate classes took. I had not factored in the time to commute (1.5 hours each way), to prepare for classes, to grade assignments, and still to attend to my daughter when I arrived home. Not to mention, I was weaning my child from breast milk, so I was exhausted.

At the end of my first semester, my teaching evaluations were not so great. My boss, a very understanding man, said that I should not worry about my first semester and that as long as I did better next time, it shouldn't be a problem.

Around the end of my first semester, my husband was offered a job at a nearby university working in academic relations. We were thrilled! We certainly needed the money. My next semester went super and my evaluations skyrocketed. I thought to myself, "I finally found my place." Of course, this thought was short-lived.

At the start of my second year, I met a research scientist in our department. He talked to me about his research in engineering education. My interest was spiked. I asked him about how he transitioned to this field, and he indicated that the first step was to get papers published in key engineering education journals and conferences. He offered to have me help him with some projects so that my name could be included in the articles. I thought to myself, "Why not?" My colleague and I spoke to our boss to see if he would approve of my added responsibilities, and he was supportive as long as it did not interfere with my teaching.

The following semester was hard. I had to teach three new classes and help my colleague with his research. I would get home, spend about an hour with my daughter, and go back to work. It is ironic. I was doing the same things that my former colleague during my postdoc had said she does to balance family and kids. Still, I wanted engineering education research experience, so I put all my efforts into that.

That year, Kayla began to suffer from separation anxiety. Every time I dropped her off at day care, she would cry for hours. She didn't want to be left alone and held on tightly whenever I would pick her up. I spoke to the childcare provider who said that she really missed me and that all she would say is "I miss mommy." My heart broke and guilt quickly seeped in. I thought to myself, "What am I doing? I am becoming what I dreaded so much."

That year was a real turning point in my life. I was having some major difficulties at work. Not many of my departmental colleagues understood that I had to leave work early to pick up my daughter from day care. I was the only female faculty member who had a child in my department. Many did not understand why I was doing research if I was a lecturer. It was a Catch-22. I realized that my role statement had to change to accommodate all my responsibilities and that I would need to better balance my time.

When I tried to get my contract modified or my title changed to better match my duties, I received a resounding "No." I didn't understand it. Why were lecturers treated as second-tier citizens if they handle and teach a bulk of the students?

That semester, my teaching evaluations slightly lowered, and while the dip was not significant, it was enough to get a call from the associate dean. I recall collecting all my teaching reports and going to his office. When I sat down, he said, "Tell me, why is it that students hate you so much?" I was thrown off by his comment. I thought to myself, "My evaluations went a bit down but not that much." I calmly showed him my teaching report. I showed him my average evaluations over the semesters and pointed to the other courses where my evaluations were "above average." I indicated to him that handling freshmen students, particularly those who are commuters and community college transfers, pose a different challenge and that it takes time to work things out. He then said something that made a huge difference in my career profession. He said, "I am not ready to give up on you. You have 1 more year to prove your worth." My educated guess was that the comment was meant to be encouraging, but to me it had a contrary effect. Not having received much mentoring throughout my education and career, this comment cemented my frustrations with engineering and reignited my desire to help other students navigate the hidden expectations of their education and career paths.

The associate dean's words helped me to understand that as a lecturer, I would always have the pressure of yearly contracts. I now understood that my performance would be judged on unclear metrics and student evaluations from rating systems that diminish the instructor's role. It clarified to me that with a small child, I simply could not take the risk of ending my contract without knowing if my job was secure. I began to think of my husband, my child, and what was most important. I realized that I wanted to have a family. I wanted my kids to love me for taking the time to be with them and not hate me for being so stressed about losing my job that I would forget them. It was at this point that I began to desire the stability that a tenure-track position has to offer even though I would have to be purposeful with how I managed my time.

My Transition to a New Career and a Tenure-Track Position

I began to once more seek positions in academia, this time seeking tenure-track positions in engineering education. I knew that I didn't want to go back to bench research. I wanted my "lab" to be my classroom. I got several interviews and job offers and ultimately chose a tenure-track position at Utah State University in engineering education within their College of Engineering.

This time, I included my family in the job negotiations. I was very clear to my department head that I had a family, that I needed childcare, and that I needed a job for my husband. This was a liberating thing for me to do. I figured that whoever was interested in my qualifications needed to know that my family was part of the deal. To my surprise, no one had any objections about this. In fact, after talking to them, they were actually appreciative that I was honest about my family in the interview and assisted me to the best of their extent.

My advice would be to not compromise your family in your job search. Your family is an important aspect of your life and employers need to be aware of it. You'd be surprised how many employers are willing to work things out if you are honest about your needs.

My husband got a position in a nearby company. Kayla got admitted into a new childcare, and I negotiated a one-semester teaching break to give me time to prepare for my new class (the best decision I could have made) and get acquainted with the new place. Since I knew what it took to teach a new class, it was not hard for me to quickly get adjusted to the culture and the educational needs of the college. Research, on the other hand, was a bit of a challenge as it was a new field for me and my exposure to engineering education research had been limited. Also, being highly underrepresented as a Latina engineering educator, I knew that I needed to be strategic about my pursuit of mentorship outside of my institution. Still, I knew that there was no other field I would rather be in.

Within my first year in the tenure track, I became pregnant with my second child. I immediately began to gather my policies and documents and spoke to a couple of moms in other departments across the campus. At the time of this writing, I was the only female engineering tenure-track faculty with two children and one of two female faculty members in our college who had gotten pregnant within the tenure

Fig. 1 Kaylee, Neil, Saidee, and Idalis (left to right) celebrating Saidee's first birthday

track. There were other moms who had their children after tenure, which is a different situation altogether.

I disclosed my pregnancy to my colleagues, my promotion and tenure committee, my ombudsman, associate deans, and dean. I became oriented on how to disclose this pregnancy within my promotion and tenure binder. For disclosing a modified teaching load within the promotion and tenure dossier, there is a bit more paperwork you have to complete, so I recommend you keep detailed records and inform your promotion and tenure committee about your adjustment to your research, teaching, and service commitments and include this in writing. I completed all of the paperwork for a parental leave after having the baby and negotiated a one-semester teaching break after the birth of my second daughter, Saidee.

Since I had informed all pertinent parties in advance, the department was able to arrange for a replacement instructor for the semester I would not teach. They even helped me by reducing my teaching hours upon my return and scheduling them between 8 am and 5 pm. The administration was wonderful about this (Figs. 1 and 2).

Fig. 2 Idalis's daughters Kayla at age 4 and Saidee at age 1

Lessons Learned as I Still Navigate My Tenure Track

Not everything in my tenure track has occurred without criticism. Ironically, the majority of the criticisms I received were from other female faculty (although some male faculty have also had their share of opinions). For some female faculty, it is "unheard of" to have a child early in the tenure track, let alone a second child! Some would look at me with pity, others with disgust, as if having a family limited my mental capabilities as a professional! Some male faculty said comments like, "Oh, my wife handles all the children stuff at work. I don't know how you are going to do this." Other comments included, "Is your husband going to stay home?" It is interesting to me that there is an expectation that my spouse had to stay home. Of course, that was not what we wanted. My husband has always been clear about wanting to work and study when finances allowed it.

When I returned to work, it was clear to me that with a second child, my schedule would need to be better managed. I had to be more organized, more purposeful, and learn to say "No" whenever needed. When Saidee was 6 months old, my husband and I decided that it would be best if one of us had a more flexible schedule. Thus, my husband decided to quit his job and go back to school full time. This has been a real challenge and a blessing at the same time. We coordinate our schedules so that one of us is free at all times. While my husband is in class, I work in the office. If I am teaching, my husband is free or waiting for classes. We wanted one of us to be available if there was an emergency. If one of our children is sick, it is easy for him to go pick them up and drop them in my office or for me to pick them up and wait until he is out of classes, so he can take them home. We tag team a lot, but it is very nice to be able to leave from work whenever you need to (except when I teach) and come back to work when you can. Also, the flexibility to work from home is nice and has certainly been a plus in working in academia. My husband since then has graduated and is employed at the institution where I work.

I work from 8:00 am to about 5:30 pm every day and come home and spend time with the girls. We eat, play, and spend quality time together before reading them a story and putting them to sleep. After they sleep, I typically spend time with my husband. Then, I go to sleep. It is as simple as that. It is true that in rare occasions, I work at home, but I tend to limit it to email replies and minor items. During days where I have to work on an important deadline, I let my husband know early so he can prepare to handle the girls. For the most part, my work stays at work. This has not deterred my productivity at all. In fact, I have been quite successful in my teaching, research, and service duties. Earlier this year I received the prestigious NSF CAREER award for my work on revealing the hidden curriculum (hidden values, norms, and beliefs) in engineering, particularly for underrepresented faculty and students. I am honored that my experiences and research will serve to help others become successful in their careers.

Reflections About My Journey

As I reflect on my journey, I realize that it has not been easy and I have not been exempt from the twists and turns of life. Although I had been clear that family was my priority, I had to rethink my professional and personal life and roles many times. I had to become strategic about collaborations. Honesty about your family is better than secrecy or omission. Despite what others may think about your career, you have to feel comfortable in your own skin about who you are as an individual and a professional. I have learned that there is no perfect time to have a child and that, when you do, you should have a plan in place to ensure you can meet all work and family responsibilities.

One of the things I still struggle with is traveling. During the tenure track, you have to travel a lot and many times, it is not conducive to have your children travel with you. I have decided to be wiser about my travel, limit and separate my travel time. This year, I am hoping my family can accompany me on some of those trips. I am slowly learning that you don't have to go to every conference and that, in the end, a focus on writing papers and grants is more important.

I have learned that maternity and paternity policies are in place in academia for a reason. Many times, faculty and academicians look down on these policies, but if they are not used, they will be gone. Many people in academia complain that the tenure process is outdated and does not fit the norms of today's society, yet they are

Fig. 3 Family picture in 2016: (left to right) Neil, Saidee, Idalis, and Kayla

unwilling to step up and make changes in the way that they live their lives. Rather, they adjust to the old ways and never push the boundaries.

If I had to do something differently, it would be that I would not have let fear control my life during my early years as an engineering student and researcher. I wish I had the strength then that I have today to be empowered by being an assistant professor in engineering, a minority, and a mom. If I had known that my attitudes had a lot to do with my happiness with my family and career, I would have taken these steps a lot earlier. Regardless, I am happy and blessed to have made the decision to have my family and keep my career. To me, they are complimentary and non-negotiable. I am a better mom because of my profession, and I am a better professional because I am a mom (Fig. 3).

> **Reflections from Idalis's Spouse**
> **Neil Rodriguez**
>
> While Idalis's professional journey hasn't been an easy one, I think Idalis has really persevered through it all. She has broken many stereotypes of what a women professional in academia should be. She has such a great passion for both her career and her family. She has and will be a great role model for our children. As her husband, I have learned to appreciate her even more as a wife and a friend. It also has cemented my role as her encourager when life gets challenging. She in turn has been my supporter throughout my career, something I am very grateful of. Her professional journey has made me love her more than ever, and our joint struggles have helped strengthen our relationship. We are each other's supporters, and I can't wait to see where this journey will take us next.

Acknowledgments I would like to first thank the monument in my life—the inner guide that reminds me every day that I possess the fire and motivation to persist despite my struggles and whose presence puts into perspective what is truly important in life. I am indebted to my wonderful husband of 12 years for the love and support he has provided throughout the years—he is my truest encourager and advocate. Nothing I have accomplished could have occurred without the inspiration I gain daily from my two beautiful daughters, Kayla and Saidee. Finally, I would like to thank people, including my parents, who, despite limited education and resources, fight daily to provide their children with a better life. Together, they teach me to live purposefully and unashamed of my walk in life and my potential to positively influence the lives of others.

About the Author

Education and Professional Career

2004	BS Chemical Engineering, University of Puerto Rico, Mayagüez, PR
2007	MS Chemical and Biological Engineering, University of Colorado, Boulder, CO
2009	PhD Chemical and Biological Engineering, University of Colorado, Boulder, CO
2009–2011	Postdoctoral Research Training Fellow, Analytical Neurobiology, National Institutes of Health, Bethesda, MD
2011–2013	Lecturer, Fischell Dept. of Bioengineering, University of Maryland, College Park, MD
2013–present	Assistant Professor of Engineering Education, Utah State University, Logan, UT
2014–present	Adjunct Professor of Bioengineering, Utah State University, Logan, UT
2016–present	Director, Steelcase Strong and Healthy Identities in Engineering (SHInE) Active Learning Center, Logan, UT

Honors and Awards (Selected)

2017	Engineering Education Undergraduate Mentor of the Year, Utah State University
2017–present	National Science Foundation CAREER Award
2015	RAND Policy Institute Fellow
2014	Research and Graduate Studies Washington, D.C., Faculty Fellow
2014	Alternate, National Science Foundation ENG IDEAS LAB
2011–2013	Chesapeake Bay Project Faculty Fellow, University of Maryland, College Park

2012–2013	Center for Teaching Excellence Faculty Fellow, University of Maryland, College Park
2010	Society for Advancement of Chicanos/Hispanics and Native Americans in Science Leadership Institute Award Recipient and Participant
2009	National Institutes of Health Incentive for Maximizing Student Development Trainee Award
2008	Honorable Mentions for Fellowship Application, Ford Foundation
2008	Invited Guest Speaker for Puerto Rico Space Grant Consortium External Advisory Board, University of Puerto Rico, Humacao
2006	Invited Guest Speaker for NASA STS-116 space shuttle launch in Kennedy Space Center, Florida
2004	Academic Excellence in Science and Engineering Dean's Award, University of Puerto Rico, Mayagüez
2003	Second Place in American Institute of Chemical Engineers Student Poster Competition, San Francisco, California

Idalis has a passion for education and student development. Through her research in engineering education and development of new and modular active learning environments, she seeks to help improve student outcomes (e.g., performance) and broaden participation in science, technology, engineering, and math.

Encounters of the Positive Kind

Michelle M. Ward

> In everyone's life, at some time, our inner fire goes out. It is then burst into flame by an encounter with another human being. We should all be thankful for those people who rekindle the inner spirit. ~Albert Schweitzer

The above quote by Albert Schweitzer was the closest succinct way to encapsulate what I am hoping is the message of this chapter. I was more than humbled when I was approached to contribute something to this fabulous idea for a book, as I feel all that has been accomplished in my life can be attributed to traits I have learned from one or more persons I have had the privilege of meeting and working with. There are too many people who have impacted my life in some way to ever do justice

M. M. Ward (✉)
Department of Chemistry, University of Pittsburgh, Pittsburgh, PA, USA
e-mail: michelle.ward@pitt.edu

to recognizing them all. What I hope to accomplish here is to layout some of the key lessons learned along the way.

Of course, my greatest encounter occurred in 1994 when I met my son, Zachery. He has given a level of beauty to my life that could not be matched by any other experience. I am proud of the different roles I have held over the years and look forward to additional ones ahead, but for me the greatest title I hold is "mom."

Just Do It

It may seem a bit cliché, but my parents taught the first key lesson learned to me. I did not come from a family where college was an expected path to be taken, let alone completing a PhD. My father was a machinist in a steel mill, and my mother was a waitress. The lessons they imparted through their example, though, were the most valuable throughout the years. From my parents I learned the value of hard work and perseverance. I think most people in the sciences know this, at least the successful ones; however, I need to give it the credit it is due, as without the work ethic my parents provided me with, I would not be where I am today.

When I decided I wanted to attend college, I found a way. I didn't really have a mentor to help me with the process, but I knew that it was something I wanted to do, and I found a way to do it. (The mentoring will come up later.) I worked two jobs to pay for my undergraduate schooling, on top of the student loans that were taken out. With the students I advise these days, and how busy they are even without outside jobs, I honestly have no idea how I pulled that off back then—aside from the innate programming to just keep moving forward and do what needed to be done.

I was married young and had my son almost 2 years after I was married. I was in the last semester of the senior year of my first undergraduate degree when he was born. I lived too far away from home to have help from family at that point, and he was too young for daycare, but I decided I would still finish that semester. Two weeks after he was born, I returned to school full-time along with him in a stroller (I was very fortunate he was a quiet newborn) and completed my courses. Again, I just kept moving forward (Fig. 1).

I had been teaching high school chemistry for 4 years when I came to the conclusion that I wanted to pursue a graduate degree. Unfortunately, likely due to the early age of the marriage, my son's father and I had been divorced for a couple of years at that point. I ended up moving closer to my parents, and I started graduate school as a single mother the year my son started kindergarten. There were multiple hurdles and key helpers throughout the experience (some of which I will touch on shortly), but I kept moving forward. One of my proudest days was when my son showed up to my dissertation defense and sat in the front row. He didn't really understand what I was talking about, but he said he was proud (Fig. 2).

I have had the opportunity to mentor and advise several young mothers in my department. I, of course, have lots to say to them, as I would love for them to have a more direct path to their goals. The one thing I keep going back to is to just keep

Fig. 1 Michelle's son Zach helping her prepare for student teaching in 1996

Fig. 2 Michelle and her son Zach at Michelle's PhD hooding ceremony in 2008

moving forward. If there is something you want and you keep working at it, even if it is just by sheer will, you can get there. In my mind, perseverance and hard work cannot be emphasized enough as a key trait for any type of success.

Embrace Serendipity

I didn't really have any opportunities to interact with a female role model in science until I transferred to the University of North Dakota. That young marriage I mentioned earlier resulted in my transferring from a very small school to this larger school with a graduate program. At this new school, I encountered my first female chemistry professor, Dr. Kathryn Thomasson, who also introduced me to the world of research.

I am a huge proponent of all types of diversity in departments. I don't think we should compromise pursuing the most qualified candidates, but to perhaps rethink how one attracts those who are qualified *and* also increase the diversity of a department. There are many books and studies out there that can provide outstanding data on this topic, and I am by no means a diversity expert. From my own experience, though, I can say having a female professor (someone who "looked like me") opened the world of academia being a possibility. I did end up pursuing teaching high school for 4 years after completing my undergraduate degrees, but I believe seeing a woman in academia allowed me to more readily see myself in graduate school when I did decide to go down that path.

Transferring to the larger school also resulted in my changing from an education major (with a focus on chemistry) to both a chemistry major and an education major. The requirement of the additional upper-level chemistry courses, and the exposure to research, resulted in an easier transition to graduate school than I would have had otherwise.

The interesting part of my undergraduate research had to do with the fact that I was to complete it while I was pregnant with my son. Kathryn facilitated this and kept me on as a work-study student after the coursework was completed, in an extraordinary way. I was drawn to the theoretical work she did, due to my circumstances and my interest in physical chemistry at the time. A good portion of my research was conducted with my son in a playpen beside my computer desk, at her recommendation. The work I did with her resulted in a couple of publications, which assisted my application process considerably for graduate school. At the time, I honestly had no idea how many doors would be opened due to the good fortune of working with this woman and her open mind regarding my situation.

Don't Be Afraid to Ask

I thought about titling this section "Science Dad," as a huge part of what I have been able to accomplish I owe to Dr. Sanford Asher (Sandy) (Fig. 3); however, there are some others who have provided very complimentary lessons. Learning to not be afraid to ask for what I want or need has changed my career—and my life. Sandy, and others, taught me that I would not get what I want in a timely manner if I don't ask for it and realize I am deserving of it.

I met Sandy when I was applying to the graduate program at the University of Pittsburgh. I was interested in his research, but was intimidated by the time requirements of the program and originally turned down my offer to attend Pitt. One evening, a few days after I sent in my refusal letter, I received a phone call at home from Sandy asking me to come back into the department and discuss my decision. We discussed my career goals, and he explained a concept that was foreign

Fig. 3 Michelle celebrating with Dr. Sanford Asher, her "Science Dad," after her PhD defense in 2008

to me—to teach at the university level, it was not about teaching training as much as it was about the quality of your research. (I certainly think that some professors could improve from some actual educational training, but that is a topic for another book.) Of course, I didn't want to limit my future career opportunities, so it made sense to attend a program like Pitt had—but there were issues regarding my situation with my son. As I mentioned earlier, I was going to start graduate school as a single mother when my son was about to start kindergarten. At the prodding of Sandy, I disclosed I was worried about things like parking on Pitt's campus (a nightmare for anyone who has experienced searching out a metered spot there) and the time I would need to spend working in the office when Zachery would be at home. As a result, Sandy arranged for the department to provide me with a parking pass and a home computer. I could arrive on campus with no issues after dropping my son off at the bus stop, and I could access online resources to do necessary desk work from home. This was the first time I was shown to ask for what I wanted, instead of assuming I would have to compromise or find a work-around. You may not always get what you ask for, but you are certainly less likely to get it if you never ask.

I owe the current position I hold at the University of Pittsburgh to interactions with two main people, Dr. George Bandik and Dr. Joe Grabowski. As I was nearing the completion of my PhD, I was only pursuing positions in industry. I had fallen in love with research over the years, and as I was geographically constrained to the Pittsburgh area I knew the type of tenure-track academic position I would want was not going to be a possibility. On multiple occasions, George would talk to me about considering applying for the lecturer and laboratory director position that was available in our department. I would talk to him and others about not wanting to leave research to "just teach." George never gave up (thank goodness for stubborn people), and Joe really made things hit home when he pointed out the fact that the interviewing process would be about me interviewing the department as much as them interviewing me. I realized there was no reason to see the position as a "lecturer box" that I would have to conform to and that I would never really know what was possible for this position until I pursued it. There are certainly compromises that have been made, but I have been able to make this position into something extremely fulfilling; and, to be honest, I feel the department is better off for my not wanting to be "put in box" and asking for what I wanted.

"Peacock-ing"

It might sound simple, but I thought that if I worked hard at something it would be recognized without me having to point it out. I also thought promoting myself and my accomplishments would be viewed in some sort of arrogant manner. In past years, I have started to read many articles and books geared toward the art of negotiation, specifically for women. It turned out I was far from alone in my thoughts that I would be recognized if I waited for my pat on the back; but more so it became very clear that recognition would not come in an expedited manner without being learning to be my own promoter.

Once again, I have to return to the guidance of my "Science Dad." When I first started attending national conferences, or even somewhat audience-limited grant progress report meetings, I was in awe of the scientific talent whose presence I was in. Sandy would make a point of inviting his graduate students to sit at the lunch table with himself and these prominent scientists. Eventually the conversation would turn to research projects going on in our group, and somehow (very often) the assumption would be made that the male graduate students sitting at the table would be the lead students on those projects. Sandy was great about very respectfully letting those scientists know that I was the one leading the project and then encourage me to discuss what the state of the research was. Through these types of experiences with him, I learned to not pause and wait for the conversation to turn to what I was doing and to not be hesitant in letting those at the table know of my accomplishments—there is a way of doing these things without being arrogant and to earn the respect of those at the table.

When I was at the point interviewing for positions toward the end of my dissertation work, I was very fortunate with job offers. I had worked hard for the credentials I had and was ecstatic, to say the least, at the thoughts of being on the verge of not living on a graduate student stipend and being able to step into my own. Sandy was my first introduction to negotiation—and the idea that I was worth more than the initial offer. In addition to learning how to ask for what I wanted or needed to increase my opportunity for success while raising my son, I learned to be willing to ask for more to benefit my career based on what I had to offer. At the time it seemed a bit foreign to me to have to point out that I was deserving of and could ask for more than what was first offered, and at times I still have to remind myself of this. Not every job offer will provide you with everything you need, but unless you are

Encounters of the Positive Kind

willing to promote yourself and confidently go after what you want, every company/institution would be happy to take you at a bargain.

In my position with the University of Pittsburgh, I am not a tenure-stream faculty member, so I go up for contract renewal on a somewhat regular basis. The first time I was up for this renewal, Dr. Adrian Michael taught me the idea of having to "peacock"; this was not his term but it succinctly grasps what he took the time to share with me. I was asked by the department to list the activities and accomplishments I had completed over the year. Being a scientist, I made a list of bullet points, for the most part—clearly everyone reading the list would see the importance and magnitude of what I had tried to accomplish in my mind. Under the guidance of Adrian, I have come to realize that is not always the case. Even if you already have your job and work with supportive colleagues, it is one's own responsibility to be sure you receive the credit that is due the hard work put in. It does not benefit your career to wait and hope that others see your worth—it is your responsibility to point that out. Reading the books mentioned earlier, it seems this comes more naturally for men, so I was very fortunate to have promoters in my corner who helped with this until it became more natural for me.

It's OK to Not Do It All

This might be the hardest thing for one to come to terms with; I know it was for me. In this day and age where more doors are open for women than ever before, we still must realize you cannot do everything you want to do 100%. There are always compromises that must be made, and one must make the choices that they can live most comfortably with.

I spent 8 years completing my graduate degrees. I had to watch some students enter the research group after me and then leave before me. It was a blow to my pride and made me question myself at times, but at some point I came to the conclusion it would not matter in the long run if it took me longer than my classmates. I was a homeroom mom, I never missed one of my son's school plays or spelling bees, I was home to pick him up off the bus almost every day, and we had family dinner every night. When I went through the graduation ceremony and my family took me out for a late lunch to celebrate, my son offered to give our name to the hostess—he put the reservation under "Dr. Mom." It really was at that point that I was able to let go of the frustrations for "taking so long" to finish my degree. I had done it in a way that meant more to me—I was a Ph.D. and a mom. I am so glad I didn't give up those opportunities to be a part of his childhood memories to finish a year or so earlier. I may have missed out on some opportunities while in graduate school, but I can live with that knowing I was there for him during that time which I could never get back.

When I was pursuing a position toward the end of my graduate work, I chose to accept the nontenure-track position I currently hold. I know that I was capable of pursuing something more prestigious; and, again, it took me some time to come to terms with that compromise. But again, I never missed one my son's soccer games, band concerts, formal dance pictures, and the like. I was able to arrange my schedule so that I was never off traveling, missing out on events I would never get a second chance to see. I was able to keep him in Pittsburgh and near my family. Now that my father has passed away, I am so very thankful I made that decision, as he and Zach were very close. I hold absolutely no judgment against those who would have chosen a more career-beneficial path successfully—I just think it needs to be said, no one should judge those who put their child's needs above their own career goals. There is nothing wrong with finding the position that allows you to be the parent you want to be, and there are ways to contribute to the scientific community outside of a research-intensive career path.

A final recommendation in this category is that you should ask for help when it is needed. Needing and asking for help doesn't make one any less of a success. I am by no means a supermom, and I readily admit it. I tried to be there for my son as much as possible, but there were times I needed outside support. My parents were always willing to step in and help with him when I couldn't pick him up from his bus for various reasons. My brother was an amazing role model for my son and would stay with him when I would be off at conference. I knew many wonderful mothers at the school he went to who would help get him to or from soccer practices when I couldn't swing both directions. There is no way anyone can do everything by themselves and hope to be successful at everything. Accepting my lack of super-powers has allowed me to take pride in what I could do and to truly appreciate those in my life all the more (Fig. 4).

Fig. 4 Michelle (far right) spending family time with her father, son, mother, and brother (left to right) in 2011

Pass It On

It might seem strange, but one of the biggest advantages in my career recently has been taking a strong role in mentoring and science outreach. When I realized a tenure-track position was not in my future, I decided I would further contribute to the scientific community by focusing the time that would have been spent on running a research group on mentoring students and being involved in science societies that emphasize science outreach and excellence. I serve as an advisor to newly declared chemistry majors and have informally advised a multitude of undergraduate and graduate students over the years. I have a personal goal of trying to help others find their way, perhaps a little more straightforwardly than I did. I certainly feel the need to "pay forward" what was done for me by those I was fortunate enough to encounter along the way.

In addition to wanting to help others, the involvement in what I would consider scholarly service has helped continue to push me forward in my own career. I started a local Women Chemists Committee in the Greater Pittsburgh Area 7 years ago, with the hopes of providing increased mentoring and networking for the generation that comes behind me. What I didn't realize at that time was that as I was focused on finding ways to help these women around me, I was being forced to focus on what steps would help one succeed—encouraging others around me to see their worth, to think outside the box, and to best prepare themselves to go after their dreams has kept this in forefront of my own mind. I love being able to work with the young women and men I have the opportunity to and find it very fulfilling, but it has also helped me to stay on top of the actions that will ultimately help me be more successful.

Final Thoughts

As far as my advice for those women considering a position in academia and successfully navigating the idea of a life-work balance, do what is right for you. Define your own path and make your own rules. Accept that nothing is perfect. Never sell yourself short.

Fig. 5 Michelle celebrating with her son Zachery at his high school graduation in 2013

Reflections from Michelle's Son Zachary Muscatello

My mother and I have always had a unique bond, and I attribute that largely to the way she raised me. Although difficult at times, watching my mother go through school taught me many valuable lessons growing up, not the least of which was perseverance. When it came to maintaining a work-life balance, my mom excelled. Incredibly, she managed to complete her schooling and research while also attending all of my school events, band concerts, and soccer matches. This impressed me growing up, but now that I am in college, I have a newfound appreciation for her level of dedication to both her work and family. Today, as a professor, she's brought that same level of dedication to the chemistry department at the University of Pittsburgh. She approaches her job with the same care and affection she has as a mother, and I've seen the positive impact she's made on students' lives firsthand. I couldn't be more proud of the woman I call Mama (Fig. 5).

Acknowledgments I would like to acknowledge those mentioned in this chapter for the lessons learned and continued support. I would also like to give special acknowledgments to my "work family" and the many students I have had the pleasure of working with during my time at Pitt. These instructors and young people give me every confidence in the future of STEM and society as a whole. It is truly an honor to have a position where the work I do never feels like a job. Finally, to my family, thank you for your support, patience, and love. I would never have the ability to do a fraction of what I do without you all. And finally, at this point in my career I have a true partner. Kevin, thank you for supporting me and the work I do. I had no idea how much that could mean until I finally found it.

About the Author

Education and Professional Career

1994	BS Chemistry, University of North Dakota, Grand Forks, ND
1996	BS Ed Secondary Education, University of North Dakota, Grand Forks, ND
1996–2000	High School Chemistry Teacher, Butler, PA
2003	MS Chemistry, University of Pittsburgh, PA
2008	PhD Analytical Chemistry, University of Pittsburgh, PA
2008–2009	Postdoctoral Research Associate, Department of Chemistry, University of Pittsburgh, PA
2009–2015	Lecturer and Analytical Laboratory Director, Department of Chemistry, University of Pittsburgh, PA
2015–present	Lecturer II and Analytical Laboratory Director, Department of Chemistry, University of Pittsburgh, PA

Honors and Awards (Selected)

2013	Choice Award from University of Pittsburgh College of General Studies Student Government
2013	Outstanding Service Award from University of Pittsburgh Department of Chemistry American Chemical Society Student Affiliates
2011	J. Kevin Scanlon Award for dedication in enhancing science education

Michelle is an active member in many associations, such as Pittcon, Society for Analytical Chemists of Pittsburgh, Spectroscopy Society of Pittsburgh, and the American Chemical Society. She is the founder of the Greater Pittsburgh Area Women Chemists Committee.

The Long and Winding Road

Gail Hartmann Webster

Photo Credit: Cavin and Stovall, Greensboro, NC

Starting the Journey

I can't remember a time during my precollege education when I didn't think that teaching would be a great profession for me. I was always the kid who enjoyed helping my classmates with their work or the children I babysat with their homework. I also remember getting very frustrated if I didn't understand something as quickly as I thought I should be able to. I can see all of these traits at play as I've

---

G. H. Webster (✉)
Department of Chemistry, Guilford College, Greensboro, NC, USA
e-mail: gwebster@guilford.edu

made my way from high school through college and graduate school and back to the classroom as a professor.

I was very fortunate to go to a great high school in Virginia. Our home was in Grafton, a small region in York County. This area is often referred to as the Tidewater area of southeastern Virginia, since it is located on a peninsula surrounded by the York and James Rivers. My teachers in the York County School System were excellent, and my classmates were too. At one point, I think I counted that there were at least five of us in my graduating class who went on to get PhDs. During high school I was continually challenged by my teachers and my classmates to push myself academically. I took two years of high school chemistry "back in the day" before Advanced Placement courses were all the rage. My high school teacher had previously worked as an industrial chemist, and I'm sure she let us do experiments in class that would never be allowed anymore. But, it was fun, it was challenging, and I enjoyed it. I liked chemistry so much, that I decided to go to college to be a pharmacist.

From Yorktown to Richmond: The College Years

I attended Virginia Commonwealth University in Richmond, Virginia, located about an hour away from my home. I completed two years of a pre-pharmacy curriculum and entered pharmacy school on VCU's medical campus in my third year of college. It wasn't what I expected. I sat in a room with about 100 other students, and the professors changed every hour or so for the next class. Occasionally, we had breaks in the day, and there were some labs, but what I remember most is sitting in the same room for most of the day and then studying like crazy at night for tests. When the summer came, I did a required externship (essentially an internship, but not supervised by the college) at a local pharmacy near my house. It was an awful experience for me. I disliked the job. My particular experience didn't give me hope that it would be a profession with much room for intellectual growth. I left pharmacy school and went back to VCU, but this time, I was an education major and a chemistry minor. It was during this time of transition that I met my husband, Jeff. He was working as a researcher at the medical campus of VCU, and he came to see me over that summer and helped me get through a very confusing time.

I took more chemistry classes than required for chemistry minor. I took the full year of physical chemistry that was not required, and I also did a year of undergraduate research, which was a major influence on my career path. My last semester of college was supposed to be spent student teaching. I say "supposed" because I was asked to leave student teaching early and head to Northern Virginia to interview for an unexpected opening for a chemistry teacher. A teacher left her position on disability, and the students were experiencing a revolving door of substitutes. I was doing well in my student teaching, so my faculty advisor recommended me for the position and suggested that I interview. I was offered the position and VCU let me leave my classes early that semester. I became a full-time chemistry (and physics) teacher. I graduated from VCU in May and finished teaching my classes in June.

My year of undergraduate research with Sarah Rutan at VCU was instrumental (pun intended) in my decision to become an analytical chemist. I thoroughly enjoyed working with her and the graduate students in the lab. Sarah encouraged me to present my work at meetings, and more importantly, she encouraged me to apply to graduate school. In the meantime, Jeff and I became engaged, and when grad school entered the picture, we decided to apply together (him for biochemistry and me for analytical chemistry), then see where we got in, get married, and move on with our lives. I think Jeff was even more insistent that I apply for graduate school than Sarah. A graduate degree was never in my original plans. Neither of my parents went to college, and just getting a BS degree was good enough for me. Looking back, I realize how lucky I was to have a professor take so much time and energy to help me realize I had the ability to get in and get through graduate school. I also know that I am even more fortunate that I found Jeff and his unwavering support.

From Richmond to Raleigh: Graduate School Years

Jeff and I entered graduate school at North Carolina State University two weeks after we were married. I'm not sure if I'd recommend that for anyone, but that's what we did. My lack of confidence resulted in my entering grad school in a master's program. Jeff knew he wanted a PhD, but I wasn't so sure. I found a great research group to join, but my advisor never had much grant money. All of the grad students in my group were teaching assistants, and for me, being a teaching assistant was the best part of graduate school. At NCSU, I had some wonderful mentors who knew I wanted to teach and supported me on my path. I remember Dr. Forrest Hentz telling me one day, "Gail, you've got to get your union card if you want to teach, and it needs to have three letters." Well, I decided that *if* I could get my MS, *then* I would stay on and get the "union card." I did, but it took me longer than most folks because I spent a good amount of time teaching, tutoring, and running a supplemental instruction program in addition to doing my research.

Meanwhile, my husband stayed focused in the lab, finished his degree, and started a postdoc at the University of North Carolina in Chapel Hill. He had some great offers for postdocs at some incredibly prestigious places, but since I wasn't finished yet, he stayed in the area for me. I am a lucky woman. At this point, I was getting anxious to start a family. I became pregnant, had a miscarriage, and became awfully discouraged. Several months later, I got pregnant again, and this time, there were no issues. I finished my last few experiments, had a baby, and wrote my dissertation. Caroline Webster entered the world the February before I graduated in May. Sometimes, I don't know how I ever finished. But then I remember the days of writing like a fiend while she was sleeping, which wasn't often those first few months, boy was she colicky. I also remember Jeff coming home from UNC, us having a quick dinner, and then him going back to NCSU with me at night, while I

Fig. 1 Gail's husband Jeff Webster, Gail, and their daughter Caroline (10 months, in her Wolfpack Red Christmas finery) at Gail's PhD graduation from North Carolina State University in December 1994

sat at the computer in the lab and worked. He would walk laps around the building with the baby, while I typed and did data analysis.

I defended my dissertation when Caroline was three months old. I nursed her just before I went to do my seminar and have my final oral exam, and she promptly threw up all over me. One of the departmental administrative assistants noticed the barf all over my shoulder and down my back and alerted me of the situation. I told her, "Well, maybe they won't keep me quite as long!" What could I do, really? (Fig. 1)

From Raleigh to Durham and Back Again: Going Where the Jobs Are

Just after Caroline was born, I saw an ad in our local newspaper for a teacher at the North Carolina School of Science and Mathematics (NCSSM). NCSSM is a public, residential high school for academically talented juniors and seniors. With Jeff in the middle of a postdoc, with pharmaceutical companies in Research Triangle Park downsizing, I was pretty limited about what I could do next. None of the big universities in the area would hire me to do anything other than teach labs, and postdocs in analytical chemistry in the area were scarce. So, I applied for the job and started teaching in the fall.

The Long and Winding Road   517

It wasn't easy being a working mom, with a postdoc husband and no family nearby. We bought a house a few years into graduate school. With a mortgage and daycare, money was pretty tight. I found a great in-home daycare for Caroline near my work place. I commuted about 45 minutes each way, so I had time in the car with her.

My job at NCSSM was a great experience. My students were engaged and fun to be with. My colleagues taught me more about running a general chemistry lab program than I ever learned in graduate school or from my teacher education program in college.

The difficult thing was what to do when either Caroline was sick or the babysitter was sick. NCSSM didn't have substitute teachers, I didn't get sick or vacation leave, and we had no family nearby to help out. Jeff was always the one who stayed home with the sick baby, and he'd go in at night or stay later to make up for his lost time. There were several times I had to take Caroline with me to school, and I often got some of the students to help me with her. I don't think my colleagues approved, but I did what I needed to do to keep my job and to help Jeff make progress in his work as well. I always kept my eyes open for other opportunities too.

I taught at NCSSM for two years. During this time, we wanted to have another child and I got pregnant again. Unfortunately, I had another miscarriage. We were sad and disheartened, but life went on. Each year that I worked at NCSSM, I applied for a summer program for high school teachers at the National Institute of Environmental Health Sciences (NIEHS), a division of the National Institutes of Health located in Research Triangle Park, NC. It was an opportunity for me to continue to be research active when I didn't have many other opportunities to be in the lab. I worked at NIEHS for two summers doing molecular biology.

In my second summer at NIEHS, my graduate research advisor called me and told me that NCSU had a job that had "Gail Webster" all over it. The chemistry department was hiring two teaching faculty to work in the general chemistry program and work toward reform in the chemistry curriculum. I wanted this job. However, I knew I was pregnant. I thought I should just let this opportunity go, and be satisfied where I was, even though I wanted to work at the college level with all of my heart. I didn't know what to do, and I remember talking to my mom about it. I clearly remember her saying, "Gail, it's 1996, women have babies and work. If you don't apply you'll never know if you can do it." So, I applied, I was interviewed while I was 12 weeks pregnant, and I was offered the job. I might add that I was offered the job even though I was told that the department chair said he would never hire an NCSU grad. As a part of the interview for the position, I also had to do a research seminar for a 100% teaching job, so my work at NIEHS was essential for me to be able to give a presentation on something other than my graduate work.

Working at NCSU was great. I was closer to home, Caroline was in daycare near Jeff's work (his advisor moved from UNC to NIEHS), and I loved teaching chemistry at the college level. Those professors at NCSU who had been great mentors during graduate school were continuing to be mentors at the beginning of my career.

When teaching schedules were being decided for the spring semester, I was in a bit of a quandary. I had to tell folks I was pregnant, since the baby was due at the beginning of the spring semester. When I told my direct supervisor, I will never forget what he did. He threw a pencil across his desk and said, "I knew this was going to happen." He told me I was responsible for finding someone to teach my classes. He was clearly angry and didn't give me any support. I was upset. Here I was, the first teaching faculty member to have a baby in the chemistry department at NCSU, in 1997. I asked to be given a night teaching assignment and told him I'd be back at work as soon as possible after the baby arrived. I was very proud of myself for keeping my cool during this conversation. As soon as our meeting was over, I went to my PhD advisor's office, walked in, shut the door, and told him my story, and I cried. He was so calm and said, "Gail, this is academia. We help each other. I'm happy to take your courses while you're gone." And so he did. I'll always be thankful to my advisor and mentor, Chuck Boss, for helping me when I needed it through graduate school and in my role as a visiting assistant professor.

Rebecca Webster was born in late January. I had two weeks at home with her, and then I started back to work teaching my evening class. I regret not having more time with her at home to this day. No woman should ever feel like they have to go back to work so quickly after giving birth. It was difficult, and Becky had jaundice soon after birth, so I was really worried about her health. It was not an easy start for us. I was fortunate that instead of teaching two classes, my load was reduced to one class that semester. I would arrive at work with Becky in a stroller, and Jeff would meet me in the lecture hall on his way home with our almost 3-year-old and take both girls home with him while I taught my class. I have to say, I had the most amazing group of students that semester. They sent me so many kind notes and emails, that I'm still touched by their kindness.

Later that spring I interviewed for the coordinator of general chemistry labs at NCSU, and I got the job. I moved into an administrative position and I wasn't teaching anymore. It wasn't my ideal job, but I needed something more than a one-year contract. Jeff was coming to the end of his postdoc and he was starting to interview for jobs. I realized that he had a greater earning potential than I did, and I was willing to move wherever we needed to be to take care of our family. At the end of that fall semester, we moved from North Carolina to Delaware for Jeff to begin his job at DuPont Pharmaceuticals (Fig. 2).

North Carolina to Delaware and Back Again

We moved in December, after I made sure the teaching assistants turned in all the grades for the many sections of general chemistry lab that I was ultimately responsible for. Jeff moved to Delaware before I did. I had to finish up the semester, get the house on the market, and get us moved. Not as easy as it sounds, but I did it. I had no job, but I was able to spend some great time with our girls. While I wasn't scouring the job listings, I was looking a bit. I found a temporary teaching assignment in

Fig. 2 Gail and her husband with their daughters Rebecca (age 5 months) and Caroline (age 3) at Corolla, NC, in the Outer Banks in 1994

Philadelphia, about 30 minutes from our home in Wilmington, DE. I worked there for a year, but with two children in daycare, the expenses didn't justify the work. I left after a year and was at home again with my girls.

The October after I left my job in Philadelphia, Bristol-Myers Pharmaceuticals bought DuPont Pharmaceuticals. Massive layoffs occurred, and my husband was out of work, and I wasn't working either. I felt awful that I did not have a full-time job to help us through such a difficult time. I found a part-time job tutoring, but I knew I needed to make sure I never put us in such a position again.

In April, Jeff started a job in High Point, North Carolina, at a small biotech company. I followed in June after Caroline finished second grade. Again, I got the house sold and got us moved back to North Carolina. I stayed at home the first year we were in North Carolina. Caroline was in third grade and Becky started kindergarten. I was able to volunteer in their school. I got in touch with some of my former colleagues when we moved back.

One of my former colleagues kept sending me job announcements and encouraged me to start applying. She sent me a job posting for teaching summer school at Guilford College, a small liberal arts college in Greensboro, very close to our home. I applied and got the job. I started teaching at Guilford College in the summer of 2003, and I've been there ever since, but it hasn't been easy.

I began working at Guilford as a visiting assistant professor in summer school. I let my colleagues in the department know I wanted more, and I was offered a job as a lab manager (part-time) and as a part-time instructor, teaching two classes the following fall. My two part-time jobs did not add to a full-time job, but I did it anyway. I did it because my colleagues in the department were awesome. They were supportive, they offered help, they listened when I had ideas, and teaching was truly important.

My teaching job in Philadelphia was the opposite. I honestly never thought I'd teach again after that year. I was told so many times at my former job that teaching was important, but when it came time to hire, the only consideration was research, which was not my strength.

After a few years as the part-time lab manager, Guilford College hired a full-time lab manager and moved me to a full-time visiting assistant professor. I worked as a visiting assistant professor until 2006. In all of that time, I never had more than a 1-year contract.

In August 2006, I began a tenure-track position. I changed my area of research from analytical chemistry to chemistry education From 2003 until 2010, Jeff was my anchor. He helped out with the girls more than anyone can imagine. He went to all of the programs at school. He took time from work when they were sick. I was never the one to stay at home with them or to take them to the doctor. He ran them around to more practices and early evening activities than most dads ever would. I missed Becky's kindergarten graduation, I missed both girls' fifth grade promotion ceremonies, and I missed Caroline's eighth grade graduation ceremony. I gave up a lot to prove that I was worthy of tenure. 2009 was a tough year for us. The biotech where Jeff worked started doing layoffs and he lost his job in January 2009. The recession was in full swing, and he didn't find a job until January 2010. The job was in Seattle, Washington.

Difficult Decisions: East and West Coast Living

Since I was waiting to hear about tenure, I didn't want to give up my job and move to Seattle. I stayed in North Carolina with the girls, now in tenth grade and seventh grade, and tried to keep things together while he started his new job at a contract research organization. It wasn't easy raising two adolescent girls, keeping up with their activities, working full-time, taking care of the house, and doing all of the professional things one needs to do as an academician. Jeff got back to North Carolina once a month to see us, and the first July he was in Seattle, the girls and I flew out and visited him for a month (Fig. 3). We never seriously considered moving across the country, although there were times I wanted to throw in the towel and move. It's a good thing we chose to tough it out. In 2013, the site he worked in was shut down, and he moved back to High Point.

Fig. 3 Gail's daughters Rebecca (left) and Caroline on a Lake Washington Cruise while visiting Seattle in 2010

Where Is the Next Journey?

In 2012, my oldest daughter, Caroline, went to college at a nearby university. Jeff moved back in 2013. Caroline went to college to study music, but she took some time off, and she eventually went back to college to study economics at Guilford. She had it especially tough when her dad moved across the country, and we still have some things to work through. My youngest daughter completed high school in 2015. She was dually enrolled in high school and college through a dual enrollment program with our local school system and Guilford College. By taking college courses in high school, Rebecca was able to take a wide variety of classes and discover what she loves: rocks (Fig. 4). Rebecca is in her junior year at Lafayette College in Easton, Pennsylvania. She is a geology and policy studies double major with a French minor. Both of my daughters work to help pay their way, and I'm proud of their hard work and their achievements in college.

Life has hit us pretty hard again with job-related issues. When my husband returned from Seattle, he found a job in Research Triangle Park, about an hour away, working with a pharmaceutical company specializing in animal health. In 2015, the company reorganized and did layoffs, and once again, we were affected. I've been the sole support for our family since then.

Reflections

I am thankful to be working at an institution that values undergraduate education. I cannot imagine having to put research above my undergraduate students. There's something very special about teaching at a small liberal arts college. It's a lot of work and there's not much administrative support, but it's a great job, and I have a

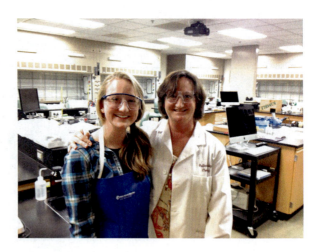

Fig. 4 Gail and her daughter Rebecca (age 17), who was taking chemistry at Guilford College as a junior in The Early College at Guilford High School, in February 2014

supportive network of colleagues at Guilford and across the country that I've met through various professional organizations.

There have been times that I thought it would be easier to walk away rather than work through the difficulties, but I'm glad I've stuck with it, even through the frustrations. Receiving one-year contracts year after year as a nontenure-track faculty member was demoralizing. Moving to the tenure track was my dream, but I didn't get tenure until I was in my late forties. I received my positive notice of tenure review while Jeff was in Seattle. It was bittersweet to read the letter and not have him with me to celebrate. I got tenure, but I did not do it alone. I was promoted to full professor in 2017. This time, everyone was living under the same roof, but unfortunately we were (and still are) experiencing another round of corporate "downsizing" and unemployment.

My husband and my daughters have been by my side this whole way, and I cannot imagine doing any of this without them. I couldn't have done it without Jeff, and there was never a time that I didn't imagine being a mom. So, while I got frustrated along the way because I wasn't getting "there" as fast as a lot of other women in chemistry, I have fulfilled two of my greatest ambitions: to be a professor and to be a mom.

Being married to a PhD biochemist and having two children limited locations where I was willing to work. I didn't want to have a long commute, so I was willing to take positions that could have been career limiting (like teaching in a high school). I made the best of the situations and learned what I could at every step along the way, and when I started my tenure-track position, I had far more experience than most, which ended up being very helpful.

My career path has made my family overly aware of issues surrounding gender equity in the workplace. My daughters will go into the workforce with their eyes far more open than mine were. I think that when my husband and I had to deal with my lack of maternity benefits at NCSU along with the negative words and actions by my supervisor, it made him keenly aware of the difficulties women scientists face in the workforce. As a group leader and manager of research scientists, he made sure that the women he supervised received positive support from him as a manager. My daughters see the long hours I've had to work over the years. They get pretty annoyed when folks outside of academia make remarks insinuating that I have less than a 40-hour work week and my "summers off." They've dealt with the nights and weekends I've spent grading, planning, going in to work to get projects finished, or traveling for work. I hope they realize that my accomplishments have only been possible because their father has been willing to be a hands-on dad, to cook, to clean, and to do anything else he could to help me along the way.

For women entering the world of academia, my only advice is to do what you love and make decisions for your life based on your circumstances. There are women in the professoriate that came before me who would consider having a child in graduate school the wrong decision. They would advise women to wait until the postdoc or even to wait until post-tenure to start a family. My obstetrician told me that you cannot plan when you're going to have a baby. It happens when it happens, and I agree. If you wait too long, the biological clock can work against you. I've seen that happen to some of my friends and colleagues. I didn't want to be a mom in my mid to late 30s, and if I had waited until post-tenure to start a family, I wouldn't be a mom at all.

My path to tenure was certainly a long one, and I've needed to step back, take a breath, and hope and pray that things would work out along the way. It hasn't been easy taking the long road, but I know I'm lucky that I have the support and love of a great family and that I get to go to work and do what I love every day.

Acknowledgments My parents, Joyce and John Hartmann, who are both deceased, worked their whole lives to give their children opportunities. I will be forever grateful for their sacrifices on my behalf. My husband and children have always been ready to do whatever is necessary to make sure I could move forward in my career. I would also like to acknowledge the Chemistry Department at Guilford College for the opportunity to teach and serve.

About the Author

Education and Professional Career

1987	BS Chemistry Education, Virginia Commonwealth University, Richmond, VA
1990	MS Analytical Chemistry, North Carolina State University, Raleigh, NC
1994	PhD Analytical Chemistry, North Carolina State University, Raleigh, NC
1994–1996	Chemistry Instructor, North Carolina School of Science and Mathematics, Durham, NC
1996–1998	Lecturer and Coordinator of General Chemistry Laboratories, Visiting Assistant Professor, North Carolina State University, Raleigh, NC

1999–2000	Visiting Assistant Professor, University of the Sciences, Philadelphia, PA
2003–2006	Visiting Assistant Professor, Guilford College, Greensboro, NC
2006–2010	Assistant Professor, Guilford College, Greensboro, NC
2010	Associate Professor of Chemistry, Guilford College, Greensboro, NC
2017	Professor of Chemistry, Guilford College, Greensboro, NC

Gail is an active member of the ACS, currently serving as the chair of the Central North Carolina Section and as a member of the Women Chemists Committee and the Division of Chemical Education. She serves as a facilitator for the Process-Oriented Guided Inquiry Learning (POGIL) Project.

I Finally Know What I Want to Be When I Grow Up

Catherine O. Welder

Photo Credit: Amy Chan Photography

What Do You Want to Be When You Grow Up?

Why is it that we are always expected to have an answer for what we want to be when we grow up? From a very young age, children are asked what they would like to become—fireman, teacher, or, my youngest son's response a few years ago, a monster truck driver. It starts with preschoolers. Then, as we hit the teenage years, career aspirations become more of a threat. My husband heard this at one point in his teenage years: "You don't want to dig ditches when you grow up, do you?" We face

---

C. O. Welder (✉)
Department of Chemistry, Dartmouth College, Hanover, NH, USA
e-mail: catherine.o.welder@dartmouth.edu

serious choices as we complete high school, and for those of us who go on to college, we are hit with another round of more ominous decisions a few years later. What, exactly, do we plan to do after graduation? A few people are fortunate to have a clear picture of where they are going next. Others seem to avoid the decision-making process as long as possible or take some time off, perhaps a gap year, as they decide where to go next. But the underlying question is always the same. What do you want to be when you grow up?

It wasn't until well into my grad school days that I discovered a love for teaching. I served as a teaching assistant for general and organic chemistry courses all the way through my graduate student years, all 5+ of them. While most of my responsibilities involved helping students navigate the labs, I also had one-on-one contact with students who needed additional help with the lecture portion of the course. It was during those office hours that I discovered I truly enjoyed helping people better understand complicated topics. I got a real thrill when a student finally had an "Ah HA!" moment.

I still remember the day I told my PhD advisor that I wanted to pursue a career in academics. I was in my third year of the program. His response was simple. "Good. I don't think you would like industry." While I have never held an industrial position, I think he was correct. I love teaching!

My dad was an industrial organic chemist. He served as the director of research at his firm. He was very excited when I took geometry in tenth grade and did well. "Great!" he said. "You will do well in chemistry." I didn't understand how he could jump to that conclusion. Now I see what he meant. The concepts you master when solving proofs in geometry are similar to the skills needed to master syntheses in organic chemistry. And being able to visualize in 3D starts with geometry and is crucial in organic. I went on to take chemistry in both my junior and senior years of high school and headed off to college with the conviction that I had found my major.

When I got to Wake Forest University, I could choose between a Bachelor of Arts (BA) or Bachelor of Science (BS) degree in chemistry. The tracks were the same for the first 2 years, so I didn't have to decide right away. Once it was necessary to make a choice, I noticed that the BS would be more work. Who wants more work? "I'll just get a BA," I thought. Well, when my dad caught wind of that, he told me to earn a BS in chemistry or pick another major! He assumed I would have difficulty finding a job with only a BA. He was probably right. The people choosing BAs in chemistry were mostly on a premed track, which was of no interest to me. So, BS it was.

Balancing School and Marriage

Among all this, Frank popped the big question. Would I marry him? We began dating as seniors in high school and went to separate colleges, seeing each other about once a month. This was back in the day before Skype, cell phones, and even e-mail. A month was an excruciatingly long period of time between visits. If we could make it through that, we could make it through anything, and I said yes. That was during the fall term of my senior year, the same term that I was applying to various graduate programs in chemistry. My dad was convinced that if I married, I wouldn't complete a graduate degree. I was convinced that if I had to choose graduate school *or* Frank, I would choose Frank, but there wasn't a need for this to be an either/or situation. We had been dating for 5 years, which seemed like an eternity, when we married. I needed his strength and companionship to have the endurance to survive a PhD program. Why not be married *and* be a graduate student?

As I began visiting chemistry departments in the spring semester of my senior year, I was thinking about the possibility of stopping at a master's degree. Again, it was about the work. A PhD program is so much more demanding, and I wasn't sure I had it in me. That's when my dad gave me another important tidbit of advice. "Don't apply to MS programs. You can always enter a PhD program and stop at a master's degree." Again, he was right. Graduate schools only have so many spots in each year's entering class. If the school has a choice between accepting a student who wishes to pursue a master's degree and one who wishes to pursue a doctorate, the doctoral student often receives the spot. I now know that many who begin a PhD program in chemistry don't actually complete it. I was told by a more senior graduate student during orientation to look to my right and to my left. Two of the three of us would not complete our PhDs. I thought, "Where are *they* going?" The prediction held true. About a third of my entering class completed PhDs in that program. However, most students left by choice, and schools vary drastically in attrition rates.

Frank and I married as soon as we had both completed our undergraduate degrees. I chose the Georgia Institute of Technology for graduate school, and, thankfully, Frank found a position as a chemist in an environmental lab nearby. Though we always wanted to have kids, we didn't want them right away. School was demanding and time-consuming and so was marriage. I remember our first year of marriage as particularly tough. We had discussed divorce in premarital counseling and had agreed that it would not be an option for us. I can't tell you how much freedom that gives us to be ourselves, knowing that despite the current situation or argument, the other will always be there.

Graduate school was certainly a challenge, but it was never overwhelming. (Then again, it's been a number of years since I was in graduate school, and perhaps I've blocked the worst of it from my memory.) I did well in my coursework and enjoyed my area of research. But I should mention a nugget of advice from my undergraduate research advisor regarding grad school selection. He advised me to select a school that had multiple research groups that I would consider joining. Excellent advice! You never know if a particular professor might be considering retirement, a position at another school, or if he or she is not accepting students due to the current size of the research group or the funding situation. I chose an advisor who worked in the area of organometallics. Historically, his group was half inorganic students and half organic students. I chose organic chemistry as my primary field of study only after another talk with my dad, during which he informed me that I would never find a job as an inorganic chemist. While there are inorganic jobs out there, it does seem that it's easier to find a teaching position as an organic chemist.

My biggest challenge in graduate school came in the form of finding balance, both within the program and outside of it. Within the program I needed to balance the time it took to study for classes and cumulative exams, to serve as a teaching assistant, and to make progress toward my research. At some point my research advisor questioned whether or not I was putting enough time into my research. I spoke with my husband and determined for myself that I was giving all I was willing and able to give. Either it was enough to successfully complete the program, or it wasn't. I couldn't give more. I needed to balance the daily grind of graduate school with extracurricular activities, too. I was very active in the church choir, handbell choir, and even the softball team. I needed the fellowship, friendship, and exercise. I needed the outlets and the support system, or I would not have successfully completed my PhD. Balance was difficult to achieve.

You're Thinking of Doing What?

One night, when my advisor was hosting his research group for dinner, his wife asked me what I hoped to do after graduate school. My advisor overheard the question. They had raised seven children of their own, and she had stayed home with the kids. I mentioned that I was thinking about being a stay-at-home mom, especially while my kids were young, and that I was considering homeschooling. The concept of obtaining a PhD degree and then being a stay-at-home mom was so foreign to my advisor that he literally didn't understand what I had said. He could not fathom earning a PhD and then not using it to launch into a wonderful, long career in chemistry. For his own clarification, he restated what he thought he heard, which wasn't even close to what I said. I did not correct him, and the topic never came up again. His reaction was very interesting to me. I remember it well as a young woman trying to figure out how pursuing a graduate degree in chemistry would fit in with raising a family. I did not see education as a waste of time or conflicting with having children and perhaps even homeschooling them. To me, it was again about balance.

Finally, after 5½ years in graduate school, I successfully defended my PhD thesis. It likely would have taken me even longer if two former group members had not remained in the department, offering me assistance as needed. The two of them served as a tremendous resource after my advisor retired a full year before I completed my degree. As his last student, I was alone in the lab for my final year. I feel that no matter how long it takes you to complete a PhD, that last year is the most significant in terms of data collection. Bitterness can set in as the degree seems so far away, yet you have invested too much to stop now. It had been a long road, and I was glad to be at the end of it.

Life After Graduate School

I had not thought much about what to do after graduate school. Frank wasn't happy in his job and was ready for a change, but he wasn't actively looking for a new position. I am thankful that I was able to find a short-term postdoctoral fellowship within the chemistry department at Georgia Tech. My new advisor was convinced that he would have more funding soon but did mention when he hired me that he only had funding for a 3-month appointment. I would be carrying out synthetic work on an NIH-funded cancer research project. Just after I was hired, the government shut down. That froze additional funding, and after 3 months my position was terminated. My advisor was devastated by the news and extremely apologetic. I, on the other hand, was slightly relieved. I had discovered a wonderful lesson. I don't like synthesis! It's tedious and slow, and I found it a bit boring. I was glad to move on to something else.

It turns out the "something else" was another postdoctoral position on the Georgia Tech campus within the Institute of Paper Science and Technology, a school designed to train graduate students in the science and engineering of papermaking. I would be studying ozone bleaching of wood pulp, a mechanistic project. After 6 months or so of pure research, I was given the opportunity to cut back on research hours and teach a survey organic chemistry lecture course to the first year graduate students in the paper science program. It was a Godsend! I taught the 10-week survey course that covered all the basics to prepare students for advanced courses in pulping and bleaching. I gained teaching experience at the collegiate level, a tremendous addition to my résumé.

As I completed my first year as a postdoctoral fellow, my husband and I began to seriously consider our next career move. He received an offer for a chemistry position at the bachelor's level in Louisiana. In the same time frame, I received an offer from my alma mater for a 1-year, nonrenewable teaching position. If we moved to Louisiana, we would try to settle down and start a family. If we moved back to North Carolina, I would have a temporary job, and Frank would work on his golf game. We weren't sure what to do, and we spent quite a lot of time in prayer and asked friends to pray for us as we made the decision. It almost seemed like a no-brainer to take the fulltime, permanent position in Louisiana, where my husband would be the primary breadwinner, and I could be the stay-at-home mom I had been thinking about all those years. But that wasn't God's plan for us. We had one weekend to decide between moving to Louisiana or North Carolina. I had been given an extension for making a decision at Wake Forest University, and Frank had only a verbal offer from his potential employer in Louisiana. We prayed that God would provide specific information to help us to make the best decision. He did. Within a few hours, Frank's written offer arrived with no mention of health benefits for his spouse or family. To me the whole point of accepting his position would be to start a family, yet dependent medical expenses wouldn't be covered? We considered this flaw to be the answer. Monday morning we accepted the position at Wake Forest.

530   C. O. Welder

A few months later, we arrived in a new role in a familiar city. I believe it was my second day in the office when a senior colleague approached me asking what Frank was doing. This professor wanted to hire Frank to set up, learn to use, and teach his research group to use a piece of equipment that Frank had never used.

Me: "You understand he's never worked on this type of equipment, right?"
Professor: "Right."
Me: "Yet you wish to make him an offer?"
Professor: "Yes. Well, I should probably talk to him first. Have him call me."

Frank started working for him that week. We hadn't finished unpacking, and Frank hadn't picked up a golf club.

The School/Marriage Balance, Again

Frank had great success working in the research lab. As a matter of fact, his boss was also the chair of the committee that determined who would be accepted into the chemistry graduate program. Once we learned that I would be at Wake Forest for the fall term, Frank applied to the graduate program in the chemistry department, but he was denied admission, likely because it was so late in the application process and all the slots had been filled by other applicants. His boss let him know that his application would be accepted for spring term (January) admission. No one had dropped out of the program. Why was there suddenly a spot available to him? (We view it as another Godsend.) So, Frank launched into his graduate studies only months after we moved to NC. He began graduate school during my first 1-year, nonrenewable contract. For each of the next 5 years, I received 1-year, nonrenewable contracts. Frank was able to complete not only his PhD in analytical chemistry but also a joint MBA from Wake Forest's business school while I taught.

Wake Forest, like many US colleges and universities, makes tenure decisions after 6 years of service. My position was not on the tenure track, so after teaching for 6 years, I knew that I would not be able to stay for a seventh. The department could not figure out how to extend my stay as a non-tenured faculty member. Frank would finish both degree programs, so we began asking the all-important question again: what do we want to be when we grow up?

Adding Kids to the Mix

By this point we had been married 13 years, and we still didn't have kids. Starting a family became more important and even somewhat urgent. As such, we decided that Frank would look for fulltime employment with the hopes that I would be home with a baby within a year or so. Frank was able to find a job in Louisville, KY. While it wasn't his dream job, he was more than qualified and accepted the position.

Soon after moving to Kentucky, I discovered I was pregnant with Luke. There were some complications during the pregnancy, and I was even hospitalized for 10 days when I was 11 weeks pregnant. After being released, I visited one doctor or another two to three times per week for the duration of the pregnancy and even beyond it. I don't know what I would have done if I were employed! For me, being pregnant with complications was a fulltime job. I think that it was another Godsend that we were in Louisville with an OB-GYN and consulting specialists who had much experience with cases such as mine.

Frank was dissatisfied with his job, and we missed friends and family terribly while in Kentucky. Once I received medical clearance, Frank began looking for a new job. He found a position that allowed him to work almost exclusively from home, and we lived in Kentucky for almost a year while trying to sell the house in preparation to move back home, North Carolina. In the meantime, friends from Wake Forest heard that we might be moving back to the area, and they contacted me to see if I might be interested in teaching there again. (Yes, you guessed it: a 1-year, nonrenewable contract.) I agreed and we moved back to the Winston-Salem area.

During my first year back, I became pregnant with our second child, Ethan. He was due in late September, and I faced another 1-year contract, so I had a decision to make. Did I want to start the fall term 8 months pregnant, take 6 weeks off for maternity leave, and then finish the fall term with a very young child in daycare? That was one of the easiest decisions I've ever made. NO! I knew from my first pregnancy that there would be many doctors' appointments in my future, especially in the third trimester. I also knew that at 6 weeks babies aren't sleeping through the night, which means parents aren't sleeping through the night. And I firmly believed that I could take better care of my 6-week-old than could even the best of daycares.

When deciding how much time to take off, I considered my own interests, as well as those of my students, co-workers, and replacement. I did not think it would be in the best interest of students to have me for the beginning and end of the 15-week term and a replacement while I was out on maternity leave. I knew a 6-week position might be difficult to fill and would not be appealing to strong applicants. So that the department would have plenty of time to carry out a search, I informed our chair during my first trimester that I planned to take the next year off. If I had it to do over, I would probably elect to wait until later in the pregnancy to announce my intentions. However, it all worked out in the end. I was able to stay home with Ethan for almost a year before accepting my final 1-year, nonrenewable contract (the eighth) at Wake Forest.

Searching for Some Sense of Longevity

During my last year at Wake, it became evident that I would always have the uncertainty and corresponding rank that a 1-year, nonrenewable contract brings with it. In the spring of that year, I reluctantly began looking at other schools. The department chair was well aware of my situation and asked me if I had seen a late ad in *Chemical and Engineering News* (C&EN), a weekly publication of the American Chemical Society, posted by Dartmouth College. I had not. They were looking for someone to direct the organic lab courses year-round and to teach the summer organic II lecture course. I had exactly the qualifications they listed in their advertisement. Most schools advertise chemistry positions early in the fall term, arrange November/December interviews, and then make offers for the following summer or fall appointments. Dartmouth placed the ad in February or March and held interviews through the end of April hoping for the candidate to start in June. If they had placed their ad the previous fall, I would not have seen it and would not have been considered for the position. God's providence was evident to us again.

So, here I am at Dartmouth. This is exactly what I want to be when I grow up—a fulltime teacher and mentor. I'm not required to write grants, though I'm certainly not discouraged from doing so. I'm not expected to carry out research, though any type of scholarly activity is a plus. I'm not facing a tenure decision, and I consider it a step in the right direction to have 3-year, renewable contracts. My official title is "senior lecturer." While not many schools have lecturer positions, they are becoming more popular, and some schools have created tenure-track teaching positions.

I teach year-round, but some schools have 9- and 10-month positions. I would prefer having summers off to enjoy with my children, now 9 and 11 (Figs. 1 and 2). Originally, Dartmouth labs were 6 hours long and began at 2 pm, which meant there were several nights in which I got home after my kids were in bed. But my colleagues were willing to help me address the concern of rarely being home in the evenings. As I negotiated the details of my offer, I asked if it would be possible for me to stay for the end of lab only two nights a week and for other organic faculty to supervise after 5 pm for additional lab sections. I never thought they would agree, but they did! Labs typically meet for 8 weeks each term, so that's only 32 weeks a year that I have what we've come to call "late nights." That's manageable. The kids completely understand that late nights typically mean mom won't be home

Fig. 1 Cathy's sons Ethan and Luke explore with magnetic stir bars and a stirring hot plate (and water, of course!) in 2012

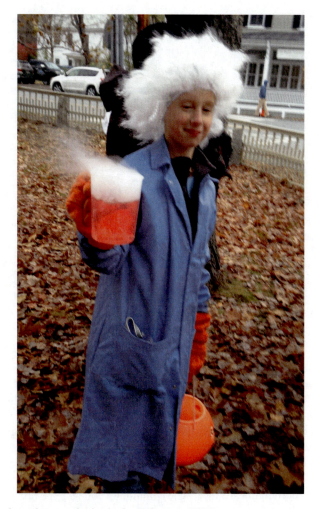

Fig. 2 Luke dressed as a mad scientist for Halloween (2016)

until after bedtime, though I must admit that I'm not too upset when students complete their experiments early enough for me to give my kids a kiss and tuck them in bed. Now the lab periods are only 4 hours long. Bedtimes are a little later as well, so I am able to spend a little time with my kids most evenings, even on late nights.

Steadfast Support from My Husband

For the most part, Frank is a stay-at-home dad. He shuttles the kids back and forth to various activities and is available as needed. Since my job requires me to be at work much of the afternoon and evening hours, Frank is limited in his job options. He

Fig. 3 Welder family in May 2013. Photo Credit: Amy Chan Photography

must be available to pick up the kids no later than 5:30 pm, the typical closing time for daycares and after-school programs in our area. A few years ago, Frank accepted a 1-year teaching position for general chemistry lecture and lab at a 4-year college located about 40 minutes from our home. That was his first teaching experience, other than serving as a teaching assistant in graduate school. He enjoyed it. The positive experience was enough to encourage him to apply to teach a general chemistry lecture course at Dartmouth College. He has since taught the course at Dartmouth a second time and is currently teaching a hazardous chemistry course in the fire science curriculum at a local state community college. The timing works well as his lecture courses are held in the mornings and my labs meet in the afternoons. That means that if the kids are out of school, one of us can be available to watch them without interrupting our work schedules too much. Since both kids are now in elementary school, I think Frank will be even more interested in intellectual pursuits. He has played a very crucial role in childcare, and I am grateful that one of us has been home for most of our children's preschool and elementary school years (Fig. 3).

So, for us, the balance of teaching and family has been rather successful. I feel that a career in academics, with a somewhat flexible teaching schedule and the possibility of having summers off, is much more attractive than a career in industry. Balancing two careers can be difficult, but Frank and I have managed each career choice so that one of us has the primary breadwinning role, while the other works and manages things on the home front. We have never felt that the pursuit of one career has come at the expense of the other, but I think that is because one of us has had an obvious career-advancing choice at a time when the other's job was stagnant (Frank) or temporary (me).

Reflections

It's tough to balance two demanding careers with family. It seems to work well when one person has flexible hours or part-time employment for the inevitable sick days or taxi service aspects of extracurricular activities. It's also tough to balance the demands of graduate school with those of being newly married. That first year of marriage, in particular, takes a great deal of work if the relationship is going to be successful for the long term. I gave all I could give toward completing a PhD and was fully aware that others may deem it insufficient. I was satisfied that I was doing my best to meet the various demands in my life.

I'm not sure that I would have done it differently, but waiting to have children has its pros and cons. A pro of waiting is that both my husband and I were able to independently complete graduate degrees without having to factor children into the mix. But, waiting to have children does have its cons. It's harder to become pregnant, remain pregnant, and to deliver a healthy baby as you age. Second, your peers likely had children years ago, meaning you can hang out with the 20 somethings who have children the age of yours, or you can interact socially with people your age who have children in middle and high school, perhaps even college when your children are in preschool and elementary grades. Again, I don't think I would change anything, but this is certainly something to carefully consider when younger. In fairness, though, there does seem to be a trend among highly educated women to postpone having children, so in the workplace there isn't such a gap between the age others began having children and when we did.

Each career decision was difficult at the time. I think it is rare to have job offers for both partners when a move is involved. Trust in and open communication with your partner are vital. I followed Frank to Kentucky hoping to either become pregnant and a stay-at-home mom or to find a collegiate teaching position. Frank followed me to New Hampshire and still hasn't found a fulltime job, but he has certainly strengthened his résumé in a new career path by teaching part-time at three local schools, including Dartmouth College (Fig. 4).

Advice

If I were advising students what to consider when selecting a program today, I would encourage them to consider various aspects of the degree program. How many classes are you expected to take? Is there a breadth requirement? (I had to be proficient in three of the five areas of chemistry in my department—analytical, organic, inorganic, physical, and biochemistry. Proficiency was achieved by either a high score on a standardized entrance exam or by taking classes in that area.) How often will you be asked to serve as a teaching assistant? (At many schools that often depends on whether or not your research advisor has external funding.) What are the additional requirements, such as cumulative exams,

Fig. 4 Cathy's sons Ethan (age 9) and Luke (age 11) visiting Dartmouth College's *Amorphophallus titanum* (corpse flower) just before it bloomed in September 2016

seminar presentations, and research proposals? You might even consider how long it takes the typical candidate to complete the degree program. Visit the school and talk with current students in the program, if you can. Of course, financials often play a factor as well. Are you selecting a field of study, such as chemistry, that supports its students with stipends? What is the cost of living in the area you are selecting? I believe several of the schools I was considering would have helped me meet my career goals.

As you begin making career choices, take comfort in the fact that there are often multiple pathways to success and happiness. Those who pursue and complete doctorate degrees often continue in a lifetime of learning. One of my most valuable lessons in graduate school was that I can now teach myself. I can read books and papers to learn more about a topic of interest. I can extend my knowledge by checking the references, and I can contact the authors for clarification. Choosing a career field does not necessarily limit future options. Instead, your degree and first job set you up for a lifetime of exploration. Collaboration with colleagues is becoming vital. You can pair your area of expertise with that of another and venture into fields currently beyond your imagination. I wish you the best as you find your niche, whether it's as a student, as a postdoctoral fellow, in a first job, or as a result of a career path change.

Fig. 5 Cathy with her husband Frank at Marsh-Billings-Rockefeller National Historic Park, Woodstock, VT in October 2016

> **Reflections from Cathy's Spouse**
> Frank says being a stay-at-home dad is the hardest job he's ever had. In the working world, we accomplish goals and are rewarded financially for our efforts. Recognition for a job well done makes us feel good. At home, there are fewer short-term goals, so the sense of accomplishment is less frequent. Children are less likely to reward a job well done with praise and admiration. They are, by nature, needy and immature. However, we feel that having a parent at home during these formative years is worth the sacrifice and effort. Frank says that he finds other stay-at-home dads, who understand his perspective and struggles, a great source of support (Fig. 5).

Acknowledgments I would like to thank three of my former Dartmouth students for reviewing this chapter and providing valuable commentary: Rebecca E. Glover ('11), Alexandra T. Geanacopoulos ('13), and Marissa H. Lynn ('13). In 2014 Rebecca completed her MSc degree in infectious diseases at the London School of Hygiene and Tropical Medicine (LSHTM) and is currently pursuing her PhD at LSHTM on the effectiveness of antimicrobial resistance interventions in the UK. Alexandra (Ally) is a third year medical student at the University of Pennsylvania and plans to pursue an academic career after a pediatrics residency. Marissa is a second year medical student at Harvard University and is interested in Internal Medicine and Pediatrics.

About the Author

Education and Professional Career

1990	BS *cum laude* with Honors in Chemistry, Wake Forest University, Winston-Salem, NC
1996	PhD Organic Chemistry, Georgia Institute of Technology, Atlanta, GA
1996–1998	Postdoctoral Fellow, Institute of Paper Science and Technology, Atlanta, GA
1998–2004	Visiting Assistant Professor, Wake Forest University, Winston-Salem, NC
2006–2007	Visiting Associate Professor, Wake Forest University, Winston-Salem, NC
2008–2009	Visiting Senior Lecturer, Wake Forest University, Winston-Salem, NC
2009–present	Senior Lecturer, Dartmouth College, Hanover, NH
2014–present	Member of the Organic Education Resources Leadership Board, www.organicers.org

Honors and Awards (Selected)

2013   Dean of Faculty Teaching Award for Visiting and Adjunct Faculty, Dartmouth College

Over the last 20 years, Cathy has taught a variety of undergraduate and graduate chemistry courses. Her academic interests include demystifying organic chemistry, developing new laboratory experiments, and training student teaching assistants as the next generation of educators.

Finding Rhythm

Leyte Winfield

Photo Courtesy of Furey Reid, Spelman College Media, 2016

Our forefathers came here as slaves and were able to create rhythm and they kept rhythm no matter what. Now, we came here as slaves, but we going out as royalty.[1]

---

[1] Miller, Percy (2016). Closing: The Chosen Ones [Recorded by Solonge Knowles featuring Percy Miller]. On: A Seat at the Table. Saint—Columbia Records, New York, NY.

L. Winfield (✉)
Department of Chemistry and Biochemistry, Spelman College, Atlanta, GA, USA
e-mail: lwinfield@spelman.edu

My Journey Begins

My path has been purposeful and represents not only my talent but my resilience. More importantly, my path has been ordained by God. It is my faith, more than anything, to which I attribute my success as a scientist and as a mother. My story is not one of perfect timing or a narrative of what I did wrong or right. This is not a story of the ideal path that should be followed by others with similar career aspirations. This is a story about my journey, my determination to be a successful mother and professor, and my path to finding rhythm.

"I was just a little girl, skinny legs a press and curl..." [1] Lauryn Hill ended the statement with, "My mamma always thought I'd be a star" (Fig. 1). But, my mom thought I would be a math or computer teacher. I was very analytical as a young person. I remember holding strategy sessions with friends over how to make the best mud pie. Everything had to be measured perfectly for texture and symmetry. In middle school, my interests ranged from developing lotions that would last longer than the walk to the bus stop, creating acne remedies that worked before the beginning of the next school term, mixing various colors to create the perfect hue of red, and evaluating the effectiveness of stain removers. All of my early projects were epic fails.

Fig. 1 Leyte, her parents and son, following her PhD graduation ceremony in New Orleans, LA 2002

Nevertheless, my emerging love of science was evident. By high school, my passion for science focused on chemistry. I attended a medical magnet high school which offered a rich and challenging math and science curriculum. The opportunity to shadow various doctors in a hospital setting was also appealing. However, during my first week of shadowing, I fainted while observing Norplant (levonorgestrel-based implants used for birth control) rods being inserted into a patient's arm. That experience killed any interest I may have had in medical school. But, I continued in the high school program, willing myself not to faint during a cesarean section, leg amputation, and motor cycle accident. During a career day presentation, a speaker described his job creating new medicines and improving old medicines. He was a chemist—a pharmaceutical chemist, to be exact. His presentation was fascinating, and I learned that there were different types of chemistry. Most importantly, I learned that chemists make consumer products like those of my earlier experiments.

Inspiration for the Journey

I can't say that I had individuals who modeled for me what it was to be a chemist. However, there were several who inspired my success. Madame C. J. Walker (born Sarah Breedlove) contributed significantly to the consumer market related to African American hair care and beauty. Like myself, Walker is a Louisiana native whose interest in science was informed by her desire to create something that could impact one's way of life and how a person presents themself to the world. It has been said that a severe case of dandruff prompted Walker to seek the training that would allow her to develop a line of shampoo and hair cream products. During my youth, I was painfully aware that many of the skin products that were within my family's budget did not provide the desired benefit. I longed for a lotion that would outlast my wait at the bus stop and an affordable lipstick that would not cause my lip to peel. Knowing Walker turned her similar concerns into a profitable industry spurred my initial interest in science.

While in high school, I joined an organization called Project Success that was hosted at Southern University in Baton Rouge, LA. The program provided development and college preparatory assistance to youth from at-risk populations. In addition to weekly tutoring and Saturday development sessions, the program offered an intensive 4-week residential summer program. It was there I met my first mentor, Mrs. Latimer.

During the program's summer session, I got in trouble for attempting to be provocative. I was terrified that my mother would be notified and I would be removed from the program. Instead, my penalty was to work with Mrs. Latimer, while others enjoyed free time in the afternoon. I did not know what to expect. Would she be angry and would her response to my behavior reflect that she expected all of the students in the program to fail at some point? Would she be disappointed, fearing that I would grow up to prove all of the negative stereotypes assigned to our community? Or, would she just be irritated by the stupid mistake I made? She was none of that. She was very firm, but patient and instructive in addressing my behavior. She acknowledged my actions and then proceeded to tell me why I was better than how I had behaved. Instead of discipline, I received encouragement from her and she expressed confidence that I would make better choices in the future.

I later discovered that she also spoke with my pastor. "Leyte," he said, "I hear you have been conducting yourself like a real lady on 'the yard' (slang for Southern University's campus). I am proud of you." With his expression of pride I realized that Mrs. Latimer's words were not empty musing, but heartfelt sentiments regarding who I was and my potential. Her confidence in me enhanced my opinion of myself and motivated me to stay on a positive trajectory toward success.

Throughout my undergraduate and graduate career, Mrs. Latimer was always a source of encouragement and affirmation. Regardless of what was happening in my life, she never addressed my problems by addressing what was wrong with the situation, but by addressing what was good in me.

One of our most meaningful conversations occurred after I became a single mother, and it was probably the shortest conversation we ever had. I was pretty distraught about the prospect of raising my son alone. She spoke in a hushed voice, ever stern, but so gentle that my tears began to dissipate. In the husky voice I had grown accustomed to, she said, "Anyone that can look at all that you are and all that you are becoming and walk away does not deserve a space on this journey." After the conversation, I have never allowed myself to be limited by the fact that I am a single mom. Mrs. Latimer continued to be there to provide guidance throughout graduate school. Without a doubt, it was her support through the years that reassured me of what I could accomplish and urged me to stay on course, even when I doubted myself.

The Road to Academia

With a goal of becoming a chemist, I set my sights on attending Dillard University in New Orleans and someday becoming a cosmetic chemist. When I met my advisor, I immediately alerted him to the fact that I was going to be a chemistry major so that I could work at the Pond's Institute. That statement marked the beginning of my journey to becoming a research chemist.

The first and second year of college courses seemed straightforward. In fact, I was able to get by on raw talent. If I went to class, I could pass the test without reading the book or studying.

Then came organic chemistry. That, of course, was a very sobering experience. Not because I did not have the talent to successfully complete the course, but because I didn't have the focus and had not cultivated the study skills needed to conquer the complexities of the subject.

My focus was further limited by the fact that I had a boyfriend at the time. He lived on one side of town, and I went to school on the other side of town. I realized too late that I could not be in both places at the same time.

Because of my lack of focus, I failed nearly every course that year and lost my scholarship. This was indeed a difficult time in my journey. I thought the failure would kill my career aspirations. Eventually, I was able to place the situation in perspective. I realized failure is part of the process and identified ways to move beyond the failures. I reminded myself of what I was trying to accomplish, got back onto the proverbial bike, and continued to push forward toward my goals.

But, before this, I had to explain to my mom how I managed to lose my scholarship. "This was my 'but God' moment. The moment when you look back and understand that God's plan for your life is perfect," I said to my mom with sincere conviction. "There is no good thing God will withhold from me. He makes the crooked way straight. He loves me so much that despite my mistakes, He still has a plan for my life. It may be a little harder. But, I didn't destroy His plan for my life." Yes, that was the speech I prepared for my mother who, consequently, wasn't interested in my testimony or my renewed faith in God. She was disappointed, but amazingly calm in her response. "I guess you better figure something out, like how are you're going to pay for college. Oh, and you can't come home."

That's when I became "army strong." I joined the army reserve and received an ROTC scholarship that covered my tuition. However, because of the choices I made, I had to negotiate the additional demands of being a reservist and an ROTC cadet while completing 21 course hours each semester so that I could graduate in 4 years.

Despite the turmoil of the failure and the hectic schedule, I was able to participate in an undergraduate research project related to white rot fungi. This was instrumental in my decision to pursue a career in research. As I shared this interest with my professors at Dillard University, they were able to help me select an appropriate career path and graduate program. Their support and encouragement led me to the University of New Orleans where I subsequently received PhD in organic chemistry.

Learning About Resilience and Determination from My Son

During my first year of graduate school, my son Acie was born. He is amazing. He is terribly handsome and has a style and personality that reflects maturity beyond his years. I'm sure all mothers feel the same (Figs. 2 and 3).

By the age of 3, he had survived eight hospitalizations related to asthma and pneumonia and an unfortunate accident that severed his Achilles tendon. He had no hearing in his left ear until he was 4 years old, which ultimately led to a speech impediment and other developmental delays. The worst was having to stand by and watch as he struggled near death to battle respiratory syncytial virus, a disease which few toddlers have survived. But, he has not been in the hospital since he was 5 years old and has not needed asthma medication since he was 7. Although the doctors said he would never be able to run or jump after his Achilles injury, he is now living a full life having played football, basketball, and lacrosse.

My son's illnesses gave way to academic struggles. He was diagnosed with dyslexia in the third grade. At my last parent-teacher conference, his principal and

Fig. 2 Leyte and her son Acie, age 8, at the Annual Meeting of National Organization for the Professional Advancement of Black Chemist and Chemical Engineers (NOBCChE), Los Angeles, CA. Photo Courtesy of NOBCChE

Fig. 3 Leyte and her son Acie, age 17, at the Spelman College 2015 Opening Convocation, where she received the Vulcan Chemical Company Award for Excellence in Teaching. Photo Courtesy of Spelman College Media and Technology 2015

a teacher spoke of his focus and his desire to always do his best. They went on to describe him as a great classmate, always willing to help, and having a genuine and kind spirit. In light of his academic difficulties, I was excited to hear that he maintains this type of rapport in an academic environment.

My friends find my son charming. He is helpful, respectful, and caring. I am impressed by his ability to engage them in various conversations. He is so much more comfortable with himself than I was at his age. When he gets into the minor mischief that boys encounter along the way, my friends remind me that he is a great kid. And, he is (minus the teenage tendencies, LOL). Despite all my son endured, he is incredibly well adjusted and undaunted by the fight. His outlook is always optimistic. Even when I'm overwhelmed, he simply says, "It's okay mommy." When I think of him, I'm reminded of the promises God has kept and His ability to bless me in spite of myself.

> **Safety as a Female Chemist**
> I would venture to say that a synthetic laboratory is not an ideal environment for an expecting mother. When I began my graduate program, I was 3 months pregnant. My heart was set on joining a group engaged in cocaine research. The group utilized raw cocaine in the synthesis of several molecules and other reagents that could have adversely affected my pregnancy and my son's health. Although I joined the research group, I did not begin research until my son was 6 months old.
>
> I believe that there are measures that will allow women to engage in research during pregnancy. Each woman has to make this choice for herself and under the advisement of a physician. When to return to the lab environment should also be considered for breastfeeding moms.
>
> Following appropriate safety guidelines should ensure that children are protected. I always wear a lab coat and gloves, and I do not wear my lab shoes home. Such precautions reduce the possibility of inadvertently exposing children to laboratory chemicals.
>
> Lastly, "bring your child to work" events are important. However, Acie would join me in conducting outreach experiments versus visiting me in the laboratory while research activities were being conducted. This allowed him to see me at work without being exposed to harmful chemicals.

Mom, Science Is Your Thing!

In his youth, my son enjoyed accompanying me to do outreach activities. He was always energetic and eager to assist with the manual tasks of packing items for demonstrations and setting up the display. He often served as my assistant during events and interacted well with the participants. Most of the activities were of interest to his age group, ranging from the creation of slime from school glue and ice cream using liquid nitrogen to crushing soda cans with steam and generating energy from bicarbonate to propel small watercrafts. This was cool stuff for kids under the age of 12.

I was certain the activities were creating in him a strong desire for science. People would often comment to him, "You have an awesome mom!" or "You must be proud of your mom," to which he would enthusiastically agree. Because of this, I was caught off guard and truly devastated when he first expressed dislike for science.

For weeks we prepared for his third grade history project. The topic was George Washington Carver, an African American scientist, of course. Together, we explored the many inventions derived from peanuts. Next, we selected a wardrobe so that he could look the part. We found a rubber apron, staying authentic to the time, and large goggles. He gave an exceptional presentation. I was so proud and filled with confident anticipation when the teacher asked, "Acie, are you going to be a scientist like your mom?" A sudden hush filled the room as he responded, "Not on her life." As one might imagine, I was crushed.

Making matters worse, he specifically hates chemistry. He just came right out and said it one day. It probably had something to do with my explanation of solids, liquids, and gases. Sometime after the presentation, he asked me to explain the concept. I rattled off some textbook-correct explanation about the rate at which particles move before he interrupted. My then third grader said, "That's not what my teacher said! She said you can hold a solid, you can pour a liquid, and a gas just fills the air." He said it as if he was disappointed by my explanation and questioned if I really knew chemistry.

On a brighter note, he recently expressed that he was proud of me, but that science was my thing and not his. As his talents come into view, I see that he is a very creative individual. I am impressed by the objectivity he displays in all areas of his life and by the clarity and commitment with which he pursues his chosen career.

Having a Family Requires Work-Life Harmony

Equilibrium is defined as a state in which opposing forces or influences are balanced. Although the academic scientific enterprise prides itself on understanding these forces in laboratory systems, it has fallen short in recognizing and remedying the imbalance between the demands of career and those of personal life. This imbalance contributes to extreme professional sacrifices and lowered retention of many talented individuals striving for successful careers in science [2].

During my graduate school years, I found myself torn between parenting, research, and military obligations. There was a sincere effort to pursue work-life balance on my part. When I was not being a mom, I was working diligently to complete my studies. My son and I enjoyed trips to the park and the children's museum as often as possible. But, graduate school did not provide much time for such leisure.

While many of my classmates and professors were very understanding of my being a mother, one classmate in particular directly supported me as a mother during this time. The individual would pick up my son from preschool so that I could have a little extra time in the lab. Others were very accommodating when Acie made an appearance at study group.

Mixing mothering with studying catalyzed my passion for mentoring and helping others succeed. I took a keen interest in the success of other African American students. Each year as the next one entered the program, I became their unsolicited mentor. One jokes that I stalked her to introduce myself and in the same setting proceeded to evaluate her 5-year plan. I believe this was a defining moment in my life as I eventually became a college professor.

My dissertation project involved a particularly difficult synthesis and purification. To allow me time to focus on the project and to complete a 3-month required military training, my son went to live with my mom and step-dad for a year. Although I was conflicted about the decision, I was able to be in the lab 12–18 hours a day while he was away. The time allowed me to work through complex compound purifications and pass my qualifying exams.

When Acie returned to New Orleans, I implemented an 8-hour workday that allowed me to balance being a mother while completing my degree. I defended my dissertation when he was 4 years old. His birth heralded some intense moments early in my career. But, I truly believe that the experiences taught me lessons that I would not have learned had I not become a mom. Without this knowledge, I would not be the scientist or professor that I am today.

My PhD advisor was not sentimental, although nurturing and encouraging. So I was humbled when he later expressed his pride in my success and the tenacity with which I pursued my studies while being a single parent. When he visited Spelman my first year as a professor, I expected my gentle-but-firm research advisor. Instead, he presented himself as a proud mentor. He reminded me of the difficulty of my dissertation work stating, "When you isolated that compound, I knew you had the talent needed to succeed." He was sincerely confident of my future success, which eased some of the anxiety I was having about tenure at that time. I will be forever grateful for that moment of encouragement.

Following graduate school, I began a postdoctoral position at Florida A&M. Moving to Florida as a single mother did not produce trepidation for me. Call it youthful naivety. My mom struggled with the fact that I would be miles away with a child without family or friends. But, I thought, "I will find an apartment, find a school for my son, go to work. How difficult could that be?" The difficulty of the transition to Florida and subsequently to Georgia did not become apparent to me until my son's 16th birthday. Reflecting I thought, God has truly been with us, keeping us and guiding our path. Prior to this revelation, however, I accepted the position at Spelman College. So with the same youthful naivety of "how difficult could that be," I packed up my son and moved to Atlanta 9 months after arriving in Florida.

I didn't enter the professorate fully aware of where the journey would lead. But, I became a quick study. The demands of seeking tenure made life pretty routine, leaving neither time nor energy for anything else. Spelman gave me the freedom to be a great researcher, a nurturing teacher, a committed mother, and nothing more. My initial schedule consisted of working on campus from 8 am to 5 pm, leaving work just in time to reach Acie's school before 6 pm, at which time aftercare ended. After a quick dinner, homework, and other nightly rituals, I was able to work at home from 9 pm until well after midnight. The schedule resulted in my productivity, no doubt, but it also produced an incredible void in my life.

My first COACh[2] meeting was amazing and a much needed intervention for balancing my life. Not only did the meeting provide insight on leadership and strategies for career progression, but it also addressed self-care and provided strategies for prioritizing one's personal life. I left the meeting with a renewed sense of self of not Leyte—the professor, but Leyte—the woman. Since this time, I have been committed to sustaining the harmony between my career and my personal life. Harmony never means an equal split in every season for all aspects of life. Nevertheless, I strive to give a solid 8 hours to work daily. I attend all of my son's events. In addition, I make time for self-care by exercising and participating in activities that are meaningful to me. I am a sister and a friend, a mother and a woman. I am very purposeful in nurturing every aspect of who I am. There are moments when working until midnight is required to finish a proposal or publication. But, for the most part, I focus and work hard for 8 hours a day. What is not accomplished during this time is rescheduled as another day's tasks. That being said, I have found that scheduling my day, allotting a specific amount of time to a given task, and sticking to the schedule helps me stay organized and productive. I believe I am more productive now that I pursue harmony between my career and my personal life than I was when I was primarily focused on my work obligations.

Inspired by My Students, Advice for Future Academicians

Once I was away from work for medical reasons. During the 2 weeks, it appeared that my students missed me terribly, or so I would like to think. Upon my return, one student greeted me with an urgent request for assistance on an assignment that was due in 15 minutes and a second with a desperate inquiry for tuition assistance. The third individual stopped by to remind me that she decided chemistry wasn't for her, but that she intended to visit me as often as she could. I was happy that she was able to identify what's best for her life without the overwhelming sense of guilt and failure that often plagues students when they decide to change their major. As she left she said energetically, "I knew you were not feeling well the last time I saw you. You just didn't look right. But, I'm glad you are back." This particular student had an old and loving spirit. The kind that is rare for someone her age. Her decision

---

[2]COACh is an organization founded by Geraldine Richmond dedicated to developing the leadership skills of female scientists.

seemed effortless and her spirit was at ease regarding her future success. It was fitting that she was one of the first individuals I saw that day. Her visit made me ponder the inspiration behind greatness, the formula for exceptional success, and whether greatness is an inherited trait or a learned behavior. I resolved that greatness is fueled by interest, talents, preparation, and opportunity.

Not only did I ponder the formula for greatness, but I also pondered the extraordinary responsibility I have been given in preparing students for the journey ahead. My students represent a range of abilities and talents. I don't take for granted my role in preparing them to use science to pursue goals and demonstrate excellence throughout their career. I have worked with students who were "gifted." These were born to fly and I simply had to show them how to use their wings. In some cases, the students were flying in the wrong air, and it was my job to direct them to the proper space to take flight. On the other hand, I have also worked with students who have yet to learn that they even have wings.

But, if the measure of my success as a teacher or mentor is only based on what I have done to help gifted students to refine their talents or find their way, then I have truly failed. For I would rather my success be measured by those who did not know they had wings, but with my help learned to fly. With that, I would say to young professors to give yourself passionately to your career. Don't just do the checklist; pursue those activities that will add meaningfully to your portfolio and development.

Fig. 4 Celebrating with students being honored by the Council of Undergraduate Research at the annual Posters on the Hill event in 2011

Don't forget to apply the same focused determination to cultivating a vibrant personal and family life. You can have it all if you recognize that you can't conquer it all in one day.

Finally, I love being in a position to nurture and influence future scientists. Working hard allows me to be the best possible example for my students. My hope is that in seeing my success, my students will be inspired to excel in their own right (Fig. 4).

Acknowledgments I would like to thank my mom who has been my biggest supporter in all things and my son who is my biggest fan. My success as a mother would not be possible without my village, those who lovingly assisted with Acie's care and nurturing, allowing me to be fully engage in my career: the Brothers of Kappa League (Bros Anderson, Brown, Gooden, and Oce), Angela Aker, Monica and Anthony Cooley, LaSynge and Richard Guyton, Marionette Holmes, the Hudsons, Marta Dark, Tasha Innis, Myrna and Edward Morris, Mary Morton, Lakesha Nelson, Florastina Payton-Stewart, Christine Washington, and Katrina and Vlad Williams. I would also like to thank Gloria Thomas for reviewing this chapter.

About the Author

Education and Professional Career

1994–2009	United States Army Reserve
1997	BS Chemistry, Dillard University, New Orleans, LA
2002	PhD Organic Chemistry, University of New Orleans, New Orleans, LA
2003	Postdoctoral Research Associate, Medicinal Chemistry, College of Pharmacy and Pharmaceutical Sciences, Florida A&M University, Tallahassee, FL
2003–2008	Assistant Professor, Department of Chemistry, Spelman College, Atlanta, GA
2010–present	Associate Professor, Department of Chemistry and Biochemistry, Spelman College, Atlanta, GA
2011–2012	Vice Chair, Department of Chemistry, Spelman College, Atlanta, GA
2012–2016	Chair, Department of Chemistry and Biochemistry, Spelman College, Atlanta, GA
2014	Post-Graduate Certification, Academic Leadership, The Chicago School for Clinical Psychology, Washington DC
2017–2018	Interim Associate Provost of Research, Spelman College, Atlanta, GA
2017–2018	Visiting Associate Professor, College of Pharmacy, University of Michigan, Ann Arbor, MI

Honors and Awards (Selected)

2018	Scholar in Residence, Rockefeller Foundation, Bellagio Center
2016	Women's International Study Center Fellow in Residence
2016	The Council of Independent Colleges and the American Academic Leadership Institute Senior Leadership Academy Fellow
2015	Vulcan Materials Company Teaching Excellence Award
2014, 2013, 2010	Scholar in Residence, Department of Chemistry, New York University
2013	National Science Foundation Opportunities for Under Represented Scholars Program Fellow
2011	American Association for Cancer Research Minority-Serving Institution Faculty Scholar in Cancer Research
2009	Presidential Award for Teaching Excellence—Junior Faculty, Spelman College
2004	Iota Sigma Pi Honor Society for Women in Chemistry

Leyte is actively involved in chemical education research and the development of curricular resources that enhance student agency and engagement in active learning environments. In addition, she leads a project for the design of small molecules that target reproductive cancers. She has served on national and local boards and has held several directorships.

References

1. Hill L (1998) Every ghetto, every city [Recorded by Hill, Lauryn]. The miseducation of Lauryn Hill. Ruffhouse Records, Philadelphia, PA
2. Richmond GL, Rohlfing CM (2013) Work-life balance. Chem Eng News 91(19):3

Constant Refinement: Learning from Life's Experiments

Kimberly A. Woznack

Photo Credit: California University of Pennsylvania

Scientists are some of the most curious people. They are curious about the way small intricate things work, and they are interested in why they work that way. They want to solve problems and make predictions. Scientists are observant, and most want to have a positive impact on students and society in general. As of 2018, I've been a chemist for the last 21 years, a professor for the last 14 years, an advocate for women in STEM for 13 years, and a mother for the last 10 years. I can say that I am very happy with all of these roles. I cannot imagine my life without any of these parts.

K. A. Woznack (✉)
Department of Chemistry and Physics, California University of Pennsylvania, California, PA, USA
e-mail: woznack@calu.edu

© Springer International Publishing AG, part of Springer Nature 2018
K. Woznack et al. (eds.), *Mom the Chemistry Professor*,
https://doi.org/10.1007/978-3-319-78972-9_41

I don't have all the answers, but I can share some of my observations and the results of my life full of learning experiments and experiences.

Preparing for Experiments

I grew up in Upstate New York. My dad is an optometrist and, while my two brothers and I were very young, my mom stayed at home with us. When my dad began his own private practice, my mom transitioned into being his full-time office manager, and she made her own schedule. (This schedule flexibility is very useful when it comes to being a parent. While the class schedule of a faculty member doesn't allow for infinite flexibility, I am often glad that I can schedule appointments for my children, during regular business hours, at a time when I don't have classes or office hours.) My parents always prioritized our education and gave us educational toys to play with. In both elementary and middle schools, I was part of some accelerated classes, and my involvement in advanced science courses set me on a path towards academic success.

In high school, I decided I wanted to become a biologist. I was fascinated by human anatomy. My high school chemistry experience wasn't as exciting. I was enrolled in the "research" chemistry course, and we designed our own experiment and wrote up a paper about it. In order to accomplish this, many regular experiments were cut. So, my love for chemistry did not yet to come to fruition.

I was not even aware at the time how fortunate I was to grow up in an upper-middle-class environment with an amazing school district. Many of my best friends from high school went on to obtain graduate degrees in science. To this day, I am grateful to have benefitted from attending a school, which offered so many opportunities.

During high school, my youngest brother was still a toddler. I have to say that watching how much work it was for my parents to balance their work and home lives made me think that I wasn't interested in having children for a very long time.

During high school, I also met the love of my life, who is now my husband, Ray (Fig. 1). We had mutual friends and similar taste in music. We were never in the same classes, but we enjoyed spending time together. Ray graduated before me and went to a local community college for their automotive repair program.

Experimenting with Majors and Observing Undergraduate Faculty Mentors

I attended Hartwick College in Oneonta, NY. Hartwick was only an hour away from home, and many of my friends from my hometown of Guilderland also enrolled at Hartwick. Like all devoted biology majors, I studied for general chemistry I and II

Fig. 1 Kim and Ray at a post-prom party in 1992

diligently. Chemistry was a far more logical science than it had appeared in high school. Studying with my lab partner for the chemistry exams really helped me see how much I liked teaching. She was really struggling with the material, and I wasn't. When I explained it to her, it reinforced my own understanding. I earned the Chemical Rubber Company (CRC) Freshmen Chemistry Award and my own copy of the CRC Handbook. My interest in chemistry was piqued. Meanwhile, my interest in biology was beginning to wane. I took courses in genetics, molecular biology, ecology, and microbiology. With the exception of perhaps developmental biology, these courses didn't seem to excite me.

For a while, I was a double major in biology and chemistry, and I took the courses needed for both. Dr. Bill Vining, who was my chemistry advisor, had an enthusiasm for chemistry that was contagious. In my junior year, I was able to attend the Spring American Chemical Society (ACS) National Meeting with the Chemistry Club. Several of my friends were presenting posters of their research. I will never forget walking into the New Orleans Convention Center at the first meeting and realizing that all of these people were chemists. I didn't know that many chemists previously and, to look through the program and see all these different names, I was floored.

I was encouraged to do a summer Research Experience for Undergraduates (REU) program and found this to be really interesting. I participated in the National Science Foundation (NSF) Solid State Chemistry REU program. This program began with a weeklong workshop on solid-state chemistry at Binghamton University, State University of New York (SUNY), and then participants dispersed around the country to complete projects. At the end of the 10-week experience, the participants returned to Binghamton University to give an oral presentation summarizing their work. My summer work was in the Department of Materials Science and

Fig. 2 Kim meeting Glenn T. Seaborg at an American Chemical Society National Meeting in 1997

Engineering at Cornell University. Completing the REU solidified my desire to go to graduate school in chemistry. With the dual major in both biology and chemistry, I was only able to earn a BA in Chemistry. If I dropped the biology courses my senior year, then I would be able to complete a BS in chemistry and enhance my chances of getting into graduate school. It was decided, I was going to be a chemist.

In my senior year, Dr. Susan Young started her tenure-track position as Hartwick's inorganic chemist after Bill left for a position at University of Massachusetts Amherst. I completed my senior research project under her supervision. I presented my results in a poster format during an ACS National Meeting in San Francisco. While at that ACS meeting, I was able to get chemist Glenn Seaborg's autograph on a T-shirt that says, "I'm in my Element." To this day, I have the signed T-shirt and a picture of myself getting it signed, framed, and mounted in my office (Fig. 2). I don't know if I will ever meet another Nobel Prize winner, but I love to share this story with my students.

I completed my senior research project under Susan's supervision and spent a lot of time in her office. This is when I got a glimpse into what the life of a faculty member is really like. She would not only have classes to prepare for, teach, and grade papers for, but I saw the committee work she was involved with and the meetings she had outside of her office. I remember sitting in her office one day and saying, "I could do this. I think I would like to be a chemistry professor." I couldn't imagine what the daily life of an industrial chemist would be like. I am not sure I would have envisioned myself so clearly in the role of "chemistry professor" if Susan had not been a female professor.

During my entire undergraduate career, Ray and I stayed together as a couple. It was a long-distance (70 miles) relationship, but we made it work. We talked on the phone, and we made many drives along Interstate-88 to visit each other on the weekends. (This was before Facebook, Skype, and FaceTime.) I even did my REU within New York State so that I wouldn't have to spend an entire summer outside of driving distance.

Fig. 3 Kim and Ray on Hartwick Graduation Day in 1997

As I was finishing my undergraduate degree, the conversation turned to what would happen when I went to graduate school (Fig. 3). Ray told me that he would move with me wherever I went to school. He was ready to leave the Capital District of New York State and see another part of the country. It was an enormous relief to me that I could look at schools without worrying about the impact this choice would have on our personal relationship. I applied to and was accepted by many graduate programs.

Experimenting with Cohabitation and Developing as an Independent Researcher

I chose to start the chemistry PhD program at the University of Wisconsin-Madison. Ray and I moved to Madison in August 1997. Very shortly after we moved, Ray found a job working for a Honda automotive dealership. They had two locations, and he would stay with them the entire 5 years we were in Madison.

I began a research project supervised by my chemistry advisor, Dr. Art Ellis, as well, as a chemical engineering professor, Dr. Tom Kuech. This project made use of the Synchrotron Radiation Center (SRC) located 15 miles away in Stoughton, WI. I am thankful for having been mentored by both Art and Tom during graduate school. They both exhibited different management styles, and I incorporate some aspects of each of their styles as I mentor undergraduate researchers.

I am really thankful for all of the connections I made to chemical educators through my membership in Art's research group. I met fabulous people from around the country who prioritized many aspects of undergraduate education and introduced me to many ways of teaching inorganic chemistry and topics related to nanotechnology.

During graduate school, I only ever had one female faculty member for any of my graduate courses. While the faculty members at UW were predominantly male, they were also predominantly parents. I met the children and families of many of the faculty, so this made it seem achievable to have a family and an academic career.

Ray and I really enjoyed our years in Madison, WI. We made many friends in the area, both chemistry graduate students and others. Ray's interests and hobbies introduced us to Wisconsin natives with common interests. We enjoyed travelling to various areas of Wisconsin, the upper peninsula of Michigan, and the sand dunes on Michigan's eastern shore. We went snowmobiling in Yellowstone National Park and in the neighboring areas of Idaho and Montana. At this point, I was nearing the point of becoming a professor but was not yet entertaining the idea of becoming a mother.

While participating in a curriculum workshop at UW, one of the project's evaluators mentioned to me that she knew someone who was looking for a postdoctoral associate (postdoc) in the realm of chemical education. She connected me with Dr. Christopher Bauer at the University of New Hampshire (UNH), who had obtained a grant to implement and study the impact of all of the NSF Chemistry Systemic Initiatives in the UNH General Chemistry Program. This was exactly the opportunity I was looking for. I knew I wanted to teach, at a predominantly undergraduate institution, but I knew I wasn't ready to begin immediately. I wanted time to learn more about pedagogy and teaching before I began.

When I graduated with my PhD, my mom insisted that I let her purchase my academic regalia for me. She had been to several college ceremonies and was always impressed by the complicated regalia worn by the faculty members. She knew that I wanted to teach and I would likely have an opportunity to wear my own regalia at least once a year as a professor. I could tell how proud both of my parents were to see me complete my PhD, and I think of them when I don my academic regalia for ceremonies each year (Fig. 4).

Shifting from Chemistry Expertise to Being a Student of Pedagogy

Ray and I moved to New Hampshire so that I could start as Chris Bauer's postdoc. Ray now had 5 years of experience working for a Honda dealership, and with his specific certifications, it was easy to find a job with another Honda dealership. We ended up renting an affordable house in a rural town about a half hour north of UNH. Ray and I commuted to work together to save money on gas, tolls, and wear and tear on our car. This half hour of time in the morning and the evening was a great time to catch up on things together.

Meanwhile, working for Chris enabled me to meet even more amazing chemical educators from around the world. Chris supported my attendance at the 2002 Biennial Conference on Chemical Education (BCCE), which occurred at the start of my postdoc. I never knew I could learn so much about teaching college chemistry in such a short period of time. Additionally, Chris supported my participation in the

Fig. 4 Kim and her parents at her PhD graduation from UW–Madison

UNH Preparing Future Faculty (PFF) program. This program was fabulous! I will be forever grateful for Chris' support in allowing me to participate while serving as his postdoc. I was able to take one or two courses a semester and learn by taking real courses about cognition and assessment. After completing my PhD at UW-Madison, I was confident in my level of knowledge about chemistry itself. Now armed with this information about education and the scholarship of teaching I felt empowered and ready enough to apply for faculty positions.

Finalizing the Best Personal and Professional Match

I applied for positions at predominantly undergraduate institutions (PUIs). I really liked to teach, and I liked doing some research, but I wanted to be at a place where the bulk of my time would be spent teaching and mentoring undergraduate students. I had two interviews in the same week at two different institutions in the greater Pittsburgh area. I liked the Pittsburgh area and California University of Pennsylvania was my top choice. Cal U was a state school, but the department was very interested in taking their General Chemistry program in a new direction. I felt my experience at UNH gave me the knowledge and desire to make positive changes to a program. Shortly after the interviews, the other Pittsburgh area school made me the job offer.

It was late in the week, and I told Ray right away that we needed to visit the Pittsburgh area that weekend to see if he liked it and would be happy living there, potentially for the rest of my academic career (at either school). When we moved to Wisconsin and New Hampshire, we knew that they were both temporary moves, and he trusted me to make the decision. I knew that now the stakes were much higher, and that he should be able to see the area and help make the decision, as the impact could be more permanent. Ray seemed surprised at the urgency and that we needed to visit immediately. It was Valentine's Day weekend, and he had wanted to take me out to dinner to our favorite restaurant on Saturday night. He talked me into going out on Friday, February 13th instead. Ray proposed to me that night, and we were now officially engaged. I guess that while I was busy interviewing at six different schools for the past few weeks, he was busy shopping for an engagement ring. We visited Pittsburgh and agreed that we would be happy to live in southwestern Pennsylvania. It was just a matter of fairly managing my job offers until I found out if Cal U was going to make me a competitive offer.

Cal U did make me an offer. It was a better offer than the others I had received, and it was the place I most wanted to go. Advice of both of my mentors, Art and Chris, indicated that it was important for me to negotiate. I had not formally negotiated any other job offer in my life, but I knew I should try. When I asked the Dean for a higher "step" on the unionized pay scale, he told me that the department was also hiring another faculty member, who had already accepted her initial offer. He asked if I still wanted him to find out if he could secure the higher rung on the ladder of the pay scale, and I said yes. He came back and agreed to the "step" that I had requested and told me that the administration was going to grant the other female faculty member the same higher step. So, I effectively negotiated for both of us! In 2004, my colleague, Dr. Yelda Hangun-Balkir, and I were the first female chemists to be hired as assistant professors on the tenure-track in the department.

Learning to Teach Through Classroom Experiments and Finding Collaborators

During these next few years, Ray and I each worked very hard. We were able to do so, in part because we weren't taking care of anyone but ourselves. We got married in 2005 (Fig. 5). During this stretch of time, Ray shifted from working at a Honda dealership to running his own Snap-on Tools franchise, to working at a small independent repair shop.

The week before we got married, I was invited to participate in the Faculty Leadership Institute of the Pennsylvania State System of Higher Education (PASSHE) Women's Consortium. I was so busy preparing for the wedding, and I naively thought, "These other women aren't chemists, what will we talk about for a whole week?" I didn't realize until I was at the institute how much the participants would all have in common as female professors, regardless of their discipline. This institute was important for my development as a female faculty member. I not only learned valuable leadership skills, but I met an amazing network of supportive colleagues throughout the state system. In 2005, I was just becoming familiar with

Fig. 5 Kim and Ray on their wedding day

this organization, and I had no idea at the time that I would eventually go on to serve this organization in various officer positions from Secretary, President-Elect, and ultimately President.

The original advertisement for the position at Cal U described the idea of a chemistry "Studio." My Dean had immediately sent me on a trip to California Polytechnic Institute to visit their chemistry "Studio" facility developed and used by Tina Bailey. I began my research on the approach and worked with the architects during my first year or two at Cal U. Then the chemistry "Studio" project seemed to slow down as the administration searched for sources of funding to carry out the project. Even though the project was slowing down, my life as a professor was plenty busy. I was teaching a full course load of 12 credits per semester, and our department was hiring new faculty members and making curriculum changes to the entire chemistry major. Ultimately, patience paid off, and the chemistry "Studio" facility renovations took place over the entire course of the 2007–2008 school year.

My colleague Yelda left our department at the end of our third year. Her husband had career opportunities in another state, so they relocated as a family. During her years at Cal U, she was a valuable role model for me, as she had arrived already a mother of a 3-year-old son. I was able to watch her experience being both a professor and mother before I ever became a parent myself.

Experimenting with Motherhood While Researching Facilities and Pedagogy

During the summer of 2007, Ray and I agreed it was time to start a family. Being one year prior to tenure it seemed like it might be the "right time," if there ever was such a thing. I knew that I was getting older and that my job situation was very stable. I liked being at Cal U. Thankfully, I became pregnant very quickly and my due date was at the end of the academic year, April 2008. So, while the facility I had worked on for several years was being built, I met weekly with the team of folks doing the renovations. I always thought that we seemed an unlikely grouping of electricians, plumbers, carpenters, and a pregnant lady looking at blueprints and talking about circuits and HVAC. I always treated all of these guys with respect, and I think I demonstrated enough mechanical and physical knowledge that the guys always treated me with respect.

I really enjoyed being pregnant. I was thankfully not experiencing morning sickness. Since I had become pregnant during the summer and my son was due in April, I felt my child should be born knowing both general chemistry I and II and inorganic chemistry because he had heard all of my lectures during the fall and spring semesters.

Cal U had no formal maternity leave program in place, so when I explained to Human Resources (HR) that I was pregnant, they explained that I could take my earned days of sick leave and, if I ran out of sick leave, I could utilize the Family Medical Leave Act (FMLA) to have unpaid time off from work. HR made no suggestions to me, about what to do with my classes that were scheduled to end a couple of weeks after my due date. I am thankful to have great colleagues who offered to cover my classes for the last two weeks of the semester. This was uncharted territory for my department since no female faculty member had ever given birth during their employment.

My goal was to work as long as I could until my baby arrived, so that I could minimize the amount of sick time I needed to use and limit the number of days of class that my colleagues needed to cover. I also didn't really want to sit home waiting for the baby to arrive. Even though I was exhausted, it seemed like it would be boring or stressful just sitting home and waiting for the baby.

The uncertainty of when he would arrive was so foreign to me. I was so used to everything being programmed into my schedule and being able to carefully time and control when things took place. Giving birth is such a major event, yet I could not dictate when it would occur. What if my water broke while I was teaching in class? In the lab? In the hallway? In my office? Like all first time mothers, I also nervous about the physical process of giving birth itself. Just 3 days before my due date, my doctor induced my labor. I was showing symptoms of preeclampsia and they wanted me to deliver as soon as possible. So, my water didn't break, I didn't go into labor, nor did I have contractions at work. On one Friday I was on campus teaching classes, and the following Monday, I was at the hospital with my labor being induced.

My son, Max, was born in April of 2008. I was just 15 days shy of my 33rd birthday. It was the end of my fourth year as a faculty member at Cal U. My colleagues, Ali Sezer, Gregg Gould, and Matt Price, only ended up covering the last 2 weeks of classes for me. During finals week I wrote my own exams and a graduate assistant even dropped papers off at the house for me to grade (Fig. 6). I am still grateful for the help of my colleagues. I didn't think it would be fair for me to use my accrued paid sick days when I was really carrying out my final exam duties (writing, grading, etc.), with the exception the actual exam proctoring. Somehow I managed to get my doctors to write a note permitting me to return to work, even though they typically won't see you or grant permission until 6 weeks postpartum.

Fig. 6 Kim at home with her son Max (age 5 days) working at the computer in 2008

Fig. 7 Kim and her son Max (age 4 months) in the new "Studio Classroom" in 2008

During that summer, I did not teach any courses. I was so happy that my position as a faculty member allowed me to stay home for a few months with my new baby. I would come in to check on the facility progress with Max in his stroller. The guys working on the room were still installing some of the finishing touches, and I think they enjoyed meeting the little baby I had been bringing in utero to the meetings we had had all year (Fig. 7).

I had never noticed parental accessibility issues until I brought an infant into a science building. We had a very large stroller, which Max's car seat snapped into, and it was too large to haul up the stairs to the second floor where my office was (we have no elevator). There is, of course, no changing station in the bathroom in my building, so I learned to manage. I would park the stroller downstairs. I would leave a towel, a few diapers and wipes in my office, as well as a spare pillow in case I needed to nurse him while we were in the building.

That summer, I was scheduled to present my work on the chemistry "Studio" at the 2008 BCCE at Indiana University. I had submitted an abstract several months before and hadn't really been concerned yet about childcare. Max was only 3 months old, and I was not ready to leave him behind. He had been nursing exclusively, and I hadn't stored enough milk to be gone that long. So, my mom and my infant son came as my companions to the conference. We stayed at the hotel right on campus where many of the sessions were held, and we drove out with our pack-n-play. It was a great way to stay professionally active and to show my mom what a chemistry conference was like as well (Fig. 8).

During the fall of 2008, Max started daycare and I returned full-time to work. I missed his company when I returned to teaching and I no longer had my companion in utero. I wanted to continue to breastfeed Max as long as I could, so I became very familiar with pumping breast milk. I had a private office so this wasn't too difficult, but I would hang a "Do Not Disturb" sign on my door so that the custodian wouldn't come in to vacuum or empty my trash while I was pumping. Despite the sign, he did

Fig. 8 Kim and her son Max (age 3 months) at a social event during the 2008 Biennial Conference on Chemical Education

Constant Refinement: Learning from Life's Experiments

come in once, when I was thankfully turned away from the door, but in retrospect I probably should have explained the scenario to him, so he would have taken more care to avoid interrupting me.

I didn't have a lot of time to dwell on returning to work with Max in daycare, because after long last, my chemistry "Studio" had been completed. I was teaching for the very first time with a different pedagogy, and my time at work was spent developing new classroom activities and experiments to be done in the new classroom.

I did sometimes remark to myself that having a baby only took 9 months, while the design and renovation of the studio chemistry facility had taken several years. I was certainly proud of both of these accomplishments.

My tenure and promotion dossier was due in November of that year. I had accomplished a lot before my son was born and was continuing to do more. I found it profoundly difficult to shift my time from "being productive" doing lots of teaching, research, and service to spending time assembling the evidence to prove that I was "being productive."

I really had wanted to breastfeed Max for a full year, but my milk supply was slowly diminishing. Most days I didn't have enough time in my schedule to pump more than once, and I was likely not staying hydrated myself since I was so busy. I was disappointed when I was only able to nurse him for about 7 months. We switched him over to formula until his first birthday.

In the summer of 2009, after my dossier was examined at all levels, I was officially awarded tenure. In 2009, I was also the first female chemist to be promoted to the rank of associate professor at Cal U.

Experimenting with Childcare

In the late summer 2009, the owner of the independent repair shop where Ray was working reported that he was going to downsize and start closing the garage so that he could retire. We considered our options, and Ray decided that he would stay home with Max. For the next 2 years, Max was home with Ray during the entire academic year. I cannot express how wonderful this was for me! The first year while Max was in daycare, there was time involved in packing his bag (extra clothes, milk bottles, later food), dropping him off, and picking him up every weekday. I was fortunate to have a daycare center located right on campus, but this daily routine amounted to at least 1 hour of time per day. With Max home with Ray, I could just wake up and sneak out early while they were still sleeping. I am so grateful that Ray was willing to try this out and that he also enjoyed it once he tried it.

Having a stay-at-home spouse made things so much easier for me at home. The job of a chemistry professor is quite demanding. I still spent loads of hours developing curricular materials, teaching, grading papers, advising the Chemistry Club, applying for grants, presenting at conferences, attending workshops and

Fig. 9 Kim, her spouse Ray, and son Max (age 2) at the Rodeo Event during the 2010 Biennial Conference on Chemical Education

seminars, and serving on many university and departmental committees. It is a time-consuming job, but it is an interesting job. No two days are identical. The personalities and talents of each student are unique. The opportunities for research and collaborations evolve. The goal of continuous improvement of the courses, degree program, department, and university is always there.

Having a family does not make the job more difficult or easier for me. I look forward to coming home to spend time with my family. It is difficult, sometimes, when I think about my work when I am home with my family or when I think about my family (like a sick child) when I am at work. I find both of these parts of my life challenging and extremely fulfilling.

I gave a presentation on the chemistry "Studio" at the 2010 BCCE in Denton, TX. This time, Ray and Max came with me (Fig. 9).

Attempting to Repeat the Motherhood Experiment

When Ray and I first talked about having children, we really liked the idea of having two. We each grew up with siblings, and we wanted Max to have a sibling as well. While Ray and I have siblings who are each 4 years younger than ourselves, we didn't really think we should wait that long to have our second child.

Constant Refinement: Learning from Life's Experiments

However, having a second child was more of a challenge than we had anticipated. I had a pregnancy that ended in miscarriage, which I was not emotionally or professionally ready to process. I had not yet told my colleagues I was pregnant, but found myself in need of someone to walk into my lab and take over for a class immediately when I knew something was wrong. I tearfully explained my situation to my most trusted departmental colleague who stepped in right away. Many months then elapsed and I had not yet become pregnant again.

My physicians assured me that I wasn't too old and that I needed to be patient. We were overjoyed when I did become pregnant and this time my due date was again in April. Being pregnant at the age of 35 meant that I was offered additional screenings during the early stages of my pregnancy. I was considered to be of "advanced maternal age." I did the same non-invasive screenings I did for my pregnancy with Max and I didn't opt for anything more invasive. I was so thankful that I was pregnant and I didn't want to take any additional risks for the purpose of screening. Thankfully, my pregnancy proceeded without major complications, and my son Samuel was born 4 days before his older brother Max's third birthday.

After Max was born, I had found out that other female faculty members on campus were able to get their colleagues officially paid for assuming responsibility for their courses when they had left for childbirth. Since I was the first female faculty member in the department to give birth, we didn't know on the department level how to go about making these arrangements when Max was born. So, while I was pregnant with Sam, I made it a point to follow my campus colleagues' recommended course of action and get an official arrangement made so that my colleagues, Ali Sezer and Min Li, could be rightly paid for the courses they took over. I could then feel a lot less guilty about them coming in to take over my class. The arrangement was rather straightforward.

Also, I wasn't sure that I wanted to work every last day until my due date the second time around. I already had Max at home and I was exhausted from not only work but also from being mom to a toddler. I decided that this time around I would choose my last day of work, instead of letting my body and baby decide. I chose April 1 as my last day of work, since I had an April 7 due date. I'm very glad that I did it this way.

The uncertainty of when my colleagues would assume my duties was gone and knowing that they were going to be paid for their time made this much easier. The experience with my work arrangements was much less stressful this time. I didn't worry at all about writing or grading my final exams.

The circumstance surrounding my son Sam's birth was somewhat similar. A few days after my due date, my doctor decided to induce my labor. With the advanced planning done at work, I was much more able to enjoy those first 3 weeks of my son's life.

Experimenting with a Role as an Administrator

Three weeks after Sam was born, I took office as the department chairperson of the Department of Chemistry & Physics at Cal U. I was the first female department chair for this department. The election for department chair occurred while I was pregnant, and some of my colleagues had previously indicated that they thought I would do a good job in that position. When hearing that I was considering running for the position, another person on campus asked me if I would be able to handle the role with small children at home. I really do think that this was meant in the best possible way, and I think that this person was genuinely concerned about the potential toll that this demanding position would take on my family life. However, if I were a male professor, whose wife was expecting a baby to be born around the same time, I'm not sure anyone would have asked me that question. I didn't really know how I felt about this. I wasn't sure if I should be disappointed that I was asked this question (Hadn't I clearly demonstrated I could cope with challenges through my competence and commitment to professional responsibilities?) or if I should be sad for the professors who are fathers who might be in a similar situation and don't get asked this type of a question. I did opt to run for the position, and I was elected (Fig. 10).

When I did take office (with a 3-week-old infant), the department secretary had her office right outside mine. Our former custodian had retired, but our secretary could also turn away any would-be interrupters during my pumping times. Thanks to a quarter-time load reduction associated with being chair, I was able to pump twice during the day and I was able to nurse Sam until he was about 9 months. I had a

Fig. 10 Kim and her son Sam at the departmental picnic the day she became chair

Fig. 11 Kim and her son Sam (age 6 months) in 2011

bigger desk, so I could just leave all of my pumping equipment and cleaning supplies there, and I didn't drag the stuff back and forth from work to home.

I had the pleasure of attending the fall ACS national meeting when my son Sam was only 4 months. I had been appointed as an associate member of the ACS Women's Chemist Committee (WCC), and I had missed the spring meeting because it was so close to my due date. I didn't want to miss the fall meeting, since I was eager to start working on the committee. Ray didn't want to travel with the full entourage to Denver, so I was diligent about pumping and freezing milk over the summer so there would be enough for a long weekend with me away. I studied the airline regulations regarding breast milk and pumps and I decided that I would "pump and dump." It felt like such a waste of a valuable resource, and I commend women who would have made the choice to freeze their milk in the hotel walk-in, but I knew this wasn't the right choice for me at this time. It was also really nice to enjoy an adult beverage, which I had not been able to do in a long time (Fig. 11).

During the same fall semester, I was headed to a Women's Consortium conference being held in the northeastern corner of Pennsylvania, about a 5-hour drive from home. I had packed my clothing bag, my toiletry bag, my backpack, and my breast pump bag, and I had even remembered to download an audiobook to listen to as I drove. An hour into my drive, my husband called and told me I had left my suitcase at home on the bed. I had made multiple trips to the car and had apparently failed to make the last trip needed. How could I forget my clothes??? I had my laptop and my work papers, I had my toiletries and my breast pump supplies in that bag, but I didn't have a stitch of clothing except what I was wearing (which was fairly casual since I knew I would be sitting in the car so many hours). Had I left behind my pump, I would have turned around and driven back to get it, even though I would have lost 2 hours' time doing so. Since it was only clothes, I went shopping. My options were rather limited driving diagonally across Pennsylvania, but I stopped at a women's clothing retailer I saw in a strip mall. It was already near 6 pm, and I knew that if I waited until I got to a bigger city or closer to my destination, stores would be closed. When the store attendant asked if there was something in particular I was looking for, I was speechless. I got tops, pants, pajamas, and undergarments, even a pair of shoes. This was an experience I will never forget, and I will also never again forget to bring my suitcase on a trip! Still, I pumped my milk and stored it in my hotel room fridge and was able to bring it back when I returned.

My 3-year term as the department chairperson ended in April 2014. I opted not to run for a subsequent term. I am glad that I did serve as the department chairperson, because it is a position you don't understand well until you experience it first-hand. I learned a lot about how the university works and I got to know the administrators on campus very well. It is a very, very, time-consuming job, if you accept the full responsibility involved. It involves solving a lot of problems and developing new policies. My colleagues indicated that I'd done a favorable job. In the end, I learned from this experiment that I really enjoy teaching and supervising undergraduate research much more than being an administrator. While I was chair, I missed being able to spend more time closely mentoring students (Figs. 12 and 13).

Fig. 12 Kim serving as the Faculty Marshall for the Honor's Convocation Ceremony at California University of Pennsylvania in 2016, carrying the University Mace

Fig. 13 Kim with a group of students, attending a monthly meeting of the Society of Analytical Chemists of Pittsburgh in spring 2017

Determining the Right Conditions for Promotion to Full Professor

I was first eligible to apply for promotion from associate professor to professor during the fall of 2012. I was very confident that I would be promoted. I had assembled a fantastic dossier of materials. I was one of the most active faculty members in the department and was serving as department chair at the time. I was active as a volunteer for a variety of professional organizations at the campus, statewide, and national levels. I had excellent teaching evaluations and several students serving as undergraduate researchers. I had overwhelmingly positive letters from my department promotion committee, the colleague who was "acting" as the department chair for the purpose of promotion (since I couldn't write my own chair's recommendation letter), and the dean of the College of Science and Technology. I still believed in the myth of the academic meritocracy. I was not promoted that year and I was devastated. I didn't understand and I took this very personally. I had not dealt with much rejection in the past. This was a very in-my-face learning experience about campus politics.

On my campus, dossiers are due by November 1. The candidates do not have access to these again until the end of the following July when decisions are announced. So, if you are unsuccessful, you have a window of 3 months to work in earnest on updating/changing your dossier. As a parent of small children, and a fan of the autumnal season, the timing of the dossier deadline really interfered with my ability to enjoy celebrating Halloween.

All told, I applied for promotion to full professor a total of four times: fall 2012, fall 2013, fall 2014, and fall 2015. Administrators and colleagues just kept encouraging me to reapply the following year. The faculty on my campus are unionized, and there is a limit on the percentage of faculty who can be at the rank of professor (30%). Some supporters suggested that perhaps that there was a "backlog" of highly qualified people who are deserving but not yet promoted. None of these hypotheses about my promotion denial were reassuring. During these years, I got to know many people on my campus in a similar situation. I wish that I had better mentorship on this process, so that I had more realistic expectations from the start of the process. I was finally notified in July of 2016 that I would be promoted effective with the start of the fall 2016 semester.

Coming Full Circle

In busy day-to-day life of balancing both motherhood and being a professor, I don't always take time to think of how some things have changed and some have stayed the same. In the spring of 2017, I had a moment when I could reflect on the big picture, and I realized just how much my career had evolved significantly over time.

My first undergraduate research presentation at a national meeting took place at the spring 1997 ACS National Meeting in San Francisco, CA. Then, exactly 20 years later, I was attending the spring national meeting, in the same city, and one of the undergraduate students whose research project I supervised was presenting her first poster at a national meeting. It was even more meaningful when I was able to introduce my own undergraduate advisor, Dr. Susan Young, to my student, Mikaylah Glenn (Fig. 14).

Grateful for My Family Collaborator

I think it would be safe to say of both parenting and professorship that, just when you have things figured out, they will change. There will be more hurdles to overcome, and life will get more hectic when both kids have school, homework, swimming lessons, and soccer practice, but working together as a couple, Ray and I have always found a way to make things work, and I'm sure we'll continue to do so. I have to give a lot of credit here to my amazing husband, Ray. When we met in high school, neither of us had any idea our life would evolve this way. He is truly a wonderful father and spouse who I'm happy to be spending my life with and he has always been immensely supportive of my career (Figs. 15 and 16).

Constant Refinement: Learning from Life's Experiments 573

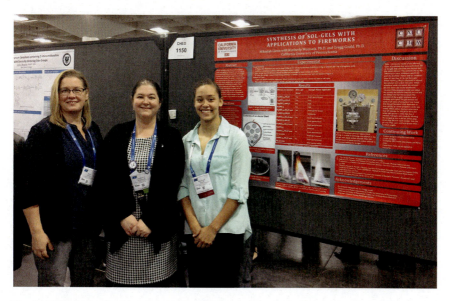

Fig. 14 Kim (center) with her undergraduate research advisor Susan Young (left) and Kim's undergraduate research student Mikaylah Glenn (right) at the American Chemical Society National Meeting held in San Francisco, spring 2017

Fig. 15 The Woznack family in Point State Park, Pittsburgh, PA, November 2017

Fig. 16 Portrait of Kim drawn by her son Max at age 6

Lessons Learned and Advice

- No one is perfect. All human beings have 24 hours in a day. It will be a challenge to any mother or parent who works full-time. Women should be honest about the multiple roles they play in many people's lives and the time those roles require. We cannot keep up the illusion of "effortless perfection," when all working moms orchestrate a delicate ballet of responsibilities every day.
- Even though we have finite hours in the day, that doesn't need to impact the quality of the time we spend at home or at work. My advice would be that as much as possible, when you are interacting with your children you should be *present*. As a faculty member, there are many hours where we cannot physically be at home, but when you are home, try to make the most of every moment by focusing on your children. I have ditched using my smartphone at home as much as possible because it kept distracting me from being "home" when I was home. Likewise, I don't stay plugged into the phone while I am at work. Ray and the elementary school have my office number and will call on the landline when there is an emergency or something important.
- Ask for support from others when you need support. This is important both at home and at work. If you need assistance with a work-related project, it is usually acceptable to ask a colleague to collaborate with you so you can split up the work. Likewise, a supportive spouse will understand if you might be delayed from your household responsibilities. At the same time, you cannot take too much advantage of a partner. You should periodically check in with your personal and professional partners to see if you both feel satisfied about the balance or distribution of responsibilities.

Constant Refinement: Learning from Life's Experiments   575

- You are your own harshest critic. Give yourself space to be human. Have realistic expectations for how much work you can complete at what level of quality in a certain period of time.
- Compliment other people (especially women) who impress you with their accomplishments. The people you are impressed by are likely their own harshest critic, and sometimes it takes an outsider to point out the obvious accomplishments. It just might make someone's day to hear that you noticed and are impressed by what they have achieved or a decision they've made.
- Surround yourself with a supportive network of both colleagues and friends. It is important to trust your "gut" or instincts when interviewing for a position. An academic position can be a work equivalent of "home" for a very long time. Try to select a supportive environment. Also find support outside of your department from other likeminded colleagues. Find support among your non-academic and non-chemist friends.
- Embrace your identity (and perhaps inner geekdom). I will be honest that when I meet people outside of the academic environment, I don't always start out with "I'm a chemistry professor" because that sometimes is a conversation killer. People will possibly report back that they hated chemistry or they did poorly and where do you go from there? I know female faculty members in other disciplines (such as psychology) have an equally difficult introduction. Come up with a response you can use that is neither derogatory nor self-deprecating. When helping out at my son's kindergarten holiday party this year, the teacher asked him to introduce his mom to the class. He said, "This is my mom and she is a chemistry teacher and her students just graduated." I was proud that he knew what I do for a living and was happy to tell his classmates.
- Be proactive about your course schedule. Even if you are not comfortable telling your colleagues that you are beginning to plan a family, you may want to think about your class schedule. If you are in a small department like mine, colleagues won't be able to cover your courses if they are all meeting during the same hour of the day. If your child is about to start preschool or kindergarten, find out when those programs start or end so that you can minimize conflicts with your course schedule. Academic schedules often get submitted over a year in advance, so if you can have a plan or back-up plan, it helps.
- If you do have the pleasure of being a mother and a chemistry professor, be a role model for your female students. I have snacks and crayons and coloring books in my desk drawers so that, in any type of emergency or when needed, my sons can come to work with me and spend time being creative. My sons love to visit my building; in many cases they associate this with vending machines and building with molecular model kits. When our campus marketing department asks if I am willing to participate in a photo shoot or video, I try to almost always say yes. I like the idea that they want a female chemist to be visible in their marketing campaign.

Acknowledgments I have to start by thanking both of my parents for being such amazing role models. They taught me to strive for excellence and to have a strong work ethic. They taught me how to be professional, and they have demonstrated how strong a partnership a marriage can be. They truly always encouraged me to succeed. I also want to thank my children Max and Sam for their unconditional love. It is a wonderful feeling to return home from work to see their smiling faces and hear them say, "Mom's home!", which they say while running up to give me a hug. I also have to thank my husband, Ray. He is patient and kind and understanding of my position. He has always supported me in my career, at every turn. I cannot express in words how grateful I am to have such a wonderful partner in life.

About the Author

Education and Professional Career

1997	BS Chemistry, Hartwick College, Oneonta, NY
2002	PhD Inorganic Chemistry, University of Wisconsin-Madison, Madison, WI
2002–2004	Postdoctoral Associate, University of New Hampshire, Durham, NH
2004–2009	Assistant Professor, California University of Pennsylvania, California, PA
2009–2016	Associate Professor, California University of Pennsylvania, California, PA
2011–2014	Department Chairperson, Department of Chemistry & Physics California University of Pennsylvania, California, PA
2014–present	Professor, California University of Pennsylvania, California, PA

Honors and Awards (Selected)

2016	"Faculty Woman of the Year," President's Commission on the Status of Women, California University of Pennsylvania
2016	Presidential Distinguished Merit Award Nominee in the Category of Research, California University of Pennsylvania
2010	Younger Chemist Committee Leadership Development Award, American Chemical Society
2010	Grant Writers Fellow, California University of Pennsylvania
2009	Faculty Professional Development Committee Faculty Merit Award in Technology

Kim has been a proud member of the Pennsylvania State System of Higher Education Women's Consortium since 2005, having served as the Secretary, Treasurer, and President. She has served on the American Chemical Society Women Chemist Committee since 2011, becoming the Chair in 2018.

Remarkable, Delightful, Awesome: It Will Change Your Life, Not Overnight But Over Time

Sherry J. Yennello

Photo Credit: Texas A&M University

She's remarkable! I said that when she was born and many times since. Stephanie was born with ten perfect little fingers and ten perfect little toes. She came out crying and all I wanted to do was soothe her and take away whatever was causing her to cry, but the nurse told me it was good for her lungs. Thus began a remarkable journey that has changed my life.

S. J. Yennello (✉)
Department of Chemistry and Cyclotron Institute, Texas A&M University, College Station, TX, USA
e-mail: yennello@chem.tamu.edu

Growing Up

Let me step back to my own childhood. I had magnificent parents and two older sisters. My mother earned a college degree, but, like many of her generation, she stopped working when my oldest sister came along. My father dropped out of school in the 6th grade to join the Merchant Marines. When my sisters and I were still quite young, he went to night school to get his GED. He didn't go back to school for career advancement; he did so to send a message to us about the importance of an education. There was never a question of whether any of his daughters would go to college, just a question of where. When my time came, I went to Rensselaer Polytechnic Institute. I got dual degrees in chemistry and physics. My father always enjoyed parents weekend, because although he had little formal education, he appreciated all the engineering demos that were always on display. My father was a very smart man.

While going through college, I didn't really have a plan for my future. In fact it wasn't until the summer after I graduated, while working at a nuclear power plant and getting ready to go to Indiana University for graduate school, that I realized I wasn't going to live the life of my mother. I was sitting by the locks in Fulton, NY, recalling a conversation I had the previous year with a friend from high school. We were talking about our futures. Even though I had worked hard through college, I said I would give it all up for the right guy—after all, that is what my mother did. My friend told me "for the right guy, you won't have to give it all up."

The Right Partner

After graduate school and a postdoc at the National Superconducting Cyclotron Laboratory at Michigan State University, I accepted a position as an Assistant Professor of Chemistry at Texas A&M University in College Station, Texas. My career as an experimental nuclear chemist was starting, and I was very excited. And, I met the "right guy." Larry was also a young faculty member in the chemistry department. He was—and is—a guy for whom I would give it all up, but for whom I don't have to. He has been tremendously supportive of my career, sometimes at the expense of his own. He has been there at every step; he went to every doctor's appointment when we were pregnant.

The Decision

In my mind, being an assistant professor and an experimental nuclear chemist presented a huge obstacle to becoming a parent. How would I balance my career and a family? How would I avoid the radiation area where I do my experiments for a

Remarkable, Delightful, Awesome: It Will Change Your Life, Not... 579

9-month pregnancy? I had great career role models in two senior nuclear scientists Ani and Jolie, but neither of them had kids. Two amazing women helped me see that being a parent and being a scholar are not mutually exclusive. Shirley Jackson, then president of my alma mater, visited my campus as part of a Women in Discovery symposium. While escorting her from one meeting to another, she explained to me that I could do an experiment without being in the radiation area as long as I had a group of people with whom to work. Additionally, Geri Richmond visited campus and also provided great encouragement about being a mother and an academic. She assured me it could be done, as long as I was willing to ask for help. These two incredible women probably don't even know how much they impacted my decision to become a mother. Now I just had to figure out the best time.

Timing

We had been married for about 5 years, and my tenure had been granted when our department made the offer to let faculty double teach—something previously much frowned upon—under certain special conditions. They were trying to get more of the tenured and tenure-track faculty into the larger service courses, because it looks better to uninformed people who don't appreciate the talent that exists in the non-tenure line faculty. So they offered to let any tenured or tenure-track faculty member have a semester off in return for teaching a larger service course in addition to a regular teaching assignment. They assumed that this would enable the faculty member to have an uninterrupted semester devoted to research. When I took the deal, the associate head assumed that this would just mean we wouldn't schedule nuclear chemistry (a small upper division course I taught once a year) the upcoming fall. I said no and that I wanted the following spring semester off. He queried "Why?" and I told him because that was what I wanted. How could I tell him I wanted to have a kid when I hadn't even discussed it with my husband yet?

So I think I shocked my husband when, over dinner at a favorite local restaurant I asked him if we wanted to have a child. The timing was perfect; I had tenure and I had arranged a semester off. Fortunately, for all of us, he agreed and we made a decision that would change our lives. We knew I would have the spring semester off, so we calculated backward and decided when we should start trying to get pregnant.

Pregnancy

Biology was good to us and our child was scheduled to arrive just after finals in December. When Larry and I were in San Francisco for an ACS meeting, I had a suspicion that we might be on our way to becoming parents, but it was too early to tell for sure. Nonetheless, despite a very nice dinner at a very nice restaurant, I passed on the wine—just in case. I didn't want to do anything that wasn't in the best interest of my future child. Additionally I gave up Twizzlers and all other junk food and ate fruit for snacks for the next 9 months. Nothing but the best for my future child.

When we first knew we were pregnant, Larry was ready to tell the world. I wasn't ready to let the people in our department know, however. There were very few women among the tenured or tenure-track faculty, and none with kids. So we told our families, but otherwise kept our news to ourselves for a number of months. I was being creative about how to avoid the radiation area, but I finally had to break my silence and tell the cyclotron laboratory director because a congressman was visiting campus as part of a science and public policy program and he wanted a tour. I had been sitting near him at lunch and did what I almost always do with people I meet, which is tell them they should come see the cyclotron. I knew his day was crammed and he wouldn't have time for the tour, so I didn't think anything about my offer. But by the time I walked back across campus, there was a phone message that his schedule had been rearranged and he was headed over for a tour. I panicked. How could I go into a radiation area to give a tour? How could I get someone else to give the tour? How could I tell the congressman that I couldn't give him the tour? So I settled on asking the cyclotron director to give him the tour; his status made it seem OK that he was to give the tour I had offered. But the price was I had to break my silence. The director promptly told his wife, and I learned that even solemnly sworn secrets came with a spousal exception. Fortunately it was still a number of months before my pregnancy became known in the department.

Although both Larry and I believe that information is good, sometimes too much information isn't a good thing. Since we were older parents, we opted to have an amniocentesis. We consulted with a neonatal specialist, who said you really shouldn't do the amnio until 16 weeks, but she did a sonogram to see if there was any reason for concern. The nuchal translucency she measured put us at a slightly higher risk of complications. I went insane. This rational scientist who spends a lot of time looking for small signals and understanding probabilities couldn't be rational. This was my future child. I needed to be certain that she was perfect. Larry convinced the doctor to do the amnio a week or so early, and we were very happy when the news arrived that we had a perfectly normal daughter on the way.

Having told my research group that I was pregnant, I would go in the cave to help set up the experiment, but once we had beam, I would confine my involvement to the counting room. My next challenge was that I was teaching the nuclear chemistry course in the fall (part of the condition for getting the spring off). Part of the course was a laboratory where we used radioactive sources. The exposure would be minimal, but I didn't want to risk the health of my child in any way. It seemed reasonable that I could teach the lab without actually being in there when the sources were being used. In order to do this, I would need a little extra help from a graduate student. I asked the department head for 1/3 of an extra teaching assistant so I could replace my physical presence during the lab with the graduate student who had taught the course with me the previous fall. Fortunately, the department head agreed.

Graduate student assistance in my nuclear chemistry lab was the only accommodation for my pregnancy that I asked for from the department—I had already "bought" my semester of teaching relief. However, I was contacted by the HR

Remarkable, Delightful, Awesome: It Will Change Your Life, Not... 581

liaison in the department to inform me about my right to take time off under the Family and Medical Leave Act (FMLA). (She hadn't felt the need to inform any of the male faculty who had recently had children.) She instructed me to get a doctor's note so I could use 6 weeks of sick leave until I had to invoke FMLA. I explained that my child wasn't due to arrive until after finals and I had arranged to have teaching relief for the following semester, so while I appreciated her advice, I didn't think I needed to invoke any of it. She took a calendar and informed me that faculty and staff were supposed to work until the 22nd of December. I assured her that there would not be a day in which she could find 50% of my faculty colleagues in their offices that I would not be accounted for. When I asked if she had done this briefing with my male colleagues who recently had kids, she told me she was just treating me like "any other pregnant female in the department." The problem was that her mental model for a pregnant female was a staff person for whom she would hire a temporary replacement while they were on maternity leave. My problem was I wasn't sure where she thought she could find a temporary replacement to run my research group. She followed up our meeting with a memo to me repeating all her instructions about getting a doctor's note and invoking FMLA. After calming down (Larry was very helpful here), I wrote the department head an email saying I hadn't asked for any leave, but I would let him know if at a later date I thought I needed to take leave. He must have asked her to step back because that was the last discussion I had with her about maternity leave.

While I was pregnant, I was informed that I had been selected as the Sigma Xi National Young Investigator, but I would have to accept the award in person in Albuquerque and give a talk at their national meeting. The meeting was scheduled 1 month before I was due. I accepted and Larry made plans to travel with me to the meeting in case I needed any help. My health insurance plan prohibited travel outside of a 90-mile range in the last month of pregnancy, and my award talk was just outside that window. But about a month before the meeting my OB said she thought my daughter would arrive about a week earlier than the original due date. Fortunately I convinced her not to move my "official" due date so I could accept the award. Larry was fabulous and even found a suit for his very pregnant wife to give a talk in.

My Daughter

Stephanie did arrive "early" as predicted and was born with jaundice. Fortunately, my last lecture had been given and my exam had already been written and printed. I called my teaching assistant who agreed to do the review for the final and give the exam. She would later deliver the final exams to my house so I could grade them. Stephanie spent the first 5 days of her life in the Pediatric Intensive Care Unit under the special lights to treat her jaundice. I took a week of sick leave. My parents, excited about their first (and still only) grandchild, flew in on the red-eye. They were amazing as they took care of Larry and I. We spent those first days driving back and

582 S. J. Yennello

forth to the hospital for feedings. Fortunately, Stephanie was able to come home before my parents had to leave.

The first weeks after Stephanie was born were eventful. Not only were we dealing with this new life, but my first PhD student, Doug, was set to walk the stage at graduation. Stephanie stayed home with her dad, but she gave me the perfect reason to leave the ceremony shortly after I had presented Doug with the hood symbolizing his degree. The following week, Larry and I had a faculty meeting to attend and we took Stephanie with us. The associate head was caught off guard. He said he didn't know that I was coming back to work—as if having a child meant the end of my career. It didn't mean the end of my career, but it did mean we needed to get some help.

Childcare

At first Larry and I thought that we might be able to arrange our schedules such that we could alternate who was home to take care of Stephi—particularly since I had a semester of teaching relief. It quickly became clear that our plan wouldn't work, regardless of a newly upgraded computer at home. So we made a plan for how one might go about finding a solution to our dilemma. Not a list of possible solutions, but a list of steps we could take that might lead to a solution. One of the first steps was to ask some of our friends who had young kids how they dealt with the childcare issue. We were told about this wonderful nanny, Dorothy, who had previously been employed at a childcare center in town, but then went on to be a nanny in Houston. Now she was coming back to town, so we got her contact info and arranged to meet her. Within 5 minutes we knew she would be perfect. Dorothy took care of Stephanie for the next 8 months, which gave us just enough time to get our bearings before the next chapter of lives started.

When Stephanie was 8 months old, we moved to Washington, DC, to work for the National Science Foundation (NSF) for 1 year. We spent 3 days driving to Arlington, Virginia. When we arrived, we put Stephi in group care for the first time. Bright Horizons, a childcare center, was on the first floor of the NSF building. Luckily a wonderful staff member at NSF, Denise, had connected me with the center, and we got on the waiting list months before we moved. Since the NSF had priority at the center, we were able to actually get Stephi a spot. We cried a lot those first times we dropped her off. The worst day was when I had to go back to College Station for an experiment the week after we moved. I flew back to Texas leaving my beautiful 8-month-old daughter in a world that had been turned upside down. The only familiarity for her was her dad, thus starting an incredible bonding between the two of them. She could only fall asleep curled up on his chest for the next year. Returning to Texas for that experiment is a decision that I would make differently if I could roll back the calendar because it took over a year for Stephi to forgive me for leaving so early on.

Remarkable, Delightful, Awesome: It Will Change Your Life, Not... 583

My sisters thought that one of the biggest changes we would have to go through with a baby was giving up going to nice restaurants, which was something Larry and I certainly enjoyed. However in October, when Stephi was 10 months old, we ended up in Hawaii and had to figure out how to feed her in a restaurant. After this went well, and we were back in DC, we proceeded to take her to all the top restaurants in DC. Many a waiter marveled at what the baby would eat. Some years later, she became quite picky about what she would eat, but just this year, she has become a vegetarian and is a most adventurous eater once again.

As the year went on, and I took trips back to College Station to run more experiments, I started checking out childcare centers again so we would be set when we returned to Texas. This was when I realized I needed a lot longer than I thought to get into a childcare center—you were supposed to get on the waiting list before you were pregnant if you wanted a spot before the child was 2. Fortunately, we were able to get Stephanie into the TAMU children's center.

Travel

At 7 weeks old, Stephi took her first airplane flight. Larry and I had a meeting to attend at Michigan State University. We had a friend there who arranged for someone to help look after her while we worked, and I was able to find an empty classroom in which I could pump. It was not much longer after that trip that I travelled alone to give a talk at Arkansas State University. I lacked the confidence to ask for time and space to pump, so I ended up sitting on the floor of a bathroom.

When Stephi was 4 months old, there was an American Chemical Society meeting in San Diego, and both Larry and I were scheduled to go. We had decided I would fly to my parents' home in Oregon with Stephi and spend a few days so everyone could get acclimated with one another. Then I would fly down to California to meet Larry. Larry was very sad the night Stephi and I left for the airport. He was about to face his first night without his daughter since we brought her home from the hospital. When my parents dropped me off at the Portland airport a few days later, I was equally sad since I was leaving her, too. I took a picture of her and left in tears. As soon as I landed in California, I sought out a 1-hour photo developer (we weren't yet in the digital age). We had that picture—if not our daughter—for the next few days. At the end of the meeting, we both flew to Portland and couldn't have been happier to have her back in our arms. We made the decision then never again to go to the same meeting without taking her with us. Now when we both go to the same conference, we creatively merge our schedules and we arrange to spend time with her, too. We just needed to coordinate our schedules. Communication and coordination are important every day. I don't agree to meetings before 8 or after 5 and never accept a travel invitation without discussing it with Larry. Seventeen years later I'm still sad every time I head to the airport without her.

As Stephanie grew older, it became more important that she knew ahead of time that one of us was going to be traveling. I would sing a refrain to her many times:

Mommy will always come back,
Mommy will always come back.
Mommy loves you very much and
Mommy will always come back.

I also left her notes to find while I was gone. Eventually, Stephi played a game where she would walk around the house pulling her small suitcase and say she was going on a business trip. She said she would call when she got to the hotel. The hotel was under the dining room table.

Professional travel is something that is always a balancing act. I travel less than I would without her and feel guilty more when I do travel. For her first birthday, I was supposed to be at a meeting in Bologna. I don't know what I was thinking when I agreed to the trip and booked the plane tickets. But while I was there, I realized what a mistake I was about to make, so I switched my plane tickets and came home early. In later years I was smart enough—and comfortable enough—to just say no to trips that conflicted with her birthday. I've also taken red-eye flights home from meetings on the west coast so that I could make it to her soccer game the next morning in Houston. I have scheduled flights to be after her games—in one case getting to the airport in just enough time to take a shower since I was on an overnight flight to Europe (and I needed to get rid of the sweat and sunscreen). My travel planning always involves several questions. Do I need to go? What is the last flight I can take and get there on time? What is the first flight I can take to get home after I have done what I needed to do? Sometimes I Skype into meetings rather than travel. Regardless I am always sad when I head to the airport. I love my daughter and don't want to miss a single day of her life.

Advice in a Nutshell

- Find a supportive partner or other support network. My husband is wonderful.
- Know your environment, or more importantly choose one that is going to be supportive. That could mean picking your thesis advisor or postdoctoral mentor. It definitely means carefully picking a department that is going to help you reach your goals.
- Timing is everything. There may never be a perfect time unless you create it.
- When you get to choose who you work with, choose wisely.
- Communication is critical. Don't assume that your partner, your child, your students, or your colleagues know what you are thinking or planning.
- If you make a decision you can't live with, unmake it.
- Be creative about travel. Do you need to go? Can you Skype in rather than attend in person?

She Has Changed My Life

My daughter is delightful. She is 17 now. I have had the privilege of watching her grow and learn, watching her tackle new challenges and become the most amazing person I know. I often wonder how she got to be so awesome. I could not imagine life without Stephanie. I try every day to appreciate the gift that is my daughter.

Acknowledgments I would like to thank my family, friends, and colleagues who have made this amazing journey possible.

About the Author

Education and Professional Career

1985	BS Chemistry, Rensselaer Polytechnic Institute, Troy, NY
1986	BS Physics, Rensselaer Polytechnic Institute, Troy, NY
1990	PhD Nuclear Chemistry, Indiana University, Bloomington, IN
1991–1992	Postdoctoral Research Associate, National Superconducting Laboratory, Michigan State University, East Lansing MI
1993–1998	Assistant Professor, Texas A&M University, College Station, TX
1998–2002	Associate Professor, Texas A&M University, College Station, TX
2000–2002	Program Director, National Science Foundation, Alexandria, VA
2002–present	Professor, Texas A&M University, College Station, TX
2004–2014, 2016–present	Associate Dean, Texas A&M University, College Station, TX
2007–present	Regents' Professor, Texas A&M University, College Station, TX
2014–present	Director, Cyclotron Institute, Texas A&M University, College Station, TX

Honors and Awards (Selected)

2014	Bright Chair in Nuclear Science, Texas A&M University
2013	Fellow, American Association for the Advancement of Science

2011	Fellow, American Chemical Society (ACS)
2011	Francis P. Garvan–John M. Olin Medal, ACS
2005	Fellow, American Physical Society
2000	Sigma Xi National Young Investigator Award
1994	National Science Foundation Young Investigator Award

Sherry is passionate about uncovering nature's deepest secrets about nuclear reactions and the origin of the chemical elements, as well as ensuring equity and access to education and professional development for all individuals regardless of any personal identifying characteristics.

Printed in the United States
By Bookmasters